Ionic Liquid Properties

Yizhak Marcus

Ionic Liquid Properties

From Molten Salts to RTILs

Springer

Yizhak Marcus
Hebrew University of Jerusalem
Jerusalem, Israel

ISBN 978-3-319-30311-6 ISBN 978-3-319-30313-0 (eBook)
DOI 10.1007/978-3-319-30313-0

Library of Congress Control Number: 2016938795

Printed on acid-free paper

This Springer imprint is published by Springer Nature
The registered company is Springer International Publishing AG Switzerland

Contents

Chapter 1
Introduction

This book deals with ionic liquids, i.e., with single substances in their liquid state of aggregation that are dissociated into ions to a major extent. Because of the enormous amount of information that is available regarding this subject, the book does not deal with mixtures of ionic liquids nor with their solution (as electrolytes) in molecular solvents. Furthermore, the information presented here is in no way exhaustive, it is rather illustrative of the subjects dealt with and deals mostly with what was published in readily accessible literature to the end of 2014. The book emphasizes the properties of the ionic liquids and does not deal with their reactions and possible applications.

The properties of ionic liquids, because they are dissociated to a major extent into ions, depend on those of the constituting cations and anions. Therefore, Chap. 2 of the book is devoted in part to the properties of isolated ions that do not interact with their environment. These properties comprise their mass and charge, their standard thermodynamics of formation, and their entropy and heat capacity under standard conditions. The sizes of ions, once they are in a condensed phase, depends on their being constrained by their neighboring particles due to the repulsion of their outer electron shells. These properties of the individual constituting ions of ionic liquids are assumed to be 'portable' when they form the ionic liquid, some of them being subject to changes induced by the mutual interactions of the ions.

Ionic liquids comprise, as the chapter headings indicate, high-temperature molten salts, network-forming ionic liquids, low-temperature molten salts, and room temperature ionic liquids (RTILs). The interest of researchers in high-temperature molten salts, dealt with in Chap. 3, peaked in the 1950s and 1960s of the last century in connection of their possible use in the metallurgical industry and in nuclear fuel reprocessing. An introductory text [1], several multi-author monographs [2–4], a handbook [5] and data compilation [6], and a series of 'advances' from vol. 1 [7] to vol. 6 [8] were published. The interest in high-temperature melting salts has waned since then, but information is still accruing, summarized in the proceedings of international conferences, e.g., that of the 17th, held in Las Vegas in 2010 or of EUCHEM conferences held regularly up to date.

Y. Marcus, *Ionic Liquid Properties*, DOI 10.1007/978-3-319-30313-0_1

Network-forming molten salts, described in Chap. 4, are based on relatively low-melting zinc halides an on molten slags, i.e., high-melting borates, silicates, and germanates. Their structures involve polyhedral anions that share corners, sides, or faces, among which small cations are dispersed. These substances in the liquid state are in thermodynamic equilibrium, but they tend to form glasses rather than to crystallize when sufficiently cooled, such glasses being outside the scope of this book.

Low-melting ionic liquids, discussed in Chap. 5, have melting points characteristically <250 °C and share many properties with the high-melting salts. They involve many substances in which either cation or anion is organic, e.g., tetraalkylammonium or carboxylate, but also purely inorganic salts. A special class of such low-melting ionic liquids are salt hydrates, in which the cation is hydrated, which on cooling from the liquid state may crystallize congruently (to a salt with the same composition as the melt) but need not do so. Salt hydrates have been suggested as thermal energy storage materials.

Room temperature ionic liquids (RTILs), the subject of Chap. 6, have become en vogue from the beginning of the present century, the number of publications dealing with them increasing exponentially since then and several books dealing with RTILs have by now been published, from an introduction [9] to further in depth discussions [10–12]. These substances are liquid at <100 °C and a large variety of them have become commercially available by now, but new ones with various functionalities are being synthesized all the time. Again, due to the enormous variety of RTILs, only a small sub-group of them is dealt with here, namely aprotic RTILs with no added functionality in their cations and with a restricted variety of anions.

Certain inorganic salts, such as HgI_2, $AsBr_3$, and $AlCl_3$, have low melting points (<250 °C) but they do not dissociate appreciable to ions in the molten state. Since they do not constitute ionic liquids, such salts are outside the scope of this book.

Once the ions are combined to form ionic liquids with whatsoever melting points, the properties of such liquids become the subject of this book, ranging from their structures to thermodynamic, transport, electrical and optical properties as well as chemical properties as appropriate. Structures are determined by diffraction methods (with x-rays and neutrons) and similar techniques (EXAFS) and computer simulations. Thermodynamic properties include the liquid range, from the melting to the (normal) boiling point, or up to the critical point when relevant, the energetic changes involved in the phase changes, heat capacities, and surface tensions. Volumetric properties, such as density, thermal expansibility, and compressibility are presented. The cohesive energy and cohesive energy density of the liquid salt are also included. The transport properties include self-diffusion, viscosity, electrical conductivity, and thermal conductivity. The electrical and optical properties include the permittivity, the refractive index, and molar refraction. The chemical properties involve those that affect the abilities of the ionic liquids to act as solvents, hence their solvation abilities, mutual solubilities with water, as the most important representative of molecular liquids, their hydrophilicity and hydrophobicity, where relevant, and the solubility of carbon dioxide and some other solutes in them.

These properties are compiled in tables and are annotated: in many cases with reference to previous critical compilations and reviews, where references to the original literature may be found, and augmented with references to more recent references or those otherwise not included in the compilations and reviews. The selection of the data included in the tables is the responsibility of the author of this book, and it was attempted to select the data critically. No attempt was made, however, to be exhaustive of the profusion of available data nor was it attempted to be comprehensive at all. The data that are presented in the Tables are discussed in terms of the physicochemical effects behind these data. Where appropriate, methods to predict such data from other information on the ionic liquids or their constituent ions or at least to make appropriate correlations with such information were presented.

1.1 List of Symbols

Chemical species and units of physical quantities are denoted by Roman type characters, whereas physical quantities that can be expressed by numerical values are denoted by Greek or *italics* characters. Mathematical symbols have their usual meaning and are not listed here. The same symbol is used for an extensive property of a system and for the molar quantity of a constituent of the system. The SI systems of physical units is used throughout, but some extra-SI ones commonly used in the physicochemical literature are also included where they simplify the notation. These include the symbols °C for centigrade temperatures (T/K – 273.15).

Principal Latin Characters

A	Helmholz energy, molar Helmholz energy (in J mol^{-1})
a_{X}	thermodynamic activity of species X
$B(T)$	Tait expression temperature-dependent parameter
B_{x}	activation energy for process x (η, κ, Λ, or D)
ce	cohesive energy (in kJ mol^{-1})
ced	cohesive energy density (in MPa)
C_{P}	molar heat capacity at constant pressure (in J·K^{-1}·mol^{-1})
C_{v}	molar heat capacity at constant volume (in J·K^{-1}·mol^{-1})
c_{X}	molar concentration of species X (in M = mol·dm^{-3})
C^{z+}	a generalized cation
D	diffusion coefficient (in m^2·s^{-1})
d	interionic distance (in nm)
E	energy, molar energy (in J·mol^{-1})
e	elementary charge (1.6022 × 10^{-19} C)
$E_{\mathrm{T}}{}^{\mathrm{N}}$	normalized polarity index
F	Faraday constant (9.6485 × 10^4 C·mol^{-1})
G	Gibbs energy, molar Gibbs energy (in J·mol^{-1})

$g(r)$	pair correlation function
GP	Gordon parameter
H	enthalpy, molar enthalpy (in $J{\cdot}mol^{-1}$)
h	solvation (hydration) number
I	radiation scattering intensity
I	moment of inertia
k_B	Boltzmann constant (1.3807×10^{-23} $J{\cdot}K^{-1}$)
K_H	Henry's law constant
m	number of segments
M_X	molar mass of species X (in $kg{\cdot}mol^{-1}$)
N_A	Avogadro's number (6.0221×10^{23} mol^{-1})
N_{co}	coordination number
n_D	refractive index at the sodium D line
P	pressure (in Pa)
p	vapor pressure (in Pa)
$P^O{}_W$	octanol/water partition constant
P_σ	parachor
R	gas constant (8.3145 $J{\cdot}K^{-1}{\cdot}mol^{-1}$)
r_{corr}	linear correlation coefficient
R_D	molar refractivity at the sodium D line (in $m^3{\cdot}mol^{-1}$)
r_X	radius of particle of species X (in nm)
S	entropy, molar entropy (in $J{\cdot}K^{-1}{\cdot}mol^{-1}$)
$S(q)$	structure factor in q (reciprocal length) space
T	temperature (in K)
t	transference number
U	Potential interaction energy in the system (in J)
u	speed of sound (in $m{\cdot}s^{-1}$)
$u(r)$	potential energy function
V	volume, molar volume (in $m^3{\cdot}mol^{-1}$)
v_X	microscopic volume of species X (in nm^3)
x_X	mole fraction of species X
y	packing fraction
Z	lattice parameter
Z_X	atomic number of element X
z_X	charge number of ionic species X (taken algebraically)

Principal Greek Characters

Δ_x	change of quantity for process x
α	Kamlet-Taft hydrogen bond donation ability of solvent
α_P	isobaric thermal expansibility (in K^{-1})
α_X	polarizability of species X (in $m^{-3)}$
β	Kamlet-Taft electron pair donation ability of solvent
δ	chemical shift of NMR signal (in ppm)

δ_H	Hildebrand solubility parameter (in $MPa^{1/2}$)
ε, ε_s,	relative static permittivity
ε_0	permittivity of empty space (8.8542×10^{-12} $C^2 \cdot J^{-1} \cdot m^{-1}$)
η	dynamic viscosity (Pa·s)
θ_D, θ_E	Debye or Einstein temperature
κ	specific conductance ($S \cdot m^{-1}$)
κ_S, κ_T	adiabatic (isentropic), isothermal compressibility (in Pa^{-1})
Λ	molar conductivity (in $S \cdot cm^2 \cdot mol^{-1}$)
λ	wavelength of radiation
λ_{th}	thermal conductivity (in $W \cdot m^{-1} \cdot K^{-1}$)
ν_I	stoichiometric coefficient (number of ions I per formula)
π^*	Kamlet-Taft polarity/polarizability of solvent
ρ	density (in $kg \cdot m^{-3}$)
σ	surface tension (in $N \cdot m^{-1}$)
σ_I	softness parameter of ion I
Φ	fluidity (reciprocal of viscosity) (in Pa^{-1} s^{-1})
χ	molar (diamagnetic) susceptibility (in $m^3 \cdot mol^{-1}$)
ω	frequency of an electromagnetic wave (in s^{-1})

Principal Subscripts

+, −	of cation, anion
0	of a unit quantity
0	characteristic temperature in VFT equation
b	of boiling (at atmospheric pressure)
c	critical
els	contribution from electrostriction
f	of formation
g	of glass transition
h	of a hole
hyd	pertaining to hydration
I	pertaining to the ion I
int	internal
intr	intrinsic value
L, latt	of lattice
m	of melting
r	reduced quantity
s	of sublimation
sd	of solid
solv	of solvation
v	of vaporization
vd	of void space, of cavity

Principal Superscripts

* characteristic quantity
o standard thermodynamic function (for $T = 298.15$ K, for $P = 0.1$ MPa)

A chemical substance or ion is generally referred to in the text by its name or formula, but in tables and as subscripts abbreviations are generally employed. The symbol $I^{z\pm}$ denotes a generalized ion, C^+ or M^{z+} denote a cation, and A^- and X^{z-} denote an anion. The abbreviations cr for crystal, g for gas, ig for ideal gas, l for liquid denote the state of the ion or substance described. The common abbreviations of alkyl chains, Me = methyl, Et = ethyl, Pr = 1-propyl, Bu = 1-butyl, Pe = 1-pentyl, Hx = 1-hexyl, Oc = 1-octyl, Dc = 1-decyl, Do = 1-dodecyl, Td = 1-tetradecyl, and also Ph = phenyl are employed.

References

1. Bloom H (1967) The chemistry of molten salts. Benjamin, New York, 184 pp
2. Blander M (ed, multi-author) (1964) Molten salt chemistry. Interscience Publication, New York, 775 pp
3. Sundheim BR (ed, multi-author) (1964) Fused salts. McGraw-Hill, New York, 464 pp
4. Inman D, Lovering DG (eds, multi-author) (1981) Ionic liquids. Plenum, New York, 459 pp
5. Janz GJ (1967) Molten salt handbook. Academic, New York, 588 pp
6. Janz GJ, Allen CB, Bansal NP, Murphy RM, Tomkins RPT (1979) Physical properties data compilations relevant to energy storage. USA NBS, Washington, DC, 442 pp
7. Braunstein J, Mamantov G, Smith GP (eds, multi-author) (1971) Advances in molten salts chemistry, Plenum Press, New York, vol 1, 250 pp
8. Mamantov G, Mamantov CB, Braunstein J (eds, multi-author) (1987) Advances in molten salts chemistry, Plenum Press, New York, vol 6, 350 pp
9. Freemantle M (ed, multi-author) (2010) An introduction to ionic liquids. Royal Society of Chemistry, Cambridge, 281 pp
10. Gaune-Escard M, Seddon KR (eds, multi-author) (2010) Molten salts and ionic liquids: never the twain? Wiley, New York, 441 pp
11. Plechkova NV, Seddon KR (eds, multi-author) (2013) Ionic liquids uncoiled: critical expert overviews. Wiley, New York, 413 pp
12. Gaune-Escard M, Haarberg, GM (eds, multi-author) (2014) Molten salts chemistry and technology. Wiley, New York, 600 pp

Chapter 2
The Properties of Ions Constituting Ionic Liquids

Ions are defined as particles of atomic or molecular size that carry electrical charges. They may exist in gaseous phases as individual ions but in condensed phases (solids and liquids) they exist as electrically neutral combinations of positively charged cations and negatively charged anions. High temperature ionic liquids may consist of monatomic ions, such as Ca^{2+} or F^-, their anions (but hardly the cations) may also consist of a few atoms, such sulfate, SO_4^{2-}. At somewhat lower temperatures molten salts may consist of quaternary ammonium or phosphonium cations with a variety of anions, e.g., Bu_4N^+ BF_4^-. A class of ionic liquids, molten hydrated salts, on the contrary, consist of cations, such as $Mg(H_2O)_6^{2+}$, in which a monatomic cation is coordinated to water molecules. Finally, room temperature ionic liquids (RTILs) consist mostly of large organic cations, such as 1-ethyl-3-methylimidazolium, and polyatomic anions, such as bis(trifluoromethylsulfonyl)amide, $(CF_3SO_2)_2N^-$. The physical properties of ions are designated in this book by having the subscript $_I$ for the ion.

2.1 The Properties of Isolated Ions

Ions in an ideal gaseous state, termed isolated or bare ions, are devoid of interactions with other particles or their surroundings and are commonly monatomic or consist of relatively few atoms. Generalized ions are designated by $I^{z\pm}$, where z_I is an integral positive or negative number. The primary characteristics of isolated ions are the amount of electrical charge they carry and their mass. The amount of charge is given in terms of $z_I e$, where $e = 1.60218 \times 10^{-19}$ C is the elementary unit of the charge. The masses of ions are generally specified as their molar mass, M_I, that is, the mass of Avogadro's number, $N_A = 6.02214 \times 10^{23}$ mol^{-1}, of ions. The units of the molar mass, are kg·mol^{-1}. The molar masses of the ions constituting the ionic liquids dealt with in this book are shown in Table 2.1, and are, of course, additive,

Y. Marcus, *Ionic Liquid Properties*, DOI 10.1007/978-3-319-30313-0_2

Table 2.1 The names, formulas, charge numbers, z_I, and molar masses, M_I, of common ions

Cations	z_I	M_I /kg·mol^{-1}
Lithium, Li^+	+1	0.006941
Sodium, Na^+	+1	0.02294
Potassium, K^+	+1	0.03910
Rubidium, Rb^+	+1	0.08547
Cesium, Cs^+	+1	0.13291
Copper(I), Cu^+	+1	0.06355
Silver, Ag^+	+1	0.10787
Thallium(I), Tl^+	+1	0.20438
Beryllium, Be^{2+}	+2	0.009012
Magnesium, Mg^{2+}	+2	0.02431
Calcium, Ca^{2+}	+2	0.04008
Strontium, Sr^{2+}	+2	0.08762
Barium, Ba^{2+}	+2	0.13733
Vanadium(II), V^{2+}	+2	0.05094
Chromium(II), Cr^{2+}	+2	0.05201
Manganese(II), Mn^{2+}	+2	0.05494
Iron(II), Fe^{2+}	+2	0.05585
Cobalt, Co^{2+}	+2	0.05893
Nickel, Ni^{2+}	+2	0.05869
Copper(II), Cu^{2+}	+2	0.06355
Zinc, Zn^{2+}	+2	0.06539
Cadmiun, Cd^{2+}	+2	0.11241
Tin(II), Sn^{2+}	+2	0.11871
Lead, Pb^{2+}	+2	0.2072
Aluminum, Al^{3+}	+3	0.02698
Scandium, Sc^{3+}	+3	0.04496
Vanadium(III), V^{3+}	+3	0.05094
Chromium(III), Cr^{3+}	+3	0.05201
Iron(III), Fe^{3+}	+3	0.05585
Gallium, Ga^{3+}	+3	0.06971
Yttrium, Y^{3+}	+3	0.08891
Indium, In^{3+}	+3	0.11482
Lanthanum, La^{3+}	+3	0.13891
Cerium(III), Ce^{3+}	+3	0.14012
Praseodymium, Pr^{3+}	+3	0.14091
Neodymium, Nd^{3+}	+3	0.14424
Promethium, Pm^{3+}	+3	0.147
Samarium(III), Sm^{3+}	+3	0.15036
Europium(III), Eu^{3+}	+3	0.15197
Gadolinium, Gd^{3+}	+3	0.15725
Terbium, Tb^{3+}	+3	0.15893
Dysprosium, Dy^{3+}	+3	0.16251

(continued)

Table 2.1 (continued)

Cations	z_I	M_I /kg·mol^{-1}
Holmium, Ho^{3+}	+3	0.164936
Erbium, Er^{3+}	+3	0.16726
Thulium, Tm^{3+}	+3	0.16893
Ytterbium(III), Yb^{3+}	+3	0.17304
Lutetium, Lu^{3+}	+3	0.17497
Thallium(III), Tl^{3+}	+3	0.20438
Bismuth, Bi^{3+}	+3	0.20898
Zirconium, Zr^{4+}	+4	0.09122
Tin(IV), Sn^{4+}	+4	0.11871
Cerium(IV), Ce^{4+}	+4	0.14012
Hafnium, Hf^{4+}	+4	0.17849
Thorium, Th^{4+}	+4	0.23204
Uranium(IV), U^{4+}	+4	0.23803
Cations of low- and room-temperature melting salts	z_I	M/kg·mol^{-1}
Ammonium, NH_4^+	+1	0.01804
Hydroxylaminium, $HONH_3^+$	+1	0.03404
Hydrazinium, $H_2NNH_3^+$	+1	0.03305
Guanidinium, $C(NH_2)_3^+$	+1	0.06008
Tetraalkylammonium, $[H(CH_2)_n]_4N^+$	+1	$0.01803 + 0.05611n$
Tetraalkylphosphonium, $[H(CH_2)_n]_4P^+$	+1	$0.03500 + 0.05611n$
1-Alkyl-3-methylimidazolium ($C_n$mim), $[H(CH_2)_n]C_4H_6N_2^+$	+1	$0.08307 + 0.01403n$
1-Alkylpyridinium (C_nPy^+), $[H(CH_2)_n]C_5H_5N^+$	+1	$0.07911 + 0.01403n$
1-Alkyl-1-methylpyrrolidinium(C_nMePyrr), $[H(CH_2)_n]C_5H_{11}N^+$	+1	$0.08506 + 0.01403n$
Tetraphenylphosphonium,Ph_4P^+	+1	0.33939
Tetraphenylarsonium, Ph_4As^+	+1	0.38334
Cations of hydrated molten salts	z_I	M/kg·mol^{-1}
$M(H_2O)_n^{z+}$ (M is a monatomic cation)	$+z$	$M_I + 0.01802n$
Anions	z_I	M/kg·mol^{-1}
Fluoride, F^-	−1	0.01899
Chloride, Cl^-	−1	0.03545
Bromide, Br^-	−1	0.07991
Iodide, I^-	−1	0.12691
Hydroxide, OH^-	−1	0.01701
Hydrosulfide,SH^-	−1	0.03307
Cyanide, CN^-	−1	0.02602
Cyanate, NCO^-	−1	0.04203
Thiocyanate, SCN^-	−1	0.05808
Azide, N_3^-	−1	0.04202
Hydrogenfluoride, HF_2^-	−1	0.03901
Triiodide, I_3^-	−1	0.38071
Metaborate, BO_2^-	−1	0.04281
Nitrite, NO_2^-	−1	0.04601
Nitrate, NO_3^-	−1	0.06201

(continued)

Table 2.1 (continued)

Cations	z_I	M_I /kg·mol^{-1}
Chlorate, ClO_3^-	−1	0.08345
Bromate, BrO_3^-	−1	0.12761
Iodate, IO_3^-	−1	0.17491
Perchlorate, ClO_4^-	−1	0.09945
Permanganate, MnO_4^-	−1	0.11894
Pertechnetate, TcO_4^-	−1	0.16301
Perrhenate, ReO_4^-	−1	0.25002
Tetrafluoroborate, BF_4^-	−1	0.08681
Trifluoromethyltrifluoroborate, $CF_3BF_3^-$	−1	0.14281
Tetracyanoborate, $B(CN)_4^-$	−1	0.11488
Formate, HCO_2^-	−1	0.04502
Acetate, $CH_3CO_2^-$	−1	0.05904
Benzoate, $PhCO_2^-$	−1	0.12112
Chloroacetate, $CH_2ClCO_2^-$	−1	0.09349
Trifluoroacetate, $CF_3CO_2^-$	−1	0.11302
Methylsulfonate, $CH_3SO_3^-$	−1	0.09510
Trifluoromethylsulfonate, $CF_3SO_3^-$	−1	0.14906
4-toluenesulfonate (tosylate) $CH_3C_6H_4SO_3^-$	−1	0.17121
Tetraphenylborate, BPh_4^-	−1	0.31923
Hexafluorophosphate, PF_6^-	−1	0.14496
Trifluoro-tris(pentafluoroethyl)phosphate (FAP), $PF_3(C_2F_5)_3$	−1	0.44503
Hexafluoroantimonate, SbF_6^-	−1	0.23574
Hydrogensulfate, HSO_4^-	−1	0.09707
Methylsulfate, $CH_3SO_4^-$	−1	0.11110
Ethylsulfate, $C_2H_5SO_4^-$	−1	0.12512
Octylsulfate, $C_8H_{17}SO_4^-$	−1	0.20931
Tetrachloroaluminate, $AlCl_4^-$	−1	0.17179
Tetrachloroferrate, $FeCl_4^-$	−1	0.20066
Dicyanamide (DCA), $N(CN)_2^-$	−1	0.06604
Difluorosulfonylamide (FSA), $N(FSO_2)_2^-$	−1	0.18014
Bis(trifluoromethylsulfonyl)amide (NTF_2), $N(CF_3SO_2)_2^-$	−1	0.28015
Oxide, O^{2-}	−2	0.01600
Sulfide, S^{2-}	−2	0.03207
Carbonate, CO_3^{2-}	−2	0.06001
Sulfite, SO_3^{2-}	−2	0.08007
Sulfate, SO_4^{2-}	−2	0.09607
Selenate, SeO_4^{2-}	−2	0.14297
Chromate, CrO_4^{2-}	−2	0.11599
Molybdate, MoO_4^{2-}	−2	0.15994
Tungstate, WO_4^{2-}	−2	0.24785
Thiosulfate, $S_2O_3^{2-}$	−2	0.11212
Hexafluorosilicate, SiF_6^{2-}	−2	0.14208
Dichromate, $Cr_2O_7^{2-}$	−2	0.21599

(continued)

Table 2.1 (continued)

Cations	z_I	M_I /kg·mol^{-1}
Phosphate, PO_4^{3-}	−3	0.09497
Hexacyanoferrate(III), $Fe(CN)_6^{3-}$	−3	0.21195
Hexacyanocobaltate(III), $Co(CN)_6^{3-}$	−3	0.21611
Hexacyanoferrate(II), $Fe(CN)_6^{4-}$	−4	0.21195

multiplied by the appropriate stoichiometric coefficients ν_I, to make up the molar mass of the ionic liquid (molten salt).

Thermodynamic quantities that pertain to isolated ions in the ideal gas state (designated as 'ig') and to their formation from the elements in their standard states are well defined. The standard molar entropy and constant-pressure heat capacity, $S°(I^{z\pm},ig)$ and $C_P°(I^{z\pm},ig)$, of isolated ions are recoded in Table 2.2 in J·K^{-1}·mol^{-1} for the standard temperature $T° = 298.15$ K and pressure $P° = 0.1$ MPa. The standard molar Gibbs energy and the enthalpy of formation, $\Delta_f G°(I^{z\pm},ig)$ and $\Delta_f H°(I^{z\pm},ig)$, in kJ·mol^{-1} of many ions are also recorded in Table 2.2, The values of the Gibbs energies having been obtained from the enthalpies and entropies: $\Delta_f G°(I^{z\pm}, ig) = \Delta_f H°(I^{z\pm}, ig) - T°[S°(I^{z\pm}, ig) - \Sigma S°(\text{elements}, ig)]$ in a manner that is thermodynamically consistent.

The standard molar entropy of isolated monatomic ions with no unpaired electrons reflects the translational entropy alone and depends only on the mass of the ion. At $T° = 298.15$ K and $P° = 0.1$ MPa the standard molar entropy is then $S°(I^{z\pm}, ig) = 108.85 + \frac{3}{2}\ln(M/M°) \text{J} \cdot \text{K}^{-1} \cdot \text{mol}^{-1}$, where $M/M°$ is the relative molar mass ($M° = 1$ kg·mol^{-1}). The standard molar heat capacity of monatomic ions with no unpaired electrons depends on the translational degrees of freedom alone, hence is common to all the monatomic ions: $C_P°(I^{z\pm}, ig) = \frac{5}{2}R = 20.79 \text{ J} \cdot \text{K}^{-1} \cdot \text{mol}^{-1}$, where $R = 8.31451$ J·K^{-1}·mol^{-1} is the gas constant. Monatomic ions with unpaired electrons have a contribution from electronic spin to the heat capacity and entropy and polyatomic ions have contributions from their rotational and vibrational modes.

The standard molar volume of an isolated ion is a trivial quantity, being the same for all ions: $V°(I^{z\pm}, ig) = RT°/P° = 0.02479 \text{ m}^3 \cdot \text{mol}^{-1}$.

Values are provided for some ions not to be found in molten salts (e.g., H^+) and for a few cations relevant to low-melting salts and RTILs, but for most of the latter no data could be found, nor for most of the polyatomic anions relevant to the RTILs.

The shape of isolated monatomic ions is spherical, but ions that consist of several atoms may have any shape. The common ones are planar (NO_3^-, CO_3^{2-}), tetrahedral (NH_4^+, BF_4^-), octahedral (PF_6^{4-}), elongated (SCN^-), or more irregular, as they are for the ions constituting RTILs that generally have low symmetry (otherwise higher melting salts result). Tetrahedral and octahedral ions approximate spherical shape for many purposes and are termed globular. The size of an isolated ion cannot be specified readily, because its outer electron shell extends indefinitely around the inner ones and the nucleus. The sizes of ions in condensed liquid phases

Table 2.2 The standard molar entropy $S^{o}(I^{z\pm},ig)$ [1] and constant-pressure molar heat capacity $C_P^{o}(I^{z\pm},ig)$ [2] and the standard molar enthalpy $\Delta_f H^{o}(I^{z\pm},ig)$ and Gibbs energy $\Delta_f G^{o}(I^{z\pm},ig)$ of formation [3] of isolated ions at the standard temperature $T^{o} = 298.15$ K and pressure $P^{o} = 0.1$ MPa

Ion	$S^{o}(I^{z\pm},ig)$/ $J{\cdot}K^{-1}{\cdot}mol^{-1}$	$C_P^{o}(I^{z\pm},ig)$/ $J{\cdot}K^{-1}{\cdot}mol^{-1}$	$\Delta_f H^{o}(I^{z\pm},ig)$/ $kJ{\cdot}mol^{-1}$	$\Delta_f G^{o}(I^{z\pm},ig)$/ $kJ{\cdot}mol^{-1}$
H^+	108.9	20.8	1536.2	1523.2
Li^+	133.0	20.8	685.78	654.8
Na^+	148.0	20.8	609.36	580.5
K^+	154.6	20.8	514.26	487.3
Rb^+	164.4	20.8	490.1	464
Cs^+	169.9	20.8	457.96	432.7
Cu^+	161.1	20.8	1089.99	1051.9
Ag^+	167.4	20.8	1021.73	984.5
Tl^+	175.3	20.8	772.2	739
Be^{2+}	136.3	20.8	2993.23	2955.4
Mg^{2+}	148.7	20.8	2348.5	2300.3
Ca^{2+}	154.9	20.8	1925.9	1892.1
Sr^{2+}	164.7	20.8	1790.54	1757
Ba^{2+}	170.4	20.8	1660.38	1628.3
Ra^{2+}	176.6	20.8	1659.79	
V^{2+}	182.1	29.6	2590.86	2545.2
Cr^{2+}	181.7	31.2	2655.71	2592
Mn^{2+}	173.8	20.8	2519.69	2477.4
Fe^{2+}	180.3	25.9	2749.93	2689.6
Co^{2+}	179.5	22.9	2844.2	2785.7
Ni^{2+}	178.1	21.7	2931.39	2873.4
Cu^{2+}	176.0	20.8	3054.07	3011.5
Zn^{2+}	161.1	20.8	2782.78	2747.2
Cd^{2+}	167.8	20.8	2623.54	2588.9
Sn^{2+}	168.5	20.8	2434.8	2399.9
Pb^{2+}	175.5	20.8	2373.33	2328.6
Al^{3+}	150.2	20.8	5483.17	5446.9
Sc^{3+}	156.4	20.8	4652.31	4616
V^{3+}	177.8	20.8	5424.6	5380.2
Cr^{3+}	179.0	20.8	5648.4	5602.1
Fe^{3+}	174.0	20.8	5712.8	5669.1
Ga^{3+}	161.9	20.8	5816.6	5780.5
Y^{3+}	164.9	20.8	4199.86	4163.9
In^{3+}	168.1	20.8	5322	5289.1
Sb^{3+}	169.0	20.8	5151	5114
La^{3+}	170.5	20.8	3904.9	3871
Ce^{3+}	185.5	20.8	3970.6	3936.8
Pr^{3+}	188.9	20.8	4005.8	3971.3

(continued)

Table 2.2 (continued)

Ion	$S°(I^{z\pm},ig)$/ $J·K^{-1}·mol^{-1}$	$C_P°(I^{z\pm},ig)$/ $J·K^{-1}·mol^{-1}$	$\Delta_f H°(I^{z\pm},ig)$/ $kJ·mol^{-1}$	$\Delta_f G°(I^{z\pm},ig)$/ $kJ·mol^{-1}$
Nd^{3+}	191.6	20.8	4050	4014.6
Pm^{3+}	191.0	20.8	4079 [4]	4044
Sm^{3+}	189.0	20.8	4100	4065.1
Eu^{3+}	181.0	20.8	4230	4199.4
Gd^{3+}	189.3	20.8	4163	4126.9
Tb3+	193.5	20.8	41967	41614.1
Dy^{3+}	195.5	20.8	4205	4169
Ho^{3+}	196.2	20.8	4243	4207
Er^{3+}	195.9	20.8	4268	4231.4
Tm^{3+}	194.3	20.8	4297	4261.2
Yb^{3+}	190.5	20.8	4367.3	4328.3
Lu^{3+}	173.4	20.8	4350	4293.5
Tl^{3+}	175.3	20.8	5639.2	5606.1
Bi^{3+}	175.9	20.8	5004	4968.1
U^{3+}	195.5	20.8	4176 [5]	4132.4
Zr^{4+}	165.2	20.8	8261 [4]	8223.4
Sn^{4+}	168.5	20.8	9320.7	9285.8
Ce^{4+}	170.6	20.8	7523	7493.6
Hf^{4+}	173.6	20.8	8192	8153.2
Th^{4+}	176.9	20.8	7021	6984.2
U^{4+}	186.4	20.8	7327 [5]	7286.4
H_3O^+	192.8	34.9	570.7 [6]	602.2
NH_4^+	186.3	34.9	630 [4]	681
$C(NH_2)_3^+$	264.5 [8]	77.9 [7]	462 [7]	
Me_4N^+	331.9 [2]	109.6	537 [9]	73.4
Et_4N^+	483	201 [10]	411 [7]	712
Pr_4N^+	641	294 [10]	307 [7]	
Bu_4N^+		386 [10]	221 [7]	
Ph_4P^+	651.0 [2]	366.3		
Ph_4As^+	650.0 [2]	368.6		
$Etmim^+$			243 [11]	
$Bumim^+$			195 [11]	
F^-	145.6	20.8	−255.39	−268.6
Cl^-	154.4	20.8	−233.13	−241.4
Br^-	163.6	20.8	−219.07	−245.1
I^-	169.4	20.8	−197	−230.2
OH^-	172.3	29.1	−143.5	−144.8
SH^-	186.2	29.1	−120 [12]	−146.6
CN^-	196.7	29.1	36 [12]	−59.2

(continued)

Table 2.2 (continued)

Ion	$S°(I^{z\pm},ig)$/ $J·K^{-1}·mol^{-1}$	$C_P°(I^{z\pm},ig)$/ $J·K^{-1}·mol^{-1}$	$\Delta_f H°(I^{z\pm},ig)$/ $kJ·mol^{-1}$	$\Delta_f G°(I^{z\pm},ig)$/ $kJ·mol^{-1}$
NCO^-	218.9	38.0	−192 [4]	−196.4
SCN^-	232.5	43.2	−49	
N_3^-	212.25	37.9	180.7	203.1
HF_2^-	211.3	34.0	−683 [4]	−666
I_3^-	334.7	61.7	−482 [4]	−529.9
BO_2^-	215.8	38.4	−667 [3]	−668
NO_2^-	236.2	37.1	−202 [12]	−182.7
NO_3^-	245.2	44.7	−320 [12]	−272.8
ClO_3^-	264.3	57.6	−200 [12]	−153.8
BrO_3^-	278.7	60.4	−145 [12]	−113.7
IO_3^-	288.2	62.0	−208 [12]	−184.9
ClO_4^-	263.0	62.0	−344 [12]	−266.8
MnO_4^-	277.8	72.4	−723.8	−674.8
ReO_4^-	284.1	74.7	−976 [2]	−930.4
BF_4^-	267.9	67.8	−1687 [4]	−1644.2
HCO_2^-	238.2	38.8	−460 [13]	−452.7
$CH_3CO_2^-$	278.2	61.4	−504.2 [13]	−464.1
$PhCO_2^-$	338.0 [8]	117.9	−400.4 [13]	−330.6
$CF_3CO_2^-$	331.0 [2]	89.3	−1194 [4]	−1138
$CF_3SO_3^-$	346.0 [2]	107.6		
BPh_4^-	656.0 [2]	363.7 [2]		
HSO_4^-	283.0	70.9	−953 [4]	−886.1
PF_6^-	299.6 [8]	104.7	−2109.9 [4]	−2005.6
SbF_6^-	345.5 [8]	124.0	−1993 [4]	−1901
$AlCl_4^-$		205 [14]	−1159 [15]	
$N(CN)_2^-$			161.5 [11][a]	
$N(CF_3SO_2)_2^-$			−2020.5 [11][a]	
O^{2-}	143.3	20.8	950 [4]	939.7
S^{2-}	152.1	20.8		
CO_3^{2-}	246.1	44.4	−321 [12]	−300.9
SiO_3^{2-}		57.6 [14]		
$Si_2O_5^{2-}$		101.9 [14]		
SO_3^{2-}	264.3	52.6		−1035.5
SO_4^{2-}	263.6	62.4	−758 [12]	−704.8
SeO_4^{2-}	281.2	73.5		
CrO_4^{2-}	281.4	74.8	−705 [12]	−659.5
MoO_4^{2-}	291.1	77.0		
WO_4^{2-}	296.6	76.6		
$S_2O_3^{2-}$	291.1	71.0		
SiF_6^{2-}	309.9	113.1	−2161 [4]	−2183.4
$Cr_2O_7^{2-}$	379.7	140.8		

(continued)

Table 2.2 (continued)

Ion	$S°(I^{z\pm},ig)$/ $J·K^{-1}·mol^{-1}$	$C_P°(I^{z\pm},ig)$/ $J·K^{-1}·mol^{-1}$	$\Delta_f H°(I^{z\pm},ig)$/ $kJ·mol^{-1}$	$\Delta_f G°(I^{z\pm},ig)$/ $kJ·mol^{-1}$
PO_4^{3-}	266.4	65.4		
$Fe(CN)_6^{3-}$	491.6 [2]	217.1		
$Co(CN)_6^{3-}$	464.8	210.2		
$Fe(CN)_6^{4-}$	469.8	210.6		

References are provided for data not found in those given in this caption
[a]When the value for $\Delta_f H°(Cl^-,ig)$ in this Table is used as a reference, but when the value for $\Delta_f H°(NO_3^-,ig)$ is used it yields values 31 kJ mol^{-1} more negative for anions, more positive for cations)

(molten salts and RTILs), however, are generally taken as being the same as those in crystals, in which they may be determined by x-ray or neutron diffraction methods. These sizes, expressed as ionic radii or ionic volumes, are dealt with below and in the following chapters.

Some other properties of isolated ions have been determined: the magnetic susceptibility, the polarizability, and the softness/hardness. These properties of isolated ions are portable and additive. This means that these properties of ions are not appreciably sensitive to the environment of the ions, whether they are isolated, in crystalline compounds, or in molten salts. The property of a compound is the sum of the stoichiometrically weighted properties of its constituting cations and anions. These values for many ions were critically selected and reported by Marcus [4].

The molar magnetic susceptibilities, $\chi_{\rm Im}$, are rather insensitive to the environment in which the ions are situated, whether isolated or in a condensed phase. Ions are diamagnetic, that is, they are repulsed out from a magnetic field, unless they have one or more unpaired electrons in their electronic shells. The consequence is that most RTILs are diamagnetic, and only a few RTILs have been prepared with paramagnetic anions, such as $FeCl_4^-$. The $\chi_{\rm Im}$ values of diamagnetic ions have the dimension of a molar volume and are negative, ranging from a few to several tens of the unit -10^{-12} $m^3·mol^{-1}$. Ions that have n unpaired electrons in their electronic shells are paramagnetic, are attracted into a magnetic field, and have positive molar susceptibilities. A paramagnetic ion with n unpaired electrons has $\chi_{\rm Im} = +1.676n(n+2) \times 10^{-9}m^3$ $·mol^{-1}$ at $T° = 298.15$ K. The values of $\chi_{\rm Im}$ for many ions are shown in Table 2.3.

The polarizibility, $\alpha_{\rm I}$, of an ion is obtained indirectly from the molar refractivity, but the molar refractivity can be obtained experimentally only for neutral species (salts), either in crystals or in dilute solutions. The molar refractivity $R_{\rm D}$ of a salt is generally determined from the refractive index at the sodium D line (589 nm), $n_{\rm D}$, according to the Lorenz-Lorentz expression: $R_{\rm D} = V\left(n_D^2 - 1\right)/\left(n_D^2 + 2\right)$, where $V = (M/\rho)$ is the molar volume and M and ρ are the molar mass and the density. The molar refractivity of a salt is not very sensitive to the environment of the ions, so that it may be ascribed to the neutral combinations of the bare cations and anions. It must then be split appropriately between the cations and the anions, but no

Table 2.3 The magnetic susceptibility, χ_{Im}, the molar refraction at the sodium D line, R_D (normalized to the value for Na^+), the polarizability α and the softness parameter σ of isolated ions

Cation	χ_{Im} 10^{-12} $m^3{\cdot}mol^{-1}$	R_D 10^{-6} $m^3{\cdot}mol^{-1}$	α 10^{-3} nm^3	σ
H^+	−6.6[a]	−0.3	−0.12	−0.30
Li^+	−3[a]	0.08	0.03	−1.32
Na^+	−2.3[a]	0.65	0.26	−0.90
K^+	−11.2[a]	2.71	1.07	−0.88
Rb^+	−20.1[a]	4.1	1.63	−0.83
Cs^+	−34[a]	6.89	2.73	−0.84
Cu^+	−12	3.1	1.23	−0.52
Ag^+	−24	5.1	2.02	−0.12
Tl^+	−34	11.5	4.56	−0.10
Be^{2+}	−6			−0.93
Mg^{2+}	−5 [17]	−0.7	−0.28	−0.71
Ca^{2+}	−8 [17]	1.59	0.63	−0.96
Sr^{2+}	−9[a]	2.65	1.05	−0.94
Ba^{2+}	−21.5[a]	5.17	2.05	−0.96
Mn^{2+}	20.7	2.2	0.87	−0.45
Fe^{2+}	19.6	2.1	0.83	−0.46
Co^{2+}	18.5	2.05	0.81	−0.41
Ni^{2+}	17.5	1.6	0.63	−0.41
Cu^{2+}	16.4	1.3	0.52	+0.08
Zn^{2+}	−10	1.39	0.55	+0.05
Cd^{2+}	−22.5	3.22	1.28	+0.28
Sn^{2+}	−20			−0.01
Pb^{2+}	−28	11.9	4.72	+0.11
Al^{3+}	−3.1	−1.18	−0.47	−0.61
Sc^{3+}	−6 [18]	1.6	0.63	−0.92
V^{3+}	10 [18]	3.5[d]	1.39	−0.59
Cr^{3+}	16			−0.40
Fe^{3+}	15.6	3.2	1.27	+0.03
Ga^{3+}	−8	5 [17]	2	−0.01
Y^{3+}	−12 [18]	2.4	0.95	−0.99
In^{3+}	−19	1.7 [17]	0.67	+0.18
Sb^{3+}	−14 [18]	8.8[d]	3.49	+0.33
La^{3+}	−20 [18]	2.74	1.09	−1.05
Ce^{3+}	20 [18]	3.4 [17]	1.35	−1.02
Pr^{3+}	20 [18]	3.3 [17]	1.31	−0.93
Nd^{3+}	20 [18]	3.1 [17]	1.23	−0.88
Sm^{3+}	20 [18]	2.9 [17]	1.15	−0.66
Eu^{3+}	20 [18]	2.7 [17]	1.07	−0.49
Gd^{3+}	20 [18]	2.6 [17]	1.03	−0.96
Tb^{3+}	19 [18]	2.5 [17]	0.99	−0.94
Dy^{3+}	19 [18]	2.4 [17]	0.95	−0.80

(continued)

Table 2.3 (continued)

Cation	χ_{Im} 10^{-12} $m^3 \cdot mol^{-1}$	R_D 10^{-6} $m^3 \cdot mol^{-1}$	α 10^{-3} nm^3	σ
Ho^{3+}	19 [18]	2.2 [17]	0.87	−0.83
Er^{3+}	18 [18]	2.1 [17]	0.83	−0.87
Tm^{3+}	18 [18]	2.0 [17]	0.79	−0.73
Yb^{3+}	18 [18]	2.0 [17]	0.79	−0.57
Lu^{3+}	−17 [18]			−0.84
Tl^{3+}	−31	2.2 [17]	0.87	+0.77
Bi^{3+}	−25	8.1	3.21	+0.52
Zr^{4+}	−12.5	1.0 [17]	0.40	−0.73
Sn^{4+}	−16	1.3 [17]	0.52	+0.26
Ce^{4+}	−21	1.9 [17]	0.75	−0.40
Hf^{4+}	−16 [18]	4.3 [19]	1.70	−0.91
Th^{4+}	−31.2	6.8 [19]	2.70	−0.97
U^{4+}	−35 [25]			−0.46
NH4+	−11.5	4.7	1.86	−0.90
$C(NH_2)_3+$		11.21 [7]	4.44	
Me_4N+	−65	22.9	9.08	+0.11
Et_4N+		43	17	
Pr_4N+		61	24	
Bu_4N+		79	31	
Ph_4As+	−229	115.3	45.7	+6.61
C_1mim^{+} [b]			10.98 [20][b]	
C_nmim^+		$21.6+4.7n$[c]		
C_nPy^+		$19.9+5.0n$[c]		
RR'_3N^+		$4.90n_{C\ total}$[c]		
$TdHx_3P^+$		158.2[c]		

Anion	χ_{Im} 10^{-12} $m^3 \cdot mol^{-1}$	R_D 10^{-6} $m^3 \cdot mol^{-1}$	$\alpha(I^{z\pm},ig)$ 10^{-3} nm^3	σ
F^-	−13[a]	2.21	0.88	−0.36
Cl^-	−28[a]	8.63	3.42	+0.21
Br^-	−39[a]	12.24	4.85	+0.47
I^-	−56.7[a]	18.95	7.51	+0.80
OH^-	−12[a]	4.65	1.84	+0.30
SH^-		12.8	5.07	+0.95
CN^-	−18	7.9	3.13	+0.71
NCO^-	−21			+1.01
SCN^-	−35	17	6.7	+1.15
N_3^-		11	4.4	+1.06
HF_2^-				−1.54
BO_2^-				−0.64
NO_2^-	−15[a]	8.7	3.45	+0.45
NO_3^-	−23[a]	10.43	4.13	+0.33
ClO_3^-	−32	12.1	4.80	+0.33
BrO_3^-	−40	15.2	6.03	

(continued)

Table 2.3 (continued)

Anion	χ_{Im} 10^{-12} $m^3 \cdot mol^{-1}$	R_D 10^{-6} $m^3 \cdot mol^{-1}$	$\alpha(I^{z\pm},ig)$ 10^{-3} nm^3	σ
IO_3^-	−50	18.85	7.47	
ClO_4^-	−34	12.77	5.06	0.00
MnO_4^-				+0.16
ReO_4^-	−60 [18]			−0.10
BF_4^-	−39	7.5[c]	2.80 [20]	0.00
PF_6^-		11.0[c]	4.18	
HCO_2^-	−21	9.43 [17]	3.74	−0.03
$CH_3CO_2^-$	−32.4	13.87	5.50	+0.08
$CF_3CO_2^-$	−50		5.23 [20]	
$CH_3SO_3^-$			6.30 [20]	
$CF_3SO_3^-$		17.7[c]	6.88 [20]	
$MePhSO_3^-$			15.55 [20]	
$N(CN)_2^-$		17.5[c]	6.11 [20]	
$B(CN)_4^-$		27.5[c]		
NTF_2^- [d]		34.8[c]	13.59 [20]	
$MeSO_4^-$ [c]			6.94[c] [20]	
BPh_4^-	−215	108.7	43.1	+7.16
HSO_4^-	−37 [17]			
O^{2-}	−12			
S^{2-}	−38			+1.39
CO_3^{2-}	−34[a]	11.45	4.54	−0.20
SO_3^{2-}	−38	12.9	5.11	−0.04
SO_4^{2-}	−40	13.79	5.47	−0.08
SeO_4^{2-}	−51	16.4	6.50	
CrO_4^{2-}	−51	27.5 [19]	10.90	
MoO_4^{2-}	−55	25.2 [19]	9.99	
WO_4^{2-}	−61	23.2 [19]	9.20	
$S_2O_3^{2-}$	−49	23.2	9.20	
SiF_6^{2-}		11.01 [19]	4.36	
SO_3^{2-}	−38	12.9	5.11	−0.04
SO_4^{2-}	−40	13.79	5.47	−0.08
SeO_4^{2-}	−51	16.4	6.50	
CrO_4^{2-}	−51	27.5 [19]	10.90	
PO_4^{3-}	−50	15.1	5.99	−0.48
$Fe(CN)_6^{3-}$		50.7 [19]	20.1	+3.52
$Co(CN)_6^{3-}$		46.7 [19]	18.5	
$Fe(CN)_6^{4-}$				+3.93

References are shown for data not selected in [4]

[a] In aqueous solutions

[b] Also for other $C_n$mim$^+$ (1-alkyl-3-methylimidazolium) cations $9.20 + 1.78n$

[c] Also for other $C_nSO_4^-$ anions $5.15 + 1.79n$

[d] bis(trifluoromethylsulfonyl)amide

[c] From values for room temperature ionic liquids, Table 6.10

theoretically valid way to do this is known. Therefore, an empirical expedient has to be resorted to, assuming the additivity of the stoichiometrically weighted values for the cations and anions, $R_D = \Sigma \nu_I R_{DI}$. The establishment of an ionic scale of polarizabilities has been made on the basis of the empirical value $R_D(Na^+) = 0.65$ cm^3 ·mol^{-1} at 25 °C. The polarizability of an isolated ion is then: $\alpha(I^{z\pm}, ig) = 3R_{ID}/4\pi N_A = R_{ID}/2.5227 \times 10^{30}$ (for R_{DI} in cm^3·mol^{-1}), yielding values of the order of 10^{-3} nm^3·particle^{-1}. The temperature coefficient of R_D is rather small, approximately +0.01 cm^3·mol^{-1}·K^{-1}. Values of R_{ID} and α are shown in Table 2.3 for many ions. Negative values of α and R_{ID} for very small ions (H^+, Mg^{2+}, and Al^{3+}) should be understood only in connection with their effects on the overall polarizability of compounds containing them.

A property of 'softness' or 'hardness' can be ascribed to an ion and is loosely related to its polarizability. The originally [16] softness parameters σ_I were based on the arbitrary assignment of zero to the hydrogen ion for cations and to the hydroxide ion for anions. A common scale for ions of both charge signs is produced when $0.3z_I$ units are subtracted/added from/to the original cation/anion values. Positive values of the softness parameter denote 'soft' ions and negative values denote 'hard' ions. The σ_I values of many ions are recorded in Table 2.3.

2.2 The Sizes of Ions in Condensed Phases

The sizes of isolated ions are ill-defined, as mentioned above, but due to the strong repulsion of the contagious electronic shells of ions in condensed phases the ions can be assigned definite sizes. The inter-ionic distances in crystals can be measured by x-ray and neutron diffraction with an uncertainty of a fraction of a pm. An early discussion of the application of diffraction methods to molten salts is that of Levy and Danford [21], a more recent review of this subject (concerning neutron diffraction) is that of Neilson and Adya [22] and still more recently also applications to RTILs have been added by Neilson et al. [23]. Diffraction measurements yield the pair distribution function, the peaks in the curves denoting the distances between pairs of atoms, as dealt with in detail in Chap. 3. However, only for monatomic and globular ions may individual ionic radii be assigned unequivocally, provided the additivity of ionic radii is assured and the radius of a key ion is assigned. These individual ionic radii, r_I, do depend on the coordination numbers of the ions and a set for the characteristic coordination in salt crystals has been established by Shannon and Prewitt [24, 25]. These radii, as selected in [4] and annotated there, are listed in Table 2.4.

In molten salts (typically, alkali metal halides) the cation-anion distances are ~5 % shorter and the anion-anion distances are ~5 % longer than in salt crystals at the melting point, but are very near the distances in the crystals at room temperature, the coordination numbers being 10 % lower in the molten salts than in the

Table 2.4 Ionic radii, r_I [4], thermochemical ionic radii, r_{Ith} [26], and ionic volumes v_I [27], from these references except where otherwise noted

Ion	r_I/pm	r_{Ith}/pm	v_I/ nm^3
Li^+	69		0.00199
Na^+	102		0.00394
K^+	138		0.00986
Rb^+	149		0.01386
Cs^+	170		0.01882
Cu^+	96		
Ag^+	115		
Tl^+	150		
Be^{2+}	35		0.0002 [29]
Mg^{2+}	72		0.00199
Ca^{2+}	100		0.00499
Sr^{2+}	113		0.00858
Ba^{2+}	136		0.01225
V^{2+}	79		0.0016 [29]
Cr^{2+}	82		0.0024 [29]
Mn^{2+}	83		0.0032 [29]
Fe^{2+}	78		0.0022 [29]
Co^{2+}	75		0.0022 [29]
Ni^{2+}	69		0.0020 [29]
Cu^{2+}	73		
Zn^{2+}	75		0.0024 [29]
Cd^{2+}	95		0.0046 [29]
Sn^{2+}	93		
Pb^{2+}	118		0.0069 [29]
Al^{3+}	53		0.0008 [29]
Sc^{3+}	75		0.0024 [29]
V^{3+}	64		0.0012 [29]
Cr^{3+}	62		0.0011 [29]
Fe^{3+}	65		0.0013 [29]
Co^{3+}	65		0.0011 [29]
Ga^{3+}	62		0.0010 [29]
Y^{3+}	90		0.0031 [29]
In^{3+}	79		0.0021 [29]
Sb^{3+}	77		0.0019 [29]
La^{3+}	105		0.0076 [29]
Ce^{3+}	101		0.0069 [29]
Pr^{3+}	100		0.0065 [29]
Nd^{3+}	99		0.0064 [29]
Pm^{3+}	97		
Sm^{3+}	96		0.0060 [29]
Eu^{3+}	95		0.0060 [29]
Gd^{3+}	94		0.0057 [29]

(continued)

Table 2.4 (continued)

Ion	r_I/pm	r_{Ith}/pm	v_I/ nm^3
Tb^{3+}	93		0.0054 [29]
Dy^{3+}	91		0.0051 [29]
Ho^{3+}	90		0.0049 [29]
Er^{3+}	89		0.0047 [29]
Tm^{3+}	88		0.0047 [29]
Yb^{3+}	87		0.0042 [29]
Lu^{3+}	86		0.0041 [29]
Tl^{3+}	88		0.0048 [29]
Bi^{3+}	102		
U^{3+}	104		
Zr^{4+}	72		0.0028 [29]
Sn^{4+}	69		0.0017 [29]
Ce^{4+}	80		0.0045 [29]
Hf^{4+}	71		0.0025 [29]
Th^{4+}	100		0.0056 [29]
U^{4+}	97		0.0049 [29]
NH_4^+	148		0.021
$C(NH_2)_3^+$	210 [7]		
Me_4N^+	280	234	0.113
Et_4N^+	337		0.199
Pr_4N^+	379		
Bu_4N^+	413		
Pe_4N^+	443		
Ph_4P^+	424		
Ph_4As^+	425		
F^-	133	126	0.025
Cl^-	181	168	0.047
Br^-	196	190	0.056
I^-	220	211	0.072
OH^-	133	152	0.032
SH^-	207	191	0.057
CN^-	191	187	0.050
NCO^-	203	193	0.054
SCN^-	213	209	0.071
N_3^-	195	180	0.058
HF_2^-	172	172	0.047
I_3^-	470	272	0.180 [29]
BO_2^-	240		
NO_2^-	192	187	0.055
NO_3^-	200	200	0.064
ClO_3^-	200	208	0.073
BrO_3^-	191	214	0.072

(continued)

Table 2.4 (continued)

Ion	r_I/pm	r_{Ith}/pm	v_I/ nm^3
IO_3^-	181	218	0.075
ClO_4^-	240	225	0.082
MnO_4^-	240	220	0.088
ReO_4^-	260	227	0.098 [29]
BF_4^-	230	205	0.073
HCO_2^-	204	200	0.056
$CH_3CO_2^-$	232	194	
BPh_4^-	421		
HSO_4^-	190	221	0.087 [29]
PF_6^-	245	242	0.109
SbF_6^-	282	252	0.121
O^{2-}	140	141	0.043
S^{2-}	184	189	0.067
CO_3^{2-}	178	189	0.061
SO_3^{2-}	200	204	0.071
SO_4^{2-}	230	218	0.091
SeO_4^{2-}	243	229	0.103
CrO_4^{2-}	240	229	0.097
MoO_4^{2-}	254	231	0.088
WO_4^{2-}	270	237	0.088
$S_2O_3^{2-}$	250	251	0.104
SiF_6^{2-}	259	248	0.112
$Cr_2O_7^{2-}$	320	292	0.167
PO_4^{3-}	238	230	0.090
AsO_4^{3-}	248	237	0.088
$Fe(CN)_6^{3-}$	440	347	0.265
$Co(CN)_6^{3-}$	430	349	0.263
$Fe(CN)_6^{4-}$	450		

crystals. Therefore the concept of an ionic radius in molten salts, equating it to that in crystals, should be used with due caution.

The ionic radius has a much more vague meaning for non-spherical ions. An approximate value of r_I is the cube root of ($3/4\pi$) times the volumes of their ellipsoids of rotation with axes a and b. For elongated rod-like ions, $a > b$ and their radii are $r_I \approx (4\pi a^2 b/3)^{1/3}$, examples being SCN^- and I_3^-. For planar ions, $a < b$ and their radii are $r_I \approx (4\pi ab^2/3)^{1/3}$, examples being NO_3^- and CO_3^{2-}. The sizes of polyatomic (non-globular) ions in crystals are also expressed by their thermochemical radii r_{Ith} according to Jenkins and co-workers [26]. The uncertainties of these radii, listed in Table 2.4, are ±20 pm for uni- and divalent anions increasing to twice this amount for trivalent ones. These values are based on the Goldschmidt radii for the alkali metal counter-ions, r_+^G, rather than the Shannon-Prewitt ones [24, 25].

A different approach to the sizes of ions is to consider their volumes rather than their radii [27]. The volume of a formula unit of the salt M_pX_q, obtained from the unit cell volume from x-ray diffraction of crystals of the salt, is assumed to be additive in the individual ionic volumes: $v(M_pX_q) = pv_+ + qv_-$. The Goldschmidt radii of the alkali metal cations were used to define $v_+ = (4\pi/3)r_+^{G\,3}$, obtaining the anion volumes by difference. Individual ionic volumes calculated on this basis by Glasser and Jenkins are also shown in Table 2.4 [28]. They differ considerably from previously reported values [17, 27] by the same authors. Protonated anions, such as HSO_4^-, are smaller than their non-protonated parent anions, such as SO_4^{2-}, because the lowered negative charge diminishes the repulsion between, the electrons, causing shrinkage of the anionic volume, whereas the added proton does not contribute appreciably to the volume.

The relationship between the ionic radii r_I and the ionic volumes v_I leads directly to $v_I = (4\pi/3)(r_I)^3$ and a set of such volumes, denoted as v_M is reported in [17]. The ionic volumes obtained from the formula unit volumes in crystals [28], denoted as v_J, is also shown in [17], with $v_J = (1.258 \pm 0.016)v_M$ having been established with a correlation coefficient of 0.995 for 55 cations. The larger size of v_J takes into account the void spaces between the (more or less spherical) ions. Another way to express this, based on the suggestion of Mukerjee [29] regarding such ions in solution, is to use the factor $k = 1.213$ multiplying the radius of univalent monatomic ions: $v_I = (4\pi/3)(kr_I)^3$. This factor is near the value, 1.159, that is geometrically required for close-packed spheres of arbitrary but comparable sizes.

The ionic volumes of the constituents of room temperature ionic liquids are derived from the densities, hence molar volumes, of these RTILs, using the ionic volumes of anions that were established for crystals [28]. These ionic volumes are presented and discussed in Chap. 6 in relationship with the densities of the RTILs [30].

References

1. Marcus Y, Loewenschuss A (1985) Standard entropies of hydration of ions. Annu Rep 81C:81–135
2. Loewenschuss A, Marcus Y (1987) Standard thermodynamic functions of gaseous polyatomic ions at 100 to 1000 K. J Phys Chem Ref Data 16:61–89
3. Wagman DD, Evans WH, Parker VB, Schumm RH, Halow I, Bailey SM, Churney L, Nuttall RL (1982) The NBS tables of chemical thermodynamic properties. J Phys Chem Ref Data 11: (Suppl. 2)
4. Marcus Y (1997) Ion properties. Dekker, New York, and references therein
5. Bratsch SB, Lagowski JL (1986) Actinide thermodynamic predictions. 3. Thermodynamics of compounds and aquo-ions of the 2+, 3+, and 4+ oxidation states and standard electrode potentials at 298.15 K. J Phys Chem 90:307–312; (1987) Predicted and experimental standard electrode potentials in liquid ammonia at 25°C. J Solution Chem 16:583–601
6. Olivella S, Urpi F, Vilarrasa J (1984) Evaluation of MNDO calculated proton affinities. J Compt Chem 5:230–236
7. Marcus Y (2012) The guanidinium ion. J ChemThermodyn 48:70–74

8. Marcus Y, Loewenschuss A (1996) Standard thermodynamic functions of some additional isolated ions at 100–1000 K. J Phys Chem Ref Data 25:1495–1507
9. Marcus Y (2014) The enthalpy of formation of gaseous tetra-*n*-propylammonium cations. J Chem Thermodyn 71:196–199
10. Abraham MH, Marcus Y (1986) The thermodynamics of solvation of ions. Part 1. The heat capacities of hydration at 298.15 K. J Chem Soc Faraday Trans 82:3255–3274
11. Verevkin SP, Emel'yanenko VN, Zaitsau DH, Heintz A, Muzny CD, Frenkel M (2010) Thermochemistry of imidazolium-based ionic liquids: experiment and first-principles Calculations. Phys Chem Chem Phys 12:14994–15000
12. Dasent WE (1982) Inorganic energetics, 2nd edn. Cambridge University Press, Cambridge, UK
13. Caldwell G, Renneboog R, Kebarle P (1989) Gas-phase acidities of aliphatic carboxylic acids, based on measurements of proton-transfer equilibria. Can J Chem 67:611–618
14. Glasser L, Jenkins HBD (2012) Single-ion heat capacities, Cp(298)ion, of solids: with a novel route to heat-capacity estimation of complex anions. Inorg Chem 51:6360–6366
15. Wood RH, D'Oratio LA (1966) Lattice energy of Na tetrachloroaluminate and the heat of formation of the tetrachloroaluminate ion. Inorg Chem 5:682–894
16. Marcus Y (1986) On enthalpies of hydration, ionization potentials, and the softness of ions. Thermochim Acta 104:389–394
17. Marcus Y, Jenkins HBD, Glasser L (2002) Ion volumes – a comparison. J Chem Soc Dalton Trans 3795–3798
18. Selwood PW (1956) Magnetochemistry, 2nd edn. Interscience, New York, p 403
19. Salzmann JJ, Jørgensen CK (1968) Molar refraction of aquo ions of metallic elements and evaluation of light refraction measurements in inorganic chemistry. Helv Chim Acta 51:1276–1293
20. Bica K, Deetlefs M, Schröder C, Seddon KR (2013) Polarisabilities of alkylimidazolium ionic liquids. Phys Chem Chem Phys 15:2703–2711
21. Levy HA, Danford MD (1964) Diffraction studies of the structure of molten salts. In: Blander M (ed) Molten salt chemistry. Interscience, New York, pp 109–125
22. Neilson GW, Adya AK (1997) Neutron diffraction studies of liquids. Annu Rep Progr Chem C 93:101–145
23. Neilson GW, Adya AK, Ansel S (2002) Neutron X-ray diffraction studies on complex liquids. Annu Rep Progr Chem C 98:273–322
24. Shannon RD, Prewitt CT (1969) Effective ionic radii in oxides and fluorides Acta Cryst. B 25:925–946; (1969) Revised values of effective ionic radii. 26:1046–1048
25. Shannon RD (1976) Revised effective ionic radii and systematic studies of interatomic distances in halides and chalcogenides. Acta Cryst V 32:751–767
26. Roobottom HK, Jenkins HDB, Passmore J, Glasser (1999) Thermochemical radii of complex ions. J Chem Educ 76:1570–1573
27. Jenkins HDB, Roobottom HK, Passmore J, Glasser L (1999) Relationships among ionic lattice energies, molecular (formula unit) volumes, and thermochemical radii. Inorg Chem 38:3609–3620
28. Glasser L, Jenkins HDB (2008) Internally consistent ion volumes and their application in volume-based thermodynamics. Inorg Chem 47:6195–6202
29. Mukerjee P (1961) Ion-solvent interactions. I. Partial molal volumes of ions in aqueous solutions. II. Internal pressure and electrostriction of aqueous solutions of electrolytes. J Phys Chem 65:740–744
30. Marcus Y (2015) Ionic and molar volumes of room temperature ionic liquids. J Mol Liq 209:289–293

Chapter 3
High-Melting Salts

Sodium chloride is often considered to be the prototype of the concept 'salt' in general as well as in the fused state of 'molten salts'. It melts without decomposition to a water-like (in appearance) liquid beyond its melting point of 801 °C and remains liquid at ambient pressure up to the normal boiling point of 1465 °C. Its critical temperature, beyond which the distinction between liquid and vapor vanishes and the fluid has no longer a surface (in a gravitational field) has been estimated to be 3100–3600 °C by various authors. Sodium chloride does *not* exhibit pre-melting solid phase transitions to a plastic crystal form, contrary to some other salts. Such a transition permits more ready movement of the ions about their equilibrium positions without destroying the long-range order of the crystal. The sodium chloride melt solidifies without significant undercooling to the usual cubic crystalline form without eventually forming a glass. In the following, 'molten salts' (synonymous with 'fused salts') are to be understood as referring to high-melting salts similar in their general properties to sodium chloride.

The structural picture of molten salts emerging from the data dealt with in Sect. 3.2, consists of intertwining quasi-lattices of cation sites and anion sites. In charge-symmetrical salts: 1:1 (e.g. NaCl) and 2:2 (e.g. CaO) a majority of the sites are occupied by the ions, with some of them empty. However, charge-unsymmetrical salts, namely 1:2 (e.g. CaF_2), 1:3 (e.g. $LaCl_3$), 2:1 (e.g. Li_2CO_3), etc., have a substantial number of unoccupied sites, whether of sizes commensurate with those of the ions or collapsed to smaller sizes. It is expedient to deal with charge-symmetrical molten salts separately from the charge-unsymmetrical ones.

3.1 The Liquid Range of High-Melting Salts

The many molten salts that behave in a similar manner to sodium chloride are discussed in the present chapter. Their liquid range thus extends from the melting point, T_m, to the critical temperature, T_c, but at ambient pressures the boiling point,

Y. Marcus, *Ionic Liquid Properties*, DOI 10.1007/978-3-319-30313-0_3

T_b, is relevant as the upper limit of this range. The melting points of alkali metal salts, and when known also their boiling points, are shown in Table 3.1, mainly taken from several compilations [1–3], converted from °C to the K-scale. Differences between the T_m values reported in the literature may exceed 10 °C in some cases, although better agreement is seen in most cases.

The alkali metal salts generally melt without decomposition, and where there are no entries in Table 3.1 there appear to be no available data. On the other hand for the univalent copper, silver and thallium salts the empty spaces in this table denote that the salts decompose on heating or are unstable in the first place (in particular Cu (I) salts). In the case of the divalent metal salts, the oxides and sulfides melt at very high temperatures, but most of the salts with polyatomic anions decompose on heating. Univalent molten salts with monatomic anions boil at ambient pressures without decomposition, hence many of their boiling points could be reported in Table 3.1. This is rarely the case for molten salts with polyatomic anions, exceptions being some of the hydroxides and cyanides.

The charge-unsymmetrical salts with the univalent halide anions generally melt without decomposition, but some of them (e.g., HgI_2 and $AlCl_3$) have low melting points (<300 °C) because they are not ionic liquids, in the sense that the bonding between the atoms is mainly covalent. Such salts are not dealt with here. The melting points, T_m, and boiling points at ambient pressures, T_b, when known, of the ionic salts are shown in Table 3.2, taken mainly from the compilation in [1] with some additional entries from [2] or as annotated, converted from °C to the K-scale.

Temperatures above ca. 1500 °C have not been obtained with accuracies better than ±2 °C, and those where '3' is the last digit reported in Table 3.2 (because of the conversion from °C to the K-scale) are generally reported to the nearest 10 °C, those ending in '73' are generally reported to the nearest 100 °C. Salts with no entries either are low melting, i.e., not characteristically ionic, or decompose on heating, for instance because the cation is reduced by bromide and iodide ions to lower valence states (e.g., $EuBr_3$).

Some alkali metal salts with divalent oxy-anions other than the carbonates and sulfates are also stable on melting, with the melting points shown in Table 3.1. The disulfates (pyrosulfates) form stable melts on heating of the corresponding hydrogensulfates and have conveniently low melting points [23–25].

The critical temperatures of high-melting salts are generally very large and practically only the alkali metal halides have been considered in this respect. Carlson et al. [50] used Eyring's significant structures theory for the estimation of the melting and boiling properties and of the critical constants of four salts. They considered the molten salts to be mixtures of solid-like and gaseous-like (dimeric molecules) domains and used parameters for such species for the calculation of the partition function and the desired data. The melting and boiling points thus obtained are generally within 10 K of the experimental values. The resulting critical constants are shown in Table 3.3. Earlier McQuarrie [51] used a different theoretical approach, considering the molten salt to be a fluid of hard spheres with definite ion sizes, to obtain the partition function, an equation of state, and the critical constants of 16 salts, shown in Table 3.3. A different approach was used by Kirshenbaum

Table 3.1 The melting points, T_m/K (upper row), and normal boiling points, T_b/K (lower row), of charge-symmetrical salts, from [1] and values in parentheses are from Barin and Knacke [2], unless otherwise annotated

	Li^+	Na^+	K^+	Rb^+	Cs^+	Cu^+	Ag^+	Tl^+
F^-	1121	1266 (1269)	1125 (1130)	1106 (1048)	976		779	659
	1946 (1966)	1977 (1983)	1775 (1663)	1683 (1666)	1524[q] (1504)		1432	1099
Cl^-	883	1074	1043 (1044)	995 (988)	915 (908)	695	728	703
	1656	1738	1680[k] (1710)	1663 (1654)	1570 (1579)	1673	1820	993
Br^-	823	1020	1007	965 (952)	909 (908)	770	707	733
	1573 (1562)	1663 (1666)	1656[k] (1671)	1613 (1625)	1573	1618	1775	1092
I^-	742	933	954	920 (913)	899 (894)	879	829	650
	1444 (1449)	1577	1596 (1618)	1573 (1577)	1553	1563	1779	1097
OH^-	744 (744)	591 (593)	679 (673)	656	616			
	1899 (1803)	1661 (1663)	1600 (1600)					
CN^-	433[a]	836 (835)	907 (895)		623	747[s]		
		(1803)	(1898)					
SCN^-	550[b]	560	446	457[m]	479[m]			507[v]
N_3^-		573	608	590	599			607[q]
BO_2^-	1122 (1117)	1239	1220[l]		1005			
NO_2^-	473[c]	557	714	722[n]	697			459[c]
NO_3^-	525	579	611 (607)	583	687		485	483
ClO_3^-	401	488	641	615[o]	661[o]		488[t]	
BrO_3^-	533[d]	654	707	703				
IO_3^-	705[e]	690[i]	833	843[p]				
ClO_4^-	509	753	798	554	523			
ReO_4^-	692[f]	693[f]	828[f]	878[f]	893[f]			
BF_4^-	577	681	843	855	828			
HCO_2^-	545[j]	531	440	443	536[r]			375[w]

(continued)

Table 3.1 (continued)

	Li^+	Na^+	K^+	Rb^+	Cs^+	Cu^+	Ag^+	Tl^+
$CH_3CO_2^-$	559	602	577	519	464			401[w]
HSO_4^-		455[j]	488[j]	479[j]	491[j]			
	O^{2-}		S^{2-}		CO_3^{2-}		SO_4^{2-}	
Be^{2+}	2850							
Mg^{2+}	3098		2499		1263		1400[u]	
	3533							
Ca^{2+}	2888		2797		1603		1733	
	3773						1673[u]	
Sr^{2+}	2804		2499		1767		1878[u]	
Ba^{2+}	2245		2502		1828		1623[u]	
Mn^{2+}	2058		1803				973[u]	
Fe^{2+}	1650		1468					
Ni^{2+}	2228		1249					
Co^{2+}	2103		1455					
Cu^{2+}	1719							
Zn^{2+}	2247		1973					
Cd^{2+}	1173[t]		2023				1273[u]	
Sn^{2+}			1154					
			1483					
Pb^{2+}	1158		1386					
	1808							

[a][5], [b][7], [c][12], [d][16], [e][17], [f][20], [g][21], [h][175], [i][18], [j][26], [k][4], [l][11], [m][8], [n][13], [o][14], [p][19], [q][10], [r][22], [s][6], [t][15], [u][3], [v][9], [w][37]

Table 3.2 The melting points, T_m/K (upper row), and normal boiling points, T_b/K (lower row), of charge-unsymmetrical salts [1] unless otherwise annotated. Values for the monovalent metal salts and for divalent metal salts with divalent anions are also included

	F^-	Cl^-	Br^-	I^-	NO_3^-	O^{2-}	S^{2-}	CO_3^{2-}	SO_4^{2-}
Li^+	1121	883	823	742	525	1843	1645	996	1132
	1946	1656	1573	1444		2836			
Na^+	1266	1074	1020	933	579	1405	1223	1131	1175
	1977	1738	1663	1577					
K^+	1125	1043	1007	954	611		1110[b]	1171	1349
	1775	1680[a]	1656[a]	1596					
Rb^+	1106	995	965	920	583		698	1146	1346
	1683	1663	1613	1573					
Cs^+	976	915	909	899	687	763		1065	1292
	1524[a]	1570	1573	1553					
Cu^+[aa]		695	770	879		1509	1402[c]		
		1673	1618	1563					
Ag^+[bb]	779	728	707	829	485		1098		933[p]
	1432	1820	1775	1779					
Tl^+[cc]	659	703	733	650	483	852	721		905
	1099	993	1092	1097			1640		
Be^{2+}	825	713	781	743		2850			
	1442	755	793	760					
Mg^{2+}	1536	981	987	923	699[d]	3098	2499	1263	1400[p]
	2500	1685				3533			
Ca^+	1691	1055	1003	1057	835[d]	2888	2797	1603	1673[p]
	2806	2208	2088			3773			
Sr^{2+}	1673	1148	916	788	843	2804	2499	1767	1878[p]
	2733	1523		2046					
Ba^{2+}	1593	1235	1123	1013	863	2245	2502	1828	1623[p]
	2533	1833	2108						
Mn^{2+}	1203	923	971	911		2058	1803		973[p]
		1463							
Fe^{2+}	1373	950	946	860		1650	1468		
		1296							
Co^{2-}	1400	1013	951	793		2103	1455		
	1673	[1323]							
Ni^{2+}	1747	1282	1236	1053		2228	1249		
Cu^{2+}	1109	770[e]	771		528	1719			
	1949	794[f]	1173						
Zn^{2+}	1145	591	671	719		1973	2247		
	1773	1005	970	898					
Cd^{2+}	1383	841	840	666	628		2023		1273[p]
	2021	1233	1117	1015					
Sn^{2+}	486	519	488	593			1154		
	1123	896	912	987			1483		

(continued)

Table 3.2 (continued)

	F^-	Cl^-	Br^-	I^-	NO_3^-	O^{2-}	S^{2-}	CO_3^{2-}	SO_4^{2-}
Pb^{2+}	1103	774	646	683	779[g]	1158	1386		1360
	1566	1224	1165	1145		1808			
UO_2^{2+}		851[y]							
Al^{3+}	2523					2326	1373		
Sc^{3+}	1788	1240	1242			2758	2048		
Cr^{3+}	1673	1425	1403			2602			
Fe^{3+}						1838			
Ga^{3+}	>1273[q]	351	395	465		2079	1363		
		474	552	613					
In^{3+}	1443	856	709	483		2185	1323		
Y^{3+}	1408[h]	953	1177	1237[i]		2711	2198		
La^{3+}	1766	1145	1061	1051		2577	2573		
						3893			
Ce^{3+}	1733	1095	1006[j]	1039		2483	2723		
Pr^{3+}	1646	1060[x]	966	1010		2456	2038		
						4033			
Nd^{3+}	1650	1031	955	1051		2506	2480		
	2573	1873	1813			4033			
Sm^{3+}	1579	955	913	1123		2542	1993		
						4053			
Eu^{3+}	1549	896				2564			
						4063			
Gd^{3+}	1504	882	1043	1198		2619			
						4173			
Tb^{3+}	1446[k]	861	1103[l]	1250		2303			
Dy^{3+}	1427	953	1152	1251		2729			
						4173			
Ho^{3+}	1416	991	1192	1267		2603			
Er^{3+}	1420	1049	1196	1287		2617	2003		
						4193			
Tm^{3+}	1431	1097	1277	1294		2614			
						4213			
Yb^{3+}	1430	1140				2638			
						4343			
Lu^{3+}	1455	1198	1298	1323		2700			
	2473					4253			
Tl^{3+}	823					1107			
Bi^{3+}	998	503	491	602		1090	1123		983[p]
	1173	720	726	895					
U^{3+}	1768[m]	1110	1000	1039					
Zr^{4+}	1176[n]	710	723	772		2950			
						4548			

(continued)

Table 3.2 (continued)

	F^-	Cl^-	Br^-	I^-	NO_3^-	O^{2-}	S^{2-}	CO_3^{2-}	SO_4^{2-}
Hf^{4+}	1298[o]	705	697	722		3047			
Th^{4+}	1383	1043	952	843		3663	2178		
						4673			
U^{4+}	1309	863	792	779		3100			
	1690	1064							

[a][4], [b][27], [c][28], [d][29], [e][30], [f][31], [g][32], [h][33], [i][34], [j][35], [k][36], [l][37], [m][37], [m][38], [n][39], [o][40], [p][2], [q][41], [x][42] a value 1060 K was reported here, the value 963 K [1] appears to be too low, [y] [43], Values for further anions: [aa][44] CuCN $T_m = 747$ K, [bb][45] $AgClO_3$ $T_m = 488$ K, [cc][46] TlSCN $T_m = 507$ K, [47] TlN_3 $T_m = 607$ K, [48] $TlNO_2$ $T_m = 459$ K, [49] $TlHCO_2$ $T_m = 375$ K, $TlCH_3CO_2$ $T_m = 401$ K

Table 3.3 Critical constants of the alkali halide salts according to various authors

	[50]			[51]			[52]			[53]	
	T_c	P_c	ρ_c	T_c	P_c	ρ_c	T_c	P_c	ρ_c	T_c	P_c
Salt	K	MPa	kg m^{-3}	K	MPa	kg m^{-3}	K	MPa	kg m^{-3}	K	MPa
LiF				6350	92.21	213	4140				
LiCl				4960	34.75	167	3080				
LiBr				4640	26.45	278	3020				
LiI				4220	18.24	324	3250				
NaF				5530	53.20	227	4270				
NaCl	3600	23.86	200	4540	24.32	175	3400	35	220	3900	25.8
NaBr	3364	18.86	300	4280	19.25	259	3200				
NaI				3950	13.88	297	3160				
KF				4780	29.79	204	3460				
KCl	3092	13.73	170	4060	15.60	161	3200	22	180	3470	18.0
KBr	3060	11.99	250	3880	12.97	223	3170				
KI				3610	9.73	252	2980				
RbF				4520	23.91	311	3280				
RbCl				3880	12.97	226	3140				
RbBr				3720	10.94	272	3130				
RbI				3490	8.41	289	3035				
CsF							2915				
CsCl							3040				
CsBr							3045				
CsI							3020				

et al. [52], based on density measurements of the molten and vapor phases of NaCl and KCl, leading by means of the rectilinear diameter law to estimated values of T_c, also shown in Table 3.3. The values for other alkali metal halides were obtained from those of NaCl and KCl according to the corresponding states principle for the entropy of vaporization. In the cases of NaCl and KCl the uncertainties in T_c were estimated as ±200 K. Pitzer [53] used an approach similar to that of Kirshenbaum et al. [52], but different values for the vapor densities, hence arrived at somewhat different values of the critical constants.

On the other hand, estimates of T_c from the extrapolation of the surface tension with increasing temperatures up to its vanishing [54, 55] yielded somewhat smaller values than in [52]: 2890 K for NaCl and 2820 K for KCl. A computer simulation study of sodium chloride [56] yielded the following critical constants: $T_c = 3300$ K, $P_c = 32.5$ MPa, and $\rho_c = 180$ kg m^{-3}, commensurate with the values in [52]. Very few salts other than the alkali metal halides had their critical constants estimated. Magnesium oxide was studied by Leu et al. [57], using the significant structure theory, and had the following constants estimated: $T_c = 5950$ K, $P_c = 34.0$ MPa, $\rho_c = 192$ kg m^{-3}. Hoch [58] studied alumina and uranium oxide, the reported critical constants were $T_c = 4950$ K and $\rho_c = 890$ kg m^{-3} for Al_2O_3 and $T_c = 7930$ K and $\rho_c = 2457$ kg m^{-3} for UO_2.

3.2 Structural Aspects from Diffraction Measurements and Computer Simulations

Some 60 years ago Zarzycki [59] studied the structure of molten sodium chloride by means of x-ray diffraction. He could conclude that the data substantiate the intuitive and theoretical picture of the structure in terms of two intertwining quasi-lattices of cation and anion sites, most of which are actually occupied by the ions. An early review of x-ray (and also some neutron diffraction) studies of molten salts is that by Levy and Danford [60], in which 25 salts are featured, most of them being molten alkali metal halides. More recently Ohno et al. [61] reviewed x-ray diffraction (XRD) studies of molten 1:1 salts, and still more recently Neilson and Adya [62] reviewed neutron diffraction studies of molten salts, and Neilson, Adya and Ansell reviewed both XRD and neutron diffraction of molten salts and also room temperature ionic liquids (RTILs) [63]. Complementary information on the structure of molten salts (excluding those consisting of light elements) can be obtained from extended x-ray absorption fine structure (EXAFS) measurements [64].

Consider an ion $I^{z\pm}$ placed at the origin of coordinates with other ions of either sign surrounding it. The number of particles of these species (subscript $_J$) in a spherical shell of thickness dr at a distance r from the center of the ion at the origin is:

$$\mathrm{d}n_{IJ}(r, \mathrm{d}r) = 4\pi r^2 g_{IJ}(r)\rho_{\mathrm{bulk}}\mathrm{d}r \quad (3.1)$$

where ρ_{bulk} is the total number density of ions of both signs in the bulk. The function $g_{IJ}(r)$, called the *pair correlation function*, is the conditional probability of finding an ion J (the same as $I^{z\pm}$ or differing from it) at the distance r from the ion

$I^{z\pm}$ at the origin. There is no correlation between the particles at large distances, hence $g_{IJ}(r \rightarrow \infty) = 1$, but at very small distances $r < d_{IJ}$, where d_{IJ} is the distance between the centers of the ions at contact, $g_{IJ}(r < d_{IJ}) = 0$, because the large repulsion of the electronic shells of the ions prevents their overlapping. Generally $4\pi r^2[g_{IJ}(r)-1]\rho_{bulk}$ has a maximum at a distance near d_{IJ} and undulates further out, has a second maximum and eventually reaches $4\pi r^2 \rho_{bulk}$ asymptotically. The *coordination number*, N_{co}, of the ion $I^{z\pm}$ is obtained from the area below the first peak in the $4\pi r^2[g_{IJ}(r)-1]\rho_{bulk}$ curve:

$$N_{co} = 4\pi\rho_{bulk} \int_0^{r'} [g_{IJ}(r) - 1]r^2 dr \tag{3.2}$$

The choice of the upper limit of the integration, r', is somewhat arbitrary: commonly it represents the distance at which $4\pi r^2[g_{IJ}(r)-1]\rho_{bulk}$ reaches its minimum after the first peak. An alternative is to assume symmetry of the peak, to use the peak distance to represent r', and to take the coordination number as twice the integral (3.2) up to this point. The second peak in the $4\pi r^2[g_{IJ}(r)-1]\rho_{bulk}$ curve represents the distance of the ion $I^{z\pm}$ from its next-nearest neighbors, and a third peak may also be observed, representing more distant correlations.

The primary information of the diffraction measurements is the intensity of the scattered beam $I(q)$ as a function of the angle φ to which the radiation is scattered, where $q = (4\pi/\lambda)\sin(\varphi/2)$ and λ is the wavelength of radiation employed, whether x-rays or neutrons. The space variable q has the dimension of a reciprocal length, its units usually being Å^{-1}. The *structure factor* $S(q)$ is obtained from the intensities of the diffracted beam, corrected for absorption and incoherent scattering and normalized to instrumental factors, in a manner dependent on the method employed, whether x-ray or neutron diffraction, as detailed in the next sections. The pair correlation function $g_{IJ}(r)$ is derived from $S(q)$ by using its Fourier transform as:

$$g_{IJ}(r) = 1 + \left(2\pi^2\rho_{bulk}r\right)^{-1} \int_0^{\infty} [S(q) - 1]q \sin(qr)dq \tag{3.3}$$

A model of the molten salt obtained by computer simulations is nowadays generally used to refine the structure, avoiding errors arising from incomplete knowledge of $S(q)$ at very small and very large angles, $q(r \rightarrow 0)$ and $q(r \rightarrow \infty)$ required for the integration (3.3).

3.2.1 X-Ray Diffraction Studies of Molten Salts

X-rays are diffracted from the electron shells of the individual ions making up the sample. The atomic scattering intensity $a(q)$ of each kind of atom in the sample depends on the electron density of the atom ρ_a and the scattering angle φ via the space variable q as:

$$a(q) = 4\pi \int_0^{ra} \rho_a \sin(qr)(qr)^{-1} dr \tag{3.4}$$

the integration extending up to the radius of the atom (ion), r_a. The structure factor for x-ray diffraction, after correction for incoherent scattering, is

$$S(q) = [I(q)/I(0)]\Sigma Z_k^{-2} n_k^{-1} \alpha_k(q)^{-2} \tag{3.5}$$

where the summation extends over all the atomic species (subscript $_k$) present, the number of each kind of atom in the sample being n_k, their atomic number being Z_k.

In a binary molten salt, consisting of only two kinds of atoms (cations and anions) I and J, the experimentally obtained structure factor $S(q)$ according to Eq. (3.5) is a linear superposition of the three *partial structure factors* for the various types of pairs of ions, I-J, I-I, and J-J, weighted by the appropriate form factors. Consequently, the overall pair correlation function $g(r)$ resulting from the Fourier transform, Eq. (3.3) represents the sum of three *partial pair correlation* functions: $g_{IJ}(r)$ for cation-anion correlation, and $g_{II}(r)$ and $g_{JJ}(r)$ for correlations of ions of the same sign. Therefore, the model of the molten salt, specifying nearest neighbors to be of opposite sign, is resorted to and yields $g_{IJ}(r)$ for cation-anion correlation up to the first peak in the $g(r)$ curve at the distance R_1 with a coordination number N_{co1}. The second peak, at a larger distance, R_2, pertains to both cation-cation and anion-anion correlations, with respect to both distances and coordination number N_{co2}.

The earlier studies of Zarzycki [59], Antonov [65], Ohno and Furukawa [66], Levy et al. [67], and Saito et al. [68] among others applied x-ray diffraction to the alkali metal halides and very few other binary salts (CaF_2, $BaCl_2$, CuBr). More recently the structures of some other 1:2 and of 1:3 salts were examined by x-ray diffraction. The first peak corresponded to the cation-anion distance d_{IJ}, which is commensurate with the sum of their ionic radii in crystals at room temperature, $r_+ + r_-$ (Table 2.4). The coordination numbers were definitely smaller than in the crystalline solids (6 for NaCl-type lattices, 8 for CsCl-type lattices), having values between 3.5 and 5.6, but they depended on the choice of r', the upper limit of integration of Eq. (3.2) as mentioned above. The second peak reflected mainly the anion-anion correlations (modified slightly by the cation-cation correlations), because the anions, having generally more electrons, dominate the scattering. The resultant inter-ionic distances and coordination numbers for the first and second peak are shown in Table 3.4.

Generally, the cation-anion distances in the melt are commensurate with the sum of their ionic radii or are slightly lower, but they are some 15 % smaller than the corresponding distance in the expanded crystal at the melting point. Notable exceptions to this in Table 3.4 are $ZnBr_2$ and $CdCl_2$, with considerably smaller

Table 3.4 Interionic distances and coordination numbers for the first and second peak of x-ray diffraction from molten salts

	T	First peak			Second peak	
Salt	K	R_1/pm	N_{co1}	$(r_+ + r_-)$/pm[a]	R_2/pm	N_{co2}
LiF	1133	195 [69]	3.7	202	300 [70]	8
LiCl		247 [67]	4.0	250	385 [67]	12.0
LiBr		268 [67]	5.2	265	412 [67]	12.8
LiI		285 [67]	5.6	289	445 [67]	11.3
NaF	1273	230 [69]	4.1	235	344 [59]	9
NaCl	1093	280 [69]	4.7	283	420 [59]	9
NaBr		305 [65]		298	456 [65]	11.7
NaI		315 [65, 67]	4.0	322	480 [67]	8.9
KF	1143	270 [69]	4.9	271	~386 [59]	9
KCl	1083	310 [59]	3.7	319		
KBr		335 [65]		334		
KI		352 [65]		358	490 [65]	7.2
RbCl		330 [60]	4.2	329		
RbBr		340 [65]		344	472 [65]	7.4
RbI		365 [65]		368	515 [65]	7.6
CsCl		353 [67]	4.6	351	487 [67]	7.1
CsBr		355 [67], 366 [65]	4.6	366	~540 [67]	8.3
CsI		385 [65, 67]	4.5	390	~550 [67]	7.2
CaF_2	1773	235 [59]	6.8	233		
$BaCl_2$	1273	320 [59]	5.8	318		
$ZnBr_2$	723	244 [72]	4.0	271	400 [72]	11.3
$CdCl_2$	923	247 [71]	3.8	276		
$LaCl_3$	1150	285 [73], 282 [74]	7.1	286	362 [73]	8.0
$LaBr_3$	1150	294 [73]	7.5	301	370 [73]	8.0
$NdCl_3$	1073	277 [75]	5.5	280	404 [75]	11.3
$ErCl_3$	1053	263 [76]	5.8	270	375 [76]	9
UCl_3	1200	285 [77]	8.0	278		

[a]In crystals at room temperature, from Table 2.4

R_1 values than their $r_+ + r_-$, for which incomplete ionization and a partial covalent bonding are responsible.

In the case of x-ray diffraction involving molten salts with polyatomic anions (nitrate, carbonate, sulfate) [69, 70] the information is obscured by the intra-anionic distances (central atom to oxygen) and little useful information on the structure of the melt is obtained. The same problem occurs with alkali metal thiocyanates [78], where the data confirm the structure of the linear SCN^- anion but does not position the cations with respect to it.

3.2.2 Neutron Diffraction Studies of Molten Salts

Neutron diffraction has several advantages over x-ray diffraction when applied to molten salts. One advantage is that the diffraction occurs from the nuclei of the ions and not from the electronic shells. Consequently the atomic scattering intensity is independent of the diffraction angle and space variable q and is given by the bound atom scattering length, averaged over all the isotopic species of the atom and spin states $<b>$. The scattering intensity $I(q)$ for neutron diffraction depends on $<b>^{-2}$, contrary to the case of x-ray diffraction, where it depends on $\alpha(q)^{-2}$. Elimination of instrumental factors and incoherent scattering from the intensities yields:

$$S(q) \approx [I(q) - I(0)]/[I(\infty) - I(0)] \quad (3.6)$$

The resulting structure factor is still a linear superposition of the three *partial structure factors* for the various types of pairs of ions, I-J, I-I, and J-J:

$$S(q) = 2n_I n_J b_I b_J (S_{IJ}(q) - 1) + n_I{}^2 b_I{}^2 (S_{II}(q) - 1) + n_J{}^2 b_J{}^2 (S_{JJ}(q) - 1) \quad (3.7)$$

Variation of b in three different experiments by varying the isotopic composition permits the three independent linear equations to be solved to yield the three partial structure factors. The Fourier transform of each of these yields the partial pair correlation functions: $g_{IJ}(r)$ for cation-anion correlation, and $g_{II}(r)$ and $g_{JJ}(r)$ for correlations of ions of the same sign. The Neutron diffraction with isotope substitution (NDIS) method permits this variation of b, because different isotopes of the elements have different b values, some of which may even be negative.

Page and Mika [79] were the first to apply this methodology, studying the structure of molten cuprous chloride. They used isotopically enriched ^{63}Cu and ^{65}Cu as well as ^{35}Cl and ^{37}Cl and derived the partial structure factor and pair correlation function for Cu-Cl. The peak in $g_{Cu\text{-}Cl}$ at 230 pm was definitely shorter than $r_+ + r_- = 277$ pm, signifying incomplete ionization of this salt (partial covalent bonding). The structure of molten sodium chloride by Edwards et al. [80] studied by the NDIS method was later refined by Biggin and Enderby [81], and many other structures of molten salts were obtained by the NDIS method as later reviewed by Neilson and Adya [62], with results shown in Table 3.5. The distance between the ions designated in the subscripts is R and the coordination number of the second ion around the first is N_{co}.

Comparison of the data in Tables 3.4 and 3.5 for salts common to the two tables shows the first peak distances obtained by the NDIS method to be ~3 % shorter on the average, and to be ~4 % shorter than the sum of the room temperature ionic radii $r_+ + r_-$. These apparent discrepancies have not been explained.

Table 3.5 Molten salt structural data from neutron diffraction with isotope substitution

Salt IJ	T/K	R_{IJ}/pm	N_{coIJ}	R_{II}/pm	N_{coII}	R_{JJ}/pm	N_{coJJ}
LiCl [82]	958	230	3.5–4.0	370		370	
LiCl [83]	958		5.5				
NaCl [80]	1148	272	5.8	389	13.0	390	13.3
NaCl [81]	1148	278	5.8	396	13.0	390	13.3
KCl [84]	1073	306	6.1	484	12	482	12.3
RbCl [85]	1023	318	6.9	486	13	480	14
CsCl [89]	968	340	5.8	495	15.4	485	16.3
CsCl [86]	973	338		385		385	
CuCl [86]	773	230	3	370		390	
CuCl [79]	713	230		330		360	
CuBr [87]	788	250	3.63			397	11.1
AgCl [88]	783	260	4.5	340	3.1		
AgCl [88]	1173	255	2.7	315	4.1	315	2.7
$MgCl_2$ [90]	998	242	4.3	381	5 ± 1	356	12 ± 1
$CaCl_2$ [91]	1093	278	5.4	360	4.2	373	7.8
$SrCl_2$ [92]	1198	290	6.9	495	13.6	380	9.3
$BaCl_2$ [93]	1298	310	7.7	490	14 ± 2	386	7 ± 1
$MnCl_2$ [90]	973	250	4			358	8.4
$NiCl_2$ [94]	1295	236	4.7	100	6	380	13.8
$NiBr_2$ [94]	1258	247	4.7	371	5.3	397	14
NiI_2 [94]	1103	260	4.2	460	5.3	410	13 ± 1
$DyCl_3$ [95]	973	265	6	420		350	

3.2.3 EXAFS Studies of Molten Salts

Extended x-ray absorption fine structure (EXAFS) is a technique to study the near-neighbor structure of liquids, including molten salts. A recent review of the application of this method by Hardacre [96] provides much information. The absorption of high energy x-rays results in the ejection of core electrons (generally the K-edge) and as the energy of the radiation is increased past the absorption edge the fine structure is revealed. The ejected electron wave is scattered by the surrounding atoms. The measured amplitude of the scattering is the sum of quantities proportional to the number of scattering atoms and their characteristic scattering amplitude, and the Fourier transform yields the pair correlation function of the back-scattering atoms around the absorbing atom. The absorption edge characteristic of a particular element identifies the atom (ion, in the case of molten salts), the local structure about which is being studied. The method is most usefully applied to atoms with relatively high atomic number Z. The resulting inter-ionic distances and the coordination numbers obtained by the EXAFS method are shown in Table 3.6.

The inter-ionic distances of the nearest neighbors in the molten salts (cation-anion distances) are obtained from the EXAFS method with the same accuracy as

Table 3.6 Molten salt structural data from EXAFS measurements

Salt IJ	T/K	d_{IJ}/pm	N_{coIJ}	d_{II}/pm	N_{coII}	d_{JJ}/pm	N_{coJJ}
KBr [97]	1025	321	3.78			507	8.79
KBr [98]	1003	~315					
RbCl [99]	1023					480 ± 90	
RbBr [98]	1003	~330					
CuBr [100]	793	244	3.0 [101]			410	
CuI [102]	903	277	2.7				
AgCl [103]	763	253	2.7				
AgBr [103]	743	258	2.8				
AgI [103]	863	273	2.2				
$SrCl_2$ [104]	1273	299	6.6 ± 0.2				
$ZnCl_2$ [99]	613	231	4.3 ± 0.2				
$ZnBr_2$ [105]	723	246	3.9 ± 0.5				
PbF_2 [106]	1130	236	3.76				
YCl_3 [107]		272	5.9 ± 0.6				
$LaCl_3$ [107]		289	7.4 ± 0.5	490	7.8 ± 0.7		
$GdCl_3$ [108]	923	274	6.5 ± 0.3				

for the x-ray and neutron diffraction methods, but the coordination numbers have lower precision. Although the application of multiple-edge EXAFS experiments has been established, permitting also cation-cation and anion-anion correlations to be measured separately, this has only rarely been done, as seen in Table 3.6. It is noteworthy that the copper(I) and silver halide melts show much shorter cation-anion distances than the sum of the ionic radii, signifying partial covalent bonding in these melts.

3.2.4 *Modeling of Molten Salts*

Models of molten salts have evolved with time and so have the inter-ionic potentials relevant to them. The restricted primitive model (RPM) considers the ions to be charged hard spheres of uniform size (radius r_I) interacting through their coulomb potentials. The potential function is then:

$$u(r) = z_+ z_- e^2/4\pi\varepsilon_0 r \text{ for } r > 2r_I \text{ and } u(r) = \infty \text{ for } r \leq 2r_I \qquad (3.8)$$

A modification introduces a relative 'dielectric constant' ε replacing $4\pi\varepsilon_0$ in the denominator of (3.8) [109, 110]. The 'restricted' is removed when the cations and anions are permitted to have different sizes, r_+ and r_-, and then the inter-ionic potential becomes similar to (3.8), but with $2r_I$ replaced by $r_+ + r_-$. The 'hard sphere' restriction is removed by the inclusion of a repulsion term [111]

$$u(r) = z_+z_-e^2/4\pi\varepsilon_0 r + b_\pm \exp[(r_+ + r_- - r)/r_0] \quad (3.9)$$

Here $b_\pm$ is a salt-dependent parameter and r_0 denotes unit radius. To this expression van der Waals interaction terms have been added [111], yielding:

$$u(r) = z_+z_-e^2/4\pi\varepsilon_0 r + b_\pm \exp[(r_+ + r_- - r)/r_0] - c_\pm r^{-6} - d_\pm r^{-8} \quad (3.10)$$

where, again, $c_\pm$ and $d_\pm$ are salt-specific parameters. This is the Born-Huggins-Mayer-Fumi-Tosi (BHMFT) potential for the rigid ion model (RIM) [56, 112] that has been widely employed for the computer simulations of molten salts (Sect. 3.2.5). A further modification [113] replaced the repulsion term $b_\pm \exp[(r_+ + r_- - r)/r_0]$ by a term $b_\pm$'$\exp[(r_+ + r_- - r)/r_0]r^{-4}$ and another modification employs perturbation theory and replaces the repulsion term by $\lambda \exp[-r/(r_+ + r_-)]$, where λ is a salt-specific parameter [114], but these modifications have not been widely used for molten salts. A polarizable ion model (PIM) [115, 116] or a soft ion model (SIM) have more recently been employed in the cases where the ions of the molten salt are highly dipole- and quadrupole-polarizable, e.g., Ag^+.

3.2.5 Computer Simulation Studies of Molten Salts

Computer simulations have been applied to studies of the structure of molten salts along two lines: one is the free standing application of the computer simulation to obtain the partial pair correlation functions, the other is the refining of x-ray and neutron diffraction and EXAFS measurements by means of a suitable model. In both cases a suitable potential function for the interactions of the ions must be employed, as discussed in Sect. 3.2.4. Such potential functions are employed in both the Monte Carlo (MC) and the molecular dynamics (MD) simulation methods. A further aspect that has been considered in the case of molten salts is the long range coulombic interaction that exceeds the limits of the periodic simulation boxes usually involved (for ~1000 ions altogether), requiring the Ewald summation that is expensive in computation time and is prone to truncation errors if not applied carefully.

Woodcock and Singer [117] were among the early persons who studied the structure of molten salts, in their case potassium chloride at 772 and 1033 °C, by means of Monte Carlo computer simulations. At about the same time Woodcock [118] reported the partial pair correlation functions for molten lithium chloride at 1000 °C obtained by molecular dynamics simulations.

Many authors have since then applied Monte Carlo (MC) and molecular dynamics (MD) simulations to molten salts. The simulations yielded the partial pair correlation functions, from which the inter-ionic distances and coordination numbers were deduced, as shown in Table 3.7. Generally the interionic distances were

Table 3.7 Molten salt structural data from computer simulations: Monte Carlo (MC) or molecular dynamics (MD)

Salt IJ	Method	*T*/*K*	R_{IJ}/pm	N_{coIJ}	R_{II}/pm	N_{coII}	R_{JJ}/pm	N_{coJJ}
LiCl [119]	MC	883	245	6.2	375	12.6	375	12.5
LiCl [126]	MD	900	225	4.05	387	11.81	376	11.86
LiI [119]	MC	742	260	4.3	445	14.1	445	14.1
LiI [125]	MC	800	270	4.8	420	13.0	420	13.4
LiI [125]	MC	1453	270	4.5	450	12.4	440	12.4
$LiNO_3$ [125]	MD	800		3.91[b]				
$LiCH_3CO_2$ [125]	MD	800		3.17[b]				
Li_2CO_3 [125]	MD	800		3.49[b]				
NaCl [119]	MC	1074	270	5.2	425	13.9	425	14.7
NaCl [126]	MD	1080	261	4.76	417	13.59	405	13.54
$NaClO_3$ [130]	MD	560	335[a]	6.6	432	8.6	518	12.2
$NaNO_3$ [125]	MD	800		4.62[b]				
Na_2CO_3 [125]	MD	800		3.76[b]				
KCl [117]	MC	1045	294	5.3	446[c]	15.7[c]	446[c]	15.7[c]
KCl [117]	MC	1306	293	5.2	461[c]	14.6[c]	461[c]	14.6[c]
KCl [126]	MD	1053	296	5.18	458	14.57	458	14.28
KBr [119]	MC	1007	320	5.5	480	15.0	480	15.0
KBr [97]	MD	1100	327	3.64			500	8.79
KI [119]	MC	954	345	5.9	490	16.0	490	16.6
K_2CO_3 [125]	MD	800		3.85[b]				
RbF [122]	MD	1155	260		400		400	
RbCl [119]	MC	995	320	5.8	460	16.5	460	16.4
RbCl [122]	MD	1359	310		480		450	
RbCl [126]	MD	1000	312	5.41	467	15.66	474	15.71
RbBr [122]	MD	1365	330		490		480	
RbI [122]	MD	1341	350		530		530	
CsF [119]	MC	976	290	5.4	440	16.9	440	16.8
CsF [123]	MD	962	270		420		410	
CsCl [123]	MD	996	330		470		460	
CsCl [126]	MD	900	330	5.70	476	16.26	490	16.34
CsBr [123]	MD	942	350		480		500	
CsI [123]	MD	871	370		520		520	
AgCl [120]	MC	728	250	5.0	380	15.3	380	15.3
AgCl [120]	MC	1500	250	4.5	400	12.9	400	13.0
$ZnCl_2$ [124]	MC	623	225		405		365	
$ZnCl_2$ [124]	MC	873	220		405		365	

[a]The Na^+ – Cl distance and coordination number, [b]Coordination number of the central atom of the anion (N or C) around the cation, [c]Only like-like values reported, not whether cations or anions

reported to the nearest 10 pm and only rarely with a better accuracy. The cation-anion $g_{IJ}(r)$ and N_{coIJ} of all the molten alkali halide salts IJ obtained by MC simulations with the rigid ion model (RIM) of pair potentials (3.8) were summarized by Baranyai et al. [121].

X-ray and neutron diffraction data (structure factors) are nowadays commonly refined by assuming a model for the molten salt and applying computer simulations to the model until full agreement with the experimental data is achieved. Examples of such applications of computer simulations are the following: x-ray diffraction data from molten UCl_3 was refined in [126], both (anomalous) x-ray diffraction and neutron scattering data of molten Cu(I) halides were refined in [127], and neutron diffraction data from molten alkaline earth chlorides according to models were refined in [128], whereas EXAFS data for molten AgBr were refined in [129]. In the case of Monte Carlo simulations a variant, caller reversed Monte Carlo (RMC) simulations, has been applied [124] to molten $ZnCl_2$. It has the advantage that no a priori interatomic potentials are required, and configurations are generated, of which the correlation functions correspond well to the experimental functions.

3.3 Static Thermochemical Data

3.3.1 Theoretical Predictions of Thermochemical Properties

Theories for the prediction of thermochemical properties of molten salts have been reviewed a number of times over the years [131–134]. The main approaches that have been used are: hole theory [135–137], corresponding states theory [138–140], perturbation theories [114, 141], and the significant structure theory [142–144]. Empirical correlation expressions have also been developed for the various properties as dealt with in the appropriate sections below.

Hole theory was first applied to molten salts by Bockris and Richards [137], regarding their compressibility. According to this approach, the main change occurring on the melting of a crystalline salt is the introduction of randomly distributed holes, accounting for the 15–25 % volume increase on melting, $\Delta_m V$. The occurrence of holes in the molten salt is manifested also in a decrease of the mean coordination numbers relative to the room temperature crystals, but not of the interionic distances, as seen in Tables 3.4, 3.5, 3.6, and 3.7. According to Fürth [136] the mean volume of a hole is $v_h = 0.68(k_B T/\sigma)^{1/2}$, where σ is the surface tension of the fluid. This quantity is directly involved in the expressions for the isothermal compressibility and the isobaric expansibility of molten salts [137]. Pandey et al. [145] applied this approach to the calculation of the surface tension of molten salts:

$$\sigma/\mathrm{mN\,m^{-1}} = 3500\left(V/\mathrm{cm^3mol^{-1}}\right)^{-1}(1 - \Delta_m V/V)^{-1} \tag{3.11}$$

The numerical coefficient has been changed from that in [145] to account for the average deviation of the calculated from the experimental values for the alkali metal halides. Cavity formation in molten salts is compatible with their restricted primitive model (RPM), Eq. (3.8) [146]. For a molten salt having N ions of mean diameter $d_{\pm} = (r_+ + r-)$ in a volume V, the size distribution of cavities with radii $r < 0.5d_{\pm}$ is:

$$C(r) = 4\pi r^2 d_{\pm}{}^3 N/V \tag{3.12}$$

The reversible work for the creation of a cavity of radius r is:

$$\Delta\mu(r) = -k_B T \ln\left[1 - \int_0^r C(r')\mathrm{d}r'\right] \tag{3.13}$$

The surface tension of an RPM fluid is therefore $\sigma = \Delta\mu(r)/4\pi r^2$, in fair agreement with the results of Monte Carlo computation with this model. The main point is that there are a good number of cavities with sizes commensurate with those of the ions in a molten salt.

The corresponding states theory was developed by Reiss, Mayer and Katz [138] to deal with molten salts. It employs the primitive model with a single distance parameter $d_{\pm}$ (=$2r_I$ in Eq. 3.8) to which all the inter-ionic distances in the entire volume of the molten salt V are proportional: $r_{IJ} = \xi_{IJ} d_{\pm}$ for all ions I and J, whether I = J or I ≠ J. Reduced thermodynamic quantities are then defined as:

$$P_r = P d_{\pm}{}^4 z_I{}^{-2}; \quad T_r = T d_{\pm} z_I{}^{-2}; \quad V_r = V d_{\pm}{}^{-3} \tag{3.14}$$

where for the charge-symmetrical salts considered $z_I = z_+ = z_-$. The distance parameter $d_{\pm}$ is *not necessarily* equal to $r_+ + r_-$, but could be so construed, or alternatively it is related to the melting point of the salt, T_m, according to:

$$d_{\pm} = \tau_m z_I{}^2 T_m{}^{-1} \tag{3.15}$$

where $\tau_m \approx 300/z_I$ K nm for uni-univalent and di-divalent salts is a universal constant. Modified reduced thermodynamic quantities then result:

$$P_r' = P z_I{}^6 T_m{}^{-4}; \quad T_r' = T T_m{}^{-1}; \quad V_r' = V T_m{}^3 z_I{}^{-6} \tag{3.16}$$

Fairly constant values of the reduced vapor pressures P_r', corresponding to deviations of <2 % in the enthalpy of vaporization, resulted at corresponding temperatures of $T = 1.30\ T_m$ and $1.55\ T_m$. Fairly constant values of the reduced surface tension of uni-univalent salts also resulted at $T = 1.00\ T_m$ and $1.10\ T_m$. Agreement with experimental data was obtained for charge-symmetrical molten salts except for the lithium halides, and for charge-unsymmetrical salts only for the alkaline earth metal fluorides, whereas for other salts of this class some degree of covalent bonding was supposed to account for the deviations.

Young and O'Connell [140] introduced a characteristic temperature $T^* = 0.4/\alpha_P$, i.e., a temperature that is inversely proportional to the isobaric expansibility α_P. The temperatures T^* for molten alkali metal halides, hydroxides and nitrates were listed. For the alkali metal halides $T^* = (1.15 \pm 0.04)T_m$, but different ratios were obtained for the nitrates (1.56 ± 0.11) and the two hydroxides tested (1.99 ± 0.003). With a set of characteristic molar volumes V^* listed for these salts a universal expression over the range $0.5 \leq T/T^* \leq 1.2$ was obtained for the reduced molar volumes:

$$V/V^* = 0.7458 + 0.1084T/T^* + 0.1458(T/T^*)^2 \tag{3.17}$$

The perturbation theories for molten salts of Harada et al. [114, 141] use a hypothetical fluid obeying the RPM as a reference, leaning on the Monte Carlo simulations of Larsen [109, 110] for such a fluid. Two state variables are employed: the reduced temperature T_r and the reduced packing fraction y_r defined in terms of the interionic distance parameter $d_\pm$ as:

$$T_r = k_B d_\pm e^{-2} T; \quad y_r = (\pi/6)(N/V){d_\pm}^3 \tag{3.18}$$

The perturbation consisted of relaxing the RPM potential (3.8) to include repulsion and multipole interactions as in the potential (3.9), but with the first term replaced by $z_+z_-e^2\xi_\pm/r$ and the second term replaced by $\lambda_\pm \exp[-r/(r_+ + r_-)]$ [114]. The two added parameters $\xi_\pm$ and $\lambda_\pm$ are salt-specific, the former being close to unity ($0.827 \leq \xi_\pm \leq 1.038$). The perturbation due to the changed potential yields an expression for the effective interionic distance parameter $d_\pm$ to be used in (3.18):

$$d_\pm/(r_+ + r_-)\zeta_\pm = 0.4069 + 0.9075\ln(\lambda_\pm/k_BT) + 6.042 \times 10^{-7}(\lambda_\pm/k_BT) \tag{3.19}$$

where $\zeta_\pm$ is a listed salt-specific parameter, close to unity ($0.929 \leq \zeta_\pm \leq 1.005$) accounting for the size differences between cation and anion. The three salt-specific parameters $\xi_\pm$, $\lambda_\pm$, and $\zeta_\pm$ together with the sum of the ionic radii $r_+ + r_-$ yield listed sets of characteristic temperatures T^* and molar volumes V^* of the salts, and a new set of reduced quantities:

$$T_r = T/T^*; \quad V_r = V/V^*; \quad p_r = pV^*/RT^* \tag{3.20}$$

where p is the vapor pressure. The melting point and the corresponding molar volume of potassium chloride are equated with its T^* and V^* and those of the other alkali metal halides are selected so that the molar volume and vapor pressure fit those for potassium chloride. Then:

$$V_r = 1.142 - 0.5570T_r + 0.4224{T_r}^2 \tag{3.21}$$

$$\log(p_r/p_0) = 4.945 - 11.905/T_r - 5.033\log T_r \qquad (3.22)$$

Expressions are also provided for the surface tension and the isothermal compressibility along the saturation (liquid-vapor coexistence) line. The average melting temperature is taken to be $T_m = 1.03\,T^*$ (except for the lithium salts) and the critical temperature is $T_c = 3.29\,T^*$ [141].

The significant structures theory of liquids [147] postulated that the liquid consists of solid-like particles or domains and gas-like ones, randomly intermingled. The partition function of the liquid is, therefore, the product of the two partition functions, as reported by Carlson et al. [50]. Subsequently, Lu et al. [142] modified this theory by noting that the gas-like particles in a molten salt are of two kinds at equilibrium: monomers and dimers. The partition function was modified accordingly. The total number of gas-like particles per mole of molten salt is $N_A\Delta_m V/V$, where, as before, V is the molar volume of the molten salt and $\Delta_m V$ is the change of the molar volume on melting. The equilibrium constant for the monomer-dimer equilibrium of the gas-like particles is subsequently disposed of, setting the number of monomeric gas-like particles approximately equal to their total number shown above. The required parameters for the calculation of the partition function, hence of the thermodynamic quantities (as well as transport quantities) are moment of inertia of the dimeric gaseous particles, I_d, and their ground state vibration frequency, ω_d, the molar sublimation energy of the solid at the melting point, E_s, the Einstein temperature, θ_E, of the solid, and the number n of nearest neighbors for the solid-like particles. The latter number is approximated by $n = Z(1 - \Delta_m V/V_m)$, where $Z = 6$, the lattice parameter for crystalline alkali metal halides and V_m is the molar volume of the molten salt at the melting point. For NaCl, NaBr, KCl, KBr, and KI, the salts tested for prediction of their properties, $4.13 \le n \le 4.35$, somewhat lower than obtained from diffraction or computer simulation studies, Tables 3.4, 3.5, 3.6, and 3.7 (except for KBr).

Vilcu and Misdolea [143] applied the significant structure theory and considered dimeric gaseous species too, obtaining acceptable values of the calculated constant volume heat capacity, but not of the constant pressure values. Contrary to these results, the constant pressure heat capacity calculated by Lu et al. [142] agreed with the experimental data, but the estimated critical temperatures were only 49–76 % of the values in Table 3.3, with opposite trends regarding the sizes of the anions. This was remedied by Cheng et al. [144] who applied a version of the significant structure theory to the estimation of thermophysical quantities for the same alkali metal halides as studied by Lu et al. [142] with acceptable success.

Although the significant structure theory specifies the Einstein temperature, θ_E, for the solid-like particles, the reference provided [148] for the values employed [142] are the Debye temperatures, θ_D. These are proportional to a function of the mean molar masses of the cations and anions of the alkali halide salts (except cesium ones) and the sum of the crystal ionic radii [149]:

$$\theta_D/\mathrm{K} = 11.7\left[(M_+ + M_-)/\mathrm{kg\ mol^{-1}}\right]^{-0.5}\left[(r_+ + r_-)/\mathrm{nm}\right]^{-1.5} \qquad (3.23)$$

The Einstein temperatures θ_E are approximately 0.75 of the Debye temperatures θ_D for the alkali metal halides [150].

All these theoretical approaches to the prediction of thermochemical properties of molten salts have dealt with charge-symmetrical salts only, and most of the calculation have been applied to alkali metal halides (mainly successful with sodium, potassium, and rubidium salts; lithium and cesium salts may show deviations). The corresponding states theory provided suggestions for corresponding temperatures at which the properties should be compared and characteristic temperatures, to be used as proportionality coefficients for reduced temperatures, have been suggested: $1.15T_m$ [139] and $0.97T_m$ [141]. A temperature somewhat above the melting point appeared to be necessary, in order to avoid problems with vestiges of the crystalline structures remaining in the melt [151]. A temperature of $1.1T_m$ was found useful [138] for the comparison of the surface tensions of molten salts. This temperature was used by Yosim and Owens [152] and Cantor et al. [153] to account for various thermochemical properties and was adopted by the present author [154–156] too for this purpose.

3.3.2 *Enthalpies of Phase Changes, Cohesive Energies, and Heat Capacities*

The molar enthalpies of melting, $\Delta_m H$, and of vaporization, $\Delta_v H$ are important quantities characterizing molten salts. The enthalpies of melting, $\Delta_m H$, shown in Table 3.8 for uni-univalent salts pertain to the melting point temperature, T_m, because most high melting salts do not supercool appreciably nor can retain crystalline states above T_m. The vapor pressures of molten salts are expressed as either the simpler or as some extended equation:

$$\ln(p/p_0) = a - b/T \tag{3.24a}$$

$$\ln(p/p_0) = a - b/T - c\ln(T/T_0) \tag{3.24b}$$

The Clausius-Clapeyron expression for the enthalpy of vaporization is:

$$\Delta_v H = -RT^2(\partial \ln p/\partial T)_P \tag{3.25}$$

resulting in $\Delta_v H = Rb$ from Eq. (3.24a) and $\Delta_v H = Rb - RcT$ from Eq. (3.24b). Although in principle the enthalpies of vaporization should depend on the temperature, for many salts only the constant values from Eq. (3.24a) have been reported. It should be noted that Eq. (3.25) is applicable only when the molar volume of the liquid is much smaller than that of the vapor, which is regarded as an ideal gas. At temperatures approaching the critical one, T_c, these postulates are no longer valid.

For all the salts with univalent cations the $\Delta_m H$ values and the $\Delta_v H$ values, as reported in [2] at the normal boiling point T_b, are reproduced in Table 3.8. Similar

Table 3.8 The molar enthalpies of fusion of univalent metal salts, $\Delta_m H$/kJ mol^{-1} (upper row), from [2], and of vaporization of the halides, $\Delta_v H$/kJ mol^{-1} (lower row) at T_m from [157], unless otherwise annotated

	Li^+	Na^+	K^+	Rb^+	Cs^+	Cu^+	Ag^+	Tl^+
F^-	27.2	33.1	28.2	26.4	21.7		16.7[r]	13.8
	231.4	228.9	195.8	186.2	167.8		204.9[s]	
Cl^-	19.8	28.2	26.3	18.4	21.8[c]	10.2	13.0	15.9
	184.9	189.5	182.8	177.0	170.3		177.8	
Br^-	17.7	26.1	25.5	15.5	20.3[c]	9.6	13.0[c]	16.3
	176.1	178.2	174.5	170.7	160.2		192.0	
I^-	14.6	23.6	24.0	12.6	25.5[c]	10.9	9.4	14.6
	161.9	163.6	166.9	162.3	160.7		144.3	
OH^-	21.0	6.4	8.6[c]	8.9[n]	4.6[n]			
	167.8		133.9					
CN^-		18.4[c]	14.6					
SCN^-		[24.1]	[12.8]		6.0[p]			
BO_2^-	33.9	33.5[c]	31.8[l]		34.8[q]			
	212.5		271.1					
NO_2^-	17.1[a]	16.5[a]	9.9[a]	11.0[a]	12.0[a]			6.9[a]
NO_3^-	26.7[b]	15.6[b]	10.0[b]	4.6[o]	14.1[o]		12.6[b]	8.8[b]
ClO_4^-	17.0[c]		20.5[h]					
ReO_4^-		32.6[h]	38.9[h]	21.8[h]	31.4[h]			
BF_4^-	14.5[d]	13.6[d]	18.0[d]	19.6[d]	19.2[d]			
HCO_2^-	16.1[e]	16.9[i]	11.9[i]					
$CH_3CO_2^-$	12.6[f]	17.4[f]	12.0[f]	8.9[f]	11.0[f]			
O^{2-}	43.1[c]	47.7		20.9[c]		64.0[c]		30.2[c]
S^{2-}		30.1[j]	16.2[c]			9.6[c]	7.9[c]	23.0[c]
CO_3^{2-}	43.9[g]	30.0[g]	32.9[g]	58.7[g]	49.2[g]			
SiO_3^{2-}	28.0[c]	52.3[c]	50.2[c]		39.7[c]			
SO_4^{2-}	7.7[m]	23.0	36.8	38.4[c]	36.4[c]		18.0	
CrO_4^{2-}		25.1[b]	31.5[b]					
MoO_4^{2-}	[47.7]	24.4[b]	38.7[b]		31.8[c]			
WO_4^{2-}	28.5[c]	31.5[b]	30.7[b]					
$S_2O_7^{2-}$		41.7[k]	21.2[k]	17.8[k]	19.5[k]			

[a][48], [b][158], [c][175], [d][159], [e][168], [f][173], [g][169], [h][162], [i][165], [j][171], [k][172], [l][164], [m][160], [n][163], [o][166], [p][174], [q][161], [r][167], [s][170]

data for the halides of multivalent metal salts are shown in Table 3.9. For several of the latter salts pre-melting solid-solid transitions occur, and the enthalpy of melting shown in these tables are from the crystalline structure stable just under the melting point. Such enthalpies may then be rather small, because a considerable amount of heat has already been absorbed during the solid-solid transition.

The cohesive energy of *molecular liquids* is related to their enthalpy of vaporization, $ce = \Delta_v H - RT$. However, because the vaporization of *molten salts* at the

Table 3.9 The molar enthalpies of melting of multivalent metal halides, $\Delta_m H$/kJ mol^{-1} (upper row), and their molar enthalpies of vaporization at T_b, $\Delta_V H$/kJ mol^{-1} (lower row), from [2] unless otherwise noted

	F^-	Cl^-	Br^-	I^-
Be^{2+}	14.6	8.7	10.0[b]	20.9
	195.4	105	100.0	96
Mg^{2+}	58.2	43.1	34.7	29.3[b]
	273.2	156.2	127.9[u,x]	
Ca^{2+}	29.7	28.5	28.9	41.8[b]
	390.8[a]	251.5[a]		
Sr^{2+}	29.7[b]	15.9	18.8	19.7
		277.2[i]		
Ba^{2+}	23.3[b]	16.7	31.4	28.5
		242.8[i]		
Mn^{2+}	25.0	37.7	33.5[b]	42
		149.0	113	
Fe^{2+}	51.9	43.1	51.9[b]	44.8
		125.5		
Co^{2+}	44.9[c]	36.4[j]	27.2[j]	
		157.3[k]		
Ni^{2+}		77.3[l]		
		223.5[m]		
Cu^{2+}	55.2[b]	200.8[m]		
	184.0			
Zn^{2+}	39.9[d]	10.3	15.6	
		119.2	98.3	
Cd^{2+}	23.0[e]	30.1	35.3[v]	20.7[v]
		110.3[m]	116.9[u]	107.2[u]
Sn^{2+}		12.8	18.0[b]	18.8[b]
		81.6		
Pb^{2+}	14.7[b]	23.8	16.4[b]	23.4[b]
	156.9	126.8	115.9	112.9
UO_2^{2+}		44.1[n]		
Fe^{3+}		43.1		
		148.4[m]		
Sc^{3+}	62.8[b]	67.4[b]		
Y^{3+}	28.0[f]	31.5[b]		
La^{3+}	55.9[j]	55.7[o]	54.4[s]	56.0[w]
		305.4[p]		
Ce^{3+}	56.5[j]	55.5[o]	52.0[w]	52.4[q]
		284.5[p]		
Pr^{3+}	57.3[j]	52.1[o]	47.3[s]	52.3[b]
		264.4[p]		
Nd^{3+}	54.8[j]	48.1[o]	45.2[s]	41.5[w]
		247.7[p]		

(continued)

Table 3.9 (continued)

	F^-	Cl^-	Br^-	I^-
Sm^{3+}	52.4[j]	44.8[q]		
Eu^{3+}	52.9[j]	45.0[r]		
Gd^{3+}	52.4[j]	40.6[o]	36.4[s]	54.0[w]
Tb^{3+}	58.4[j]	25.3[b]		57.4[w]
Dy^{3+}	58.4[j]	25.5[q]		
Ho^{3+}	56.8[j]	29.3[s]	50.1[w]	
Er^{3+}	27.5[j]	32.6[s]		
Tm^{3+}	28.9[j]	35.6[o]		
Yb^{3+}	29.7[j]	28.8[q]		
Lu^{3+}	29.3[j]			
Bi^{3+}	21.6[h]	23.7[t]	21.7[v]	39.1[b]
Th^{4+}	43.9[b]	61.5[b]	54.4[b]	48[u]
		152.7		
U^{4+}	30.7[b]	44.8	55.2	23.6[b]
		141.4		

[a][195], [b][175], [c][185], [d][188], [e][189], [f][187], [g][180], [h][186], [i][191], [j][177], [k][179], [l][183], [m][31], [n][43], [o][193], [p][176], [q][194], [r][196], [s][184], [t][182], [u][192], [v][181], [w][190], [178]

corresponding temperature $1.1T_m$ is to a large extent to (monomeric or dimeric) molecular species of the salt rather than to individual ions at infinite separation, $ce \neq \Delta_v H - RT$. Instead, the cohesive energies of molten salts at $1.1T_m$ are calculated from Eq. (3.26) (Figs. 3.1, 3.2, and 3.3):

$$ce = -U_L - [(H_{298} - H_0) - (H_{1.1Tm} - H_{298}) - 1.1\nu RT_m] \qquad (3.26)$$

as shown schematically in Fig. 3.3. The lattice potential energies $-U_L$ are from the compilation by Jenkins and Roobottom in [1]. Relatively small corrections to $-U_L$ to yield ce pertain to the raising of the temperature from 0 to 298.15 K as $(H_{298} - H_0)$ with data from [197], and from 298.15 K through the melting process up to $1.1T_m$ as $(H_{1.1Tm} - H_{298})$ with data from [2]. The term $\nu R(1.1T_m)$ is the correction from the enthalpies to the energies, where v is the number of ions per unit formula of the salt. The sum of corrections $(H_{298} - H_0) + (H_{1.1Tm} - H_{298})$, if not available in [197] and [2], may be approximated by $2\nu RT_m$. Some lattice energies have been obtained from [198–200] as noted. In those cases where the $-U_L$ values were not available in [1] nor in the latter three references, they can be estimated according to Jenkins et al. [201] from the microscopic volumes, v_m, which are additive for the constituent ions (Table 2.4) times their number per unit formula of the salt, and the expression:

$$-U_L/\text{kJ}\cdot\text{mol}^{-1} = \nu z_+|z_-|\left(138.7/\left(v_m/\text{nm}^3\right)^{1/3} + 26.7\right) \qquad (3.27)$$

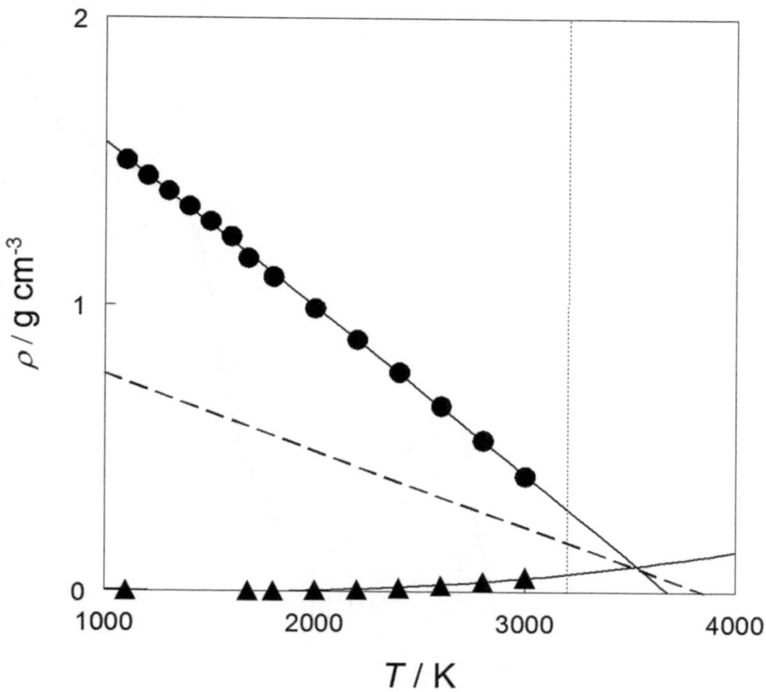

Fig. 3.1 Molten KCl: liquid (*circles*) and vapor (*triangles*) densities and the rectilinear diameter (*dashed line*) from data of Kirschenbaum et al. [52]. Their meeting point should represent the upper limit of T_c, the vertical dotted line is the reported T_c, which is supposed to be 200 K below this upper limit, but is actually ~400 K below it

The derived cohesive energies at the corresponding temperature $1.1T_m$ are shown in Table 3.10. Included also in this table are the cohesive energy densities, *ced*, at the same temperature, using the molar volumes reported in Sect. 3.3.3 in Tables 3.13 and 3.14. Where no entries for *ced* are shown in Table 3.10 the reason is lack of density data for the molten salt.

The most widely used method for the measurement of the constant pressure molar heat capacity of molten salts, C_P, was drop calorimetry, in which a sample preheated to some definite temperature imparted its heat to a fluid in the calorimeter, raising its temperature. The expected accuracy of the measurements was $\pm 1\,\%$. An alternative method introduced more recently is differential scanning calorimetry, but it has a somewhat lower accuracy (uncertainty $\geq 2\,\%$), due to the smallness of the samples employed. It turned out that C_P of molten salts varied generally hardly at all with the temperature as it was reported in [2], [175], and [202] and shown in Table 3.11.

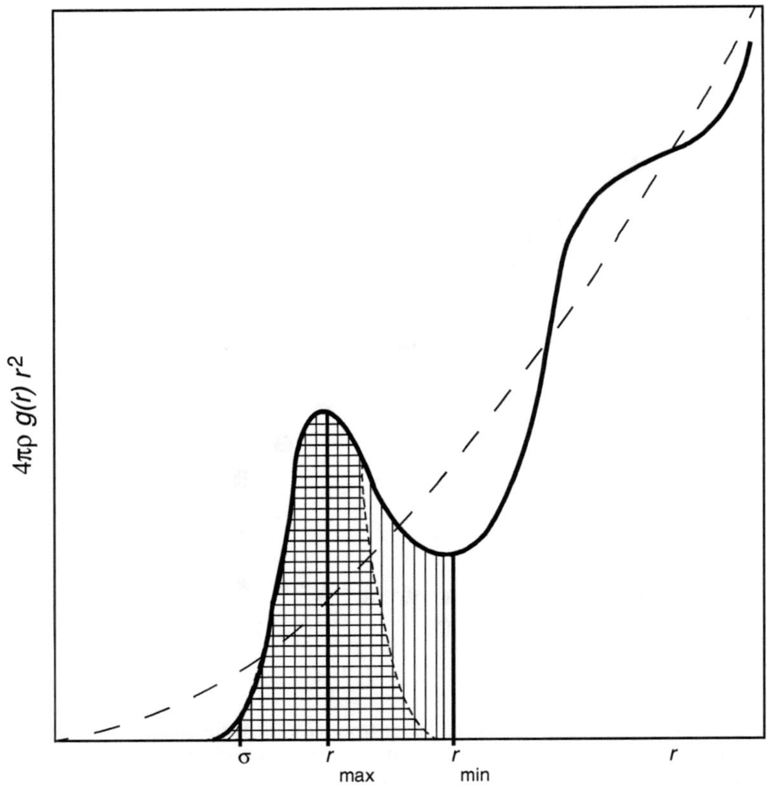

Fig. 3.2 The number of neighboring particles in a fluid that a particle at the origin has, thick curve, as a function of the distance r. It reaches asymptotically the dashed line that signifies no correlation between the particles. Two maxima are seen, representing the first and second coordination spheres of the particle at the origin. The vertically hatched area represents the coordination number N_{co} in the first coordination sphere up to the first minimum in the curve, r_{min}; the cross-hatched area represents the coordination number N_{co} in the first coordination sphere assuming symmetry around the peak, r_{max}. (From: Y. Marcus, Introduction to Liquid State Chemistry, Wiley, Chichester, 1977, by permission of the publisher)

The constant volume molar heat capacity of molten salts, C_V, is not measured directly but is derived from C_P and requires volumetric data:

$$C_V = C_P - TV{\alpha_P}^2/\kappa_T \tag{3.28}$$

where V is the molar volume, α_P is the isobaric expansibility, and κ_T is the isothermal compressibility (see Tables 3.13, 3.14, 3.15, and 3.16 for such data).

The C_P values of symmetrical molten salts with univalent monatomic ions (i.e., practically the alkali metal halides) are approximated well [152] by:

$$C_P = 3R + 2RT\alpha_P\left(1 + y + y^2\right)/(1 - y)^3 \tag{3.29}$$

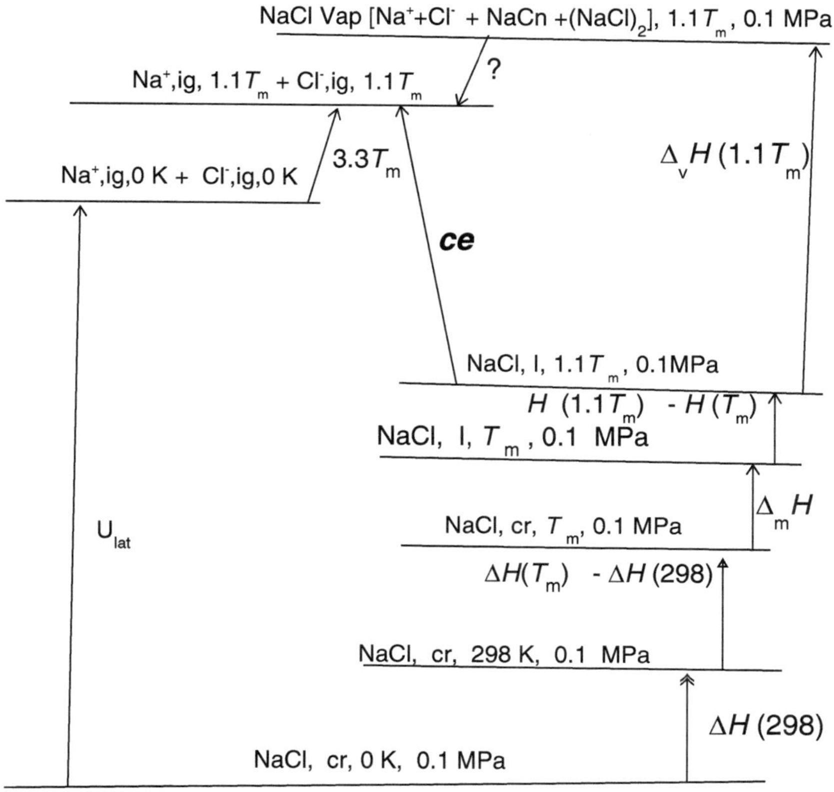

Fig. 3.3 A schematic representation of the energetic steps leading from the crystalline salt (NaCl) at 0 K to its ions in the ideal gas state (Na^+,ig and Cl^-,ig) at $1.1\,T_m$ at the standard pressure, 0.1 MPa. On the left hand side the lattice energy U_{lat} is invested and the ions are heated to $1.1\,T_m$. On the right hand side The crystal is heated to 298 K, then to the melting point T_m, the enthalpy of melting is invested, $\Delta_m H$, and the liquid salt is heated to $1.1\,T_m$, then vaporized by investment of $\Delta_v H$ to produce the vapor at $1.1\,T_m$ and 0.1 MPa. The vapor is not an ideal gas, and some energy is regained on expansion to this state. The cohesive energy *ce* of the molten salt at $1.1\,T_m$ is calculated from the energies of the ideal gas ions and that of the liquid at this temperature

where $y = \pi N_A d^3/V$ is the packing fraction and $d = r_+ + r_-$ is the sum of the cation and anion radii (Table 2.4). Computer simulations were applied in [118] to charge-symmetrical salts to obtain the constant volume heat capacities C_V of some molten salts. On the whole, for the uni-univalent salts with monatomic ions, Table 3.11, C_P varies little with the nature of the ions, being $(8 \pm 1)R$, but is linearly correlated with the sum d of the cation and anion radii [202]:

$$C_P/\mathrm{J\,K^{-1}mol^{-1}} = (54.4\ \pm\ 4.8) + (45.4\ \pm 16.0)(d/\mathrm{nm}) \qquad (3.30)$$

Table 3.10 The cohesive energy *ce* and its density *ced* of molten salts at $1.1T_m$

Salt	$1.1T_m$/K	*ce*/kJ mol^{-1}	*ced*/MPa	Salt	$1.1T_m$/K	*ce*/kJ mol^{-1}	*ced*/MPa
LiOH	820	1035	58.1	BeF_2	908	3487	147.8
LiF	1232	956	64.6	$BeCl_2$	784	3024	53.5
LiCl	971	818	28.4	MgF_2	1690	2918	110.1
LiBr	902	735	20.9	$MgCl_2$	1057	2467	42.9
LiI	816	691	15.5	$MgBr_2$	1086	2433	33.9
$LiNO_3$	578	796	20.1	MgI_2	1015	2318	24.8
$LiClO_3$	441		16.0[b]	CaF_2	1860	2630	82.7
$LiClO_4$	560	700	13.1	$CaCl_2$	1151	2316	42.5
$LiBF_4$	635	710	13.9	$CaBr_2$	1103	1702	26.1
$LiCH_3CO_2$	615	851	13.0	CaI_2	1163	1978	22.6
Li_2CO_3	1109	2428	60.4	$Ca(NO_3)_2$	906	2291	33.5[b]
Li_2SO_4	1245	2158	38.4	SrF_2	1925	2472	67.0
Li_2MoO_4	1076	1999	32.4[b]	$SrCl_2$	1261	2134	35.8
Li_2WO_4	1117	2000	32.2[b]	$SrBr_2$	1008	2020	29.7
NaOH	656	825	36.2	SrI_2	867	2293	27.1
NaF	1395	883	39.4	$Sr(NO_3)_2$	966	2200	27.4[b]
NaCl	1181	690	17.7	BaF_2	1805	2330	53.7
NaBr	1122	665	14.6	$BaCl_2$	1359	2035	30.2
NaI	1026	610	10.8	$BaBr_2$	1265	1958	25.4
$NaNO_2$	599	758	19.4	BaI_2	1114	2166	22.9
$NaNO_3$	638	682	15.3	$Ba(NO_3)_2$	955	2086	22.5[b]
$NaClO_3$	573	780	14.0[b]	$MnCl_2$	1015	2535	46.6
$NaBF_4$	749	669	11.4	$FeCl_2$	1045	2595	
$NaCH_3CO_2$	662	766	11.6	$CoCl_2$	1114	2735	49.0
Na_2CO_3	1244	2274	41.2	$NiCl_2$	1309	2786	57.8
Na_2SO_4	1273	1758	24.9	$CuCl_2$	847	2793	57.8
Na_2CrO_4	1177	1866	25.3	$ZnCl_2$	650	2549	46.6
Na_2MoO_4	1058	1983	26.4[b]	$ZnBr_2$	734	2637	39.9
Na_2WO_4	1064	1987	25.7[b]	ZnI_2	791	2601	30.8
Na_3AlF_6	1407	3809	36.0	$CdCl_2$	925	2567	46.5
KOH	696	776	23.6	$CdBr_2$	924	2493	36.5
KF	1244	725	22.9	CdI_2	726	2641	31.1
KCl	1154	622	12.2	$SnCl_2$	572	2316	40.7
KBr	1108	594	10.2	$PbCl_2$	851	2291	39.8
KI	1049	558	7.9	$PbBr_2$	711	2229	34.1
KSCN	495	583	10.7	PbI_2	751	2182	26.4
KNO_2	761	711[a]	14.1[b]	GaI_3	512	4628	36.6
KNO_3	668	617	11.1	$InCl_3$	945	4764	42.3
$KClO_3$	705	724[a]	11.3[b]	$InBr_3$	780	4560	38.7
KBF_4	927	609	8.2	InI_3	531	4260	32.2

(continued)

Table 3.10 (continued)

Salt	$1.1T_m$/K	ce/kJ mol^{-1}	ced/MPa	Salt	$1.1T_m$/K	ce/kJ mol^{-1}	ced/MPa
KCH_3CO_2	635	684	9.3	YCl_3	1048	4541	57.7
K_2CO_3	1289	2045	27.3	LaF_3	1943	4747	108.4
K_2SO_4	1476	1635	17.0	$LaCl_3$	1260	4305	54.6
K_2MoO_4	1319	1941	20.5[b]	$CeCl_3$	1205	4434	56.5
K_2WO_4	1323	1908	18.1	$PrCl_3$	1165	4357	56.9
$K_2Cr_2O_7$	738	1748	13.3[b]	$NdCl_3$	1134	4381	56.1
RbOH	722	715	17.8	$SmCl_3$	1051	4411	57.2
RbF	1217	707	18.6	$EuCl_3$	986	4426	
RbCl	1087	614	11.0	$GdCl_3$	970	4438	58.9
RbBr	1051	589	9.3	$DyCl_3$	1048	4516	59.7
RbI	1012	560	7.4	$HoCl_3$	1090	4534[c]	61.4
$RbNO_3$	641	613	10.1	$ErCl_3$	1154	4565	61.2
$RbBF_4$	941	593	7.2	$YbCl_3$	1263	4693	63.1
$RbCH_3CO_2$	571	724	9.9	$LuCl_3$	1318	4609[c]	55.0
Rb_2CO_3	1221	2030	24.3	$BiCl_3$	556	4707	57.4
Rb_2SO_4	1456	1576	14.5	$BiBr_3$	540	3311	34.1
CsF	1074	674	15.6	UCl_3	1221	4204[c]	48.3
CsCl	1011	590	9.4				
CsBr	1000	563	8.0				
CsI	989	531	6.3				
$CsNO_3$	756	661	7.9				
$CsBF_4$	911	574	6.2				
$CsCH_3CO_2$	514	691	8.6				
Cs_2CO_3	1173	1949	20.1[b]				
Cs_2SO_4	1406	1537	9.7				
CuCl	765	1005	36.8				
AgCl	801	917	27.2				
AgBr	778	904	23.7				
AgI	912	890	18.7				
$AgNO_3$	534	829	16.2				
Ag_2SO_4	1031	2130	32.3				
TlCl	773	748	14.1				
TlBr	806	732	15.6				
TlI	715	699	13.0				
$TlNO_3$	531	680	12.3[b]				

[a][198], [b]Evaluated in [199] from estimated lattice energies and molar volumes, [c][200], [d][199]

For ions with univalent cations and polyatomic anions C_P is very roughly $(3 \pm 1)\nu bR$, where ν is the number of ions per formula unit of the salt and b is the number of bonds in the anion. Much fewer data are available for the C_P of molten salts with multivalent cations, practically limited to those with divalent cations and lanthanides, shown in Table 3.12. Correlation expressions similar to Eq. (3.30) were presented in [202] for molten salts with polyvalent cations.

Table 3.11 The molar heat capacities, C_P/J K^{-1} mol^{-1}, of molten univalent metal salts, from Barin and Knacke [2] (Roman font) and from Marcus [202] (*italics* font), unless otherwise annotated

	Li^+	Na^+	K^+	Rb^+	Cs^+	Cu^+	Ag^+	Tl^+
F^-	64.2	70.6	66.9	58.8	74.1			67.3
Cl^-	64.2[a]	68.9[a]	73.6	64.0	77.4[a]	*66.9*	66.9	59.4
Br^-	65.3	62.3	69.9	66.9	77.4	*66.9*	62.3	75.1[a]
I^-	63.2	64.9	72.4	67	67.4[a]	*70.8*	58.6	72.0
OH^-	86.8	85.7[a]	83.1	*83.7*	*81.6*			
CN^-		*79.5*	*75.3*					
BO_2^-	144.6	146.4[c]	126[d]					
NO_3^-	*152.7*	155.6	123.4		*136.0*		*128.0*	*130.3*
ClO_3^-	*122.2*	*133.9*						
ClO_4^-	161.1							
BF_4^-		*165.4*	*167.2*					
HCO_2^-	137.2[b]	*142.3*						
$CH_3CO_2^-$	187.9[b]	*226.4*	181.3[e]					
O^{2-}	97.1[c]	104.6		95.8[c]		100.4[c]		95.0[c]
S^{2-}		77.9	*101.0*			89.1[c]	*93.1*	99.6[c]
CO_3^{2-}	185.4	*185.9*	209.2[c]					
SiO_3^{2-}	*167.4*	*179.1*	167.4[c]		194.6[c]			
SO_4^{2-}	*207.9*	197.4	200.0	*207.9*	*207.1*		205.0	*205.0*
CrO_4^{2-}	*200.0*	*204.6*	*209.2*		*210.9*			
MoO_4^{2-}	*215.1*	*213.0*			*210.0*			
WO_4^{2-}	*205.0*	*209.2*	*213.4*	*213*	*214.2*			
$S_2O_7^{2-}$		*244.8*	*267*	*272.2*	*222*			
$Cr_2O_7^{2-}$			*415.9*					
$B_4O_7^{2-}$	*453.6*	*444.9*	*464.6*					

[a]When so marked the heat capacity is temperature dependent and the entry pertains to $T = 1.1T_m$, [b]From [168], [c][175], [d]From [167], [e]From [203]

3.3.3 Volumetric Properties

A great deal of experimental information on the densities of molten salts as functions of the temperature is available in the Molten Salts Handbook by Janz [3] and subsequent compilations by Janz and coworkers [211–214] and recently by Marcus [215]. A discussion of the molar volumes of the various salts should be based on the concept of corresponding states, and in view of the large range of melting points of the salts, a corresponding temperature should be selected. As in other sections of this Chapter, $1.1T_m$, where T_m/K is the melting point of the salt, is used as the corresponding temperature.

The densities of molten salts [3, 211–215] can be measured accurately as functions of the temperature (and pressure) by a variety of methods [215]. At the elevated temperatures needed the archimedean sinker method is widely used and

Table 3.12 The molar heat capacities, C_P/K K^{-1} mol^{-1}, of multivalent metal salts, from [2] and [197] and as noted

	F^-	Cl^-	Br^-	I^-	O^{2-}	Other
Be^{2+}	88	121.4	113.0	113.0	64.9[i]	
Mg^{2+}	94.4	92.5	97.3	100.4	60.7	159.0[m]
Ca^+	100.0	103.3	114.6	117.2	83.7[i]	444.8[n]
Sr^{2+}	100.4	104.6	108.8	110.0		
Ba^{2+}	99.8	104.4	104.9	113.0	66.9[i]	
Mn^{2+}	98.3	94.6[a]	100	109	60.7	71[o]
Fe^{2+}	98.3	102.1	106.7	108.8	68.2	72.8[p]
Co^{2+}		100.4				
Ni^{2+}		100.6[b]				76.8[q]
Zn^{2+}	94.1	100.8	113.8			62.0[q]
Cu^{2+}	100.4	100[c]				
Cd^{2+}		110.0[d]	101.7[d]	102.1[d]		
Sn^{2+}		92.0	99.6	94.6		59.3[q]
Pb^{2+}	109.2	104.2	115.5	135.6	64.3	70.1[q]
Fe^{3+}		133.9				
Sc^{3+}	88.9	143.4				
Y^{3+}	133.7	177.6				
La^{3+}	152.7	157.7[e]	151.1[j]	151.8		
Ce^{3+}	125.0	145.2	152.7	152.7[e]		
Pr^{3+}	130.8	155.3[f]	154.8	143.1[e]		
Nd^{3+}	133.9	148.1[e]	148.5	151.9[e]		
Sm^{3+}		145.3[g]				
Eu^{3+}		156.0[g]				
Gd^{3+}	127.8	139.9[f]	139.3	155.9		
Tb^{3+}		139.3[h]	145.0[k]			
Dy^{3+}	156.9	159.4[f]				
Ho^{3+}		147.7				
Er^{3+}	139.1	148.5				
Tm^{3+}	140.3	148.5[g]				
Yb^{3+}		121.3[i]				
Bi^{3+}	184.6	143.5	157.7[d]	157.7	152.7[i]	188.3[q]
U^{3+}		129.7[i]				
Zr^{4+}					87.9[i]	
Th^{4+}	152.7	163.2	171.5	177.4		
U^{4+}	165.3	162.5	163.2	165.7	136.0[l]	

[a][201], [b][204], [c][202], [d][205], [e][206], [f][207], [g][208], [h][209], [i][175], [j][211], [k][210], [l][212], [m]$MgSO_4$, [n]For CaB_4O_7 near T_m, [o]MnS, [p]FeS, [q]For the sulfides at $1.1T_m$ from [203]

the densities are known to better than 0.1 %. They are generally measured at ambient pressures and are linear with the temperature over a wide temperature range above the melting point, up to and much beyond $1.1T_m$. This is expressed by Eq. (3.31):

$$\rho/(\mathrm{g\cdot cm^{-3}}) = a - 10^3 b(T/\mathrm{K}) \tag{3.31}$$

Given this dependence and the coefficients a and b, the following quantities for a large number of molten salts at $1.1T_m$ are then readily calculated: the densities, $\rho = a - 1.1 \times 10^3 b(T_m/\mathrm{K})$, the isobaric expansibilities, $10^3\alpha_P/\mathrm{K}^{-1} = b/\rho$, and the molar volumes, $V/(\mathrm{cm^{-3}\cdot mol^{-1}}) = M/\rho$, where the $M/(\mathrm{g\cdot mol^{-1}})$ are the molar masses. These quantities are shown in Tables 3.13 and 3.14. The densities and molar volumes at other temperatures are obtained from the values listed in Tables 3.13 and 3.14 as: $\rho(T) = \rho_{1.1Tm}[1 - \alpha_P(T - 1.1T_m)]$ and $V(T) = V_{1.1Tm}[1 + \alpha_P(T - 1.1T_m)]$.

The molar volumes at $1.1T_m$ of uni-univalent molten salts are related to the radii of the ions as

$$V_{1.1Tm}/\mathrm{cm^3\ mol^{-1}} = (4\pi N_A/3)\left[3.3(r_+/\mathrm{nm})^3 + 1.78(r_-/\mathrm{nm})^3\right] \tag{3.32}$$

within $\pm 3\ \mathrm{cm^3\ mol^{-1}}$ [216]. Salts with considerable covalent bonding, such as CuCl, have molar volumes smaller than those corresponding to Eq. (3.32). For nonsymmetrical salts the correlations take into account the presence of vacancies in the quasi-lattice of the melts and other expressions have been obtained [215, 216].

The expansibilities of molten salts at $1.1T_m$ correlate well with the inverse of their cohesive energies ce (listed in Table 3.10):

$$10^3\alpha_P/\mathrm{K}^{-1} = (245 \pm 35)z_+z_-ce^{-1} \tag{3.33}$$

where the z are the numerical charge numbers of the ions. This relationship is rationalized by noting that the larger the cohesive energy, the more difficult is for thermal agitation to expand the volume of the molten salt. Silver halides and salts of divalent post-transition-metals do not conform to this relationship.

The adiabatic compressibilities of molten salts are obtained from measurements of the ultrasound velocities in them, u, with a probable uncertainty of ± 1 %, and of their densities, ρ: $\kappa_S = u^{-2}\rho^{-1}$. The isothermal compressibilities can be obtained from measured thermal pressure coefficients $(\partial P/\partial T)V$ as $\kappa_T = \alpha_P/(\partial P/\partial T)V$ (with a probable uncertainty of ± 1.5 %). Otherwise the isothermal compressibilities are obtained from the adiabatic ones as $\kappa_T = \kappa_S + TV\alpha_P^2/C_P$. The available values of κ_T and κ_S are shown in Tables 3.15 and 3.16, as listed also in [215].

The isothermal compressibilities of halides of the alkali metals, alkaline earth metals, and divalent transition- and post-transition-metals are inversely correlated with their cohesive energies [221] in analogy with the expansibilities with a similar rationalization:

$$\kappa_T/\mathrm{GPa}^{-1} = (4.37 \pm 0.80)z_+^2ce^{-1} \tag{3.34}$$

Table 3.13 The densities, ρ/g·cm^{-3}, the expansibilities, α_P/K^{-1}, and the molar volumes, V/cm^3·mol^{-1}, of molten univalent metal salts at the corresponding temperatures 1.1(T_m/K) from [214] and [216] or as noted

Salt	$1.1T_m$	ρ	$10^3\alpha_P$	V
LiF	1233	1.7536	0.278	14.8
LiCl	971	1.4706	0.287	28.8
LiBr	905	2.4759	0.264	35.1
LiI	816	3.0415	0.303	44.6
LiOH	820	1.3429	0.339	17.8
$LiNO_3$	578	1.7410	0.236	39.6
$LiBF_4$	635	1.8392	0.249	50.9
$LiClO_3$[a]	441	2.0545	0.087	44.0
$LiClO_4$	560	1.9944	0.307	53.3
Li_2CO_3	980	1.8372	0.203	40.2
Li_2SO_4	1245	1.9573	0.208	56.2
Li_2MoO_4	1071	2.804	0.181	61 0.7
Li_2WO_4[b]	1114	4.0084	0.805	65.3
NaF	1395	1.8731	0.302	22.4
NaCl	1181	1.4977	0.362	39.0
NaBr	1122	2.2582	0.361	45.6
NaI	1026	2.6544	0.356	56.5
NaOH	650	1.7570	0.273	22.8
$NaNO_2$	613	1.7684	0.422	39.0
$NaNO_3$	637	1.9070	0.324	44.6
$NaClO_3$	537	2.0718	0.466	51.4
$NaPO_3$	991	2.2101	0.164	46.1
$NaBF_4$	749	1.9055	0.394	57.6
$NaCH_3CO_2$	662	1.2363	0.487	66.4
Na_2CO_3	1244	1.9213	0.234	55.2
Na_2SO_4	1273	2.0128	0.241	70.6
Na_2CrO_4[b]	1177	2.193	0.221	73.8
Na_2MoO_4	1056	2.7427	0.230	75.1
Na_2WO_4[b]	1068	3.5515	0.907	82.8
$Na_2S_2O_7$	781	1.7329	0.328	128.2
Na_3AlF_6	1407	1.9834	0.495	105.8
KF	1238	1.8399	0.353	31.6
KCl	1147	1.4670	0.395	50.8
KBr	1108	2.0440	0.403	58.3
KI	1049	2.3571	0.407	70.4
KOH	696	1.7069	0.258	32.9
KSCN	495	1.5623	0.512	54.5
KNO_2	781	1.6419	0.425	51.8
KNO_3	672	1.8235	0.400	55.4
KHF_2	563	1.9213	0.296	40.7
KBF_4	927	1.6950	0.481	74.3

(continued)

Table 3.13 (continued)

Salt	$1.1T_m$	ρ	$10^3\alpha_P$	V
KCH_3CO_2	635	1.3357	0.480	72.7
K_2CO_3	1289	1.8456	0.238	74.9
K_2SO_4	1484	1.8121	0.267	96.2
K_2MoO_4	1319	2.5147	0.113	94.7
K_2WO_4[b]	1323	2.8825	0.727	105.6
$K_2S_2O_7$	795	1.5633	0.410	160.4
$K_2Cr_2O_7$	738	2.2395	0.313	131.4
RbF	1217	2.7526	0.371	38.0
RbCl	1095	2.1537	0.410	56.1
RbBr	1062	2.6008	0.412	63.6
RbI	1012	2.7927	0.408	76.1
RbOH	722	2.5476	0.306	40.2
$RbNO_3$	641	2.4261	0.400	60.8
$RbBF_4$	941	2.1003	0.495	82.0
$RbCH_3CO_2$	571	1.9460	0.494	73.4
Rb_2CO_3	1221	2.7689	0.245	83.4
Rb_2SO_4	1481	2.4571	0.271	108.7
$Rb_2S_2O_7$	814	2.1132	0.400	164.2
CsF	1074	3.5227	0.363	43.1
CsCl	1010	2.6932	0.396	62.5
CsBr	1000	3.0216	0.406	70.4
CsI	989	3.0709	0.386	84.8
$CsNO_3$	756	2.3415	0.500	83.2
$CsBF_4$	911	2.3928	0.497	91.8
$CsCH_3CO_2$	514	2.3723	0.414	80.5
Cs_2CO_3	1172	3.3734	0.254	96.6
Cs_2SO_4	1421	2.2833	0.257	158.5
$Cs_2S_2O_7$	850	2.1513	0.422	205.4
CuCl	765	3.6215	0.218	27.3
AgCl	801	4.2525	0.221	33.7
AgBr	778	4.9330	0.211	38.1
AgI	912	4.9424	0.204	47.5
$AgNO_3$	534	3.3185	0.325	51.2
Ag_2SO_4	1031	4.734	0.156	65.9
InCl	548	2.8755	0.505	52.2
TlCl	773	4.5192	0.398	53.1
TlBr	806	5.8840	0.326	48.3
TlI	715	6.1421	0.304	53.9
$TlNO_2$	505	4.8238	0.325	51.9
$TlNO_3$	531	4.8158	0.363	55.3
Tl_2SO_4	996	5.5029	0.237	91.7

[a][217], [b][218]

Table 3.14 The densities, ρ/g·cm^{-3}, the expansibilities, α_P/K^{-1}, and the molar volumes, V/cm^3 ·mol^{-1}, of molten multivalent metal salts at the corresponding temperatures 1.1(T_m/K) from Janz [214] or as noted

Salt	$1.1T_m$	ρ	$10^3\alpha_P$	V
BeF_2	908	1.9899	0.26	23.6
$BeCl_2$	784	1.4139	0.778	56.5
MgF_2	1690	2.3494	0.223	26.5
$MgCl_2$	1079	1.6601	0.193	57.4
$MgBr_2$	1086	2.5689	0.186	71.7
MgI_2	1015	2.9808	0.218	93.3
CaF_2	1860	2.4517	0.160	31.8
$CaCl_2$	1161	2.0357	0.209	54.5
$CaBr_2$	1103	3.0660	0.163	65.2
CaI_2	1163	3.3598	0.223	87.5
SrF_2	1840	3.4022	0.221	36.9
$SrCl_2$	1263	2.6598	0.218	59.6
$SrBr_2$	1008	3.6393	0.205	68.0
SrI_2	867	4.0351	0.219	84.6
BaF_2	1752	4.0405	0.247	43.4
$BaCl_2$	1359	3.0895	0.220	67.4
$BaBr_2$	1235	3.8567	0.223	77.1
BaI_2	1114	4.1338	0.236	94.6
$MnCl_2$	1015	2.3128	0.438	54.4
$FeCl_2$ [a]	1045	2.282	0.288	55.5
$CoCl_2$ [b]	1114	2.3251	0.335	55.8
$NiCl_2$	1309	2.6349	0.251	49.2
$ZnCl_2$	650	2.4917	0.186	54.7
$ZnBr_2$	738	3.4091	0.280	66.1
ZnI_2	791	3.7805	0.360	84.4
$CdCl_2$	925	3.3221	0.250	55.2
$CdBr_2$	924	3.9849	0.271	68.3
CdI_2	726	4.3221	0.258	84.7
$SnCl_2$	571	3.3301	0.360	56.9
$PbCl_2$	851	4.8350	0.310	57.5
$PbBr_2$	711	5.6159	0.296	65.4
PbI_2	751	5.5827	0.285	82.6
GaI_3	512	3.5610	0.668	126.5
$InCl_3$	945	1.9600	1.059	112.5
$InBr_3$	780	3.0128	0.498	117.7
InI_3	531	3.7455	0.398	132.3
YCl_3	1048	2.4800	0.209	78.7
LaF_3	1943	4.4679	0.153	43.8
$LaCl_3$	1260	3.1101	0.251	78.9
$LaBr_3$	1167	4.1180	0.243	91.9
LaI_3	1156	4.1739	0.266	124.5

(continued)

Table 3.14 (continued)

Salt	$1.1T_m$	ρ	$10^3\alpha_P$	V
CeF_3	1906	4.4690	0.209	44.7
$CeCl_3$	1205	3.1396	0.293	78.5
$PrCl_3$	1059	3.266	0.230	76.6
$NdCl_3$	1134	3.2094	0.290	78.1
NdI_3	1156	4.1699	0.257	125.9
$SmCl_3$[a]	1051	3.352	0.218	76.6
$EuCl_3$[a]	986	3.495	0.237	73.9
$GdCl_3$	970	3.4978	0.192	75.4
$DyCl_3$	1048	3.5520	0.192	75.7
$HoCl_3$[a]	1090	3.523	0.176	77.0
$ErCl_3$[a]	1154	3.521	0.145	77.7
$YbCl_3$[a]	1263	3.573	0.186	78.2
$LuCl_3$[a]	1318	3.541	0.193	79.4
$BiCl_3$	553	3.8453	0.598	82.0
$BiBr_3$	540	4.6205	0.523	97.1
BiI_3	750	4.5500	0.484	129.6
UCl_3	1221	3.9536	0.508	87.1
ThF_4	1521	5.3595	0.127	51.7
$ThCl_4$	1147	3.2172	0.435	116.2
UF_4	1440	6.3555	0.156	49.4
UCl_4	949	3.454	0.663	110.0

[a][220], [b][219]

Other molten salts have somewhat different correlation expressions [221]. Isothermal compressibilities of molten salts may be estimated from the scaled particle theory if no experimental data are available [222]:

$$\kappa_T = (V/RT)(1-y)^4/(1+2y)^2 \quad (3.35)$$

where, as defined above, $y = \pi N_A d^3/V$ is the packing fraction and $d = r_+ + r_-$ is the sum of the cation and anion radii.

Densities of molten salts at high pressures P are expressed by means of the Tait equation:

$$\rho(P,T) = \rho(P_0,T)/[1 - 0.1\ln\{1 + P/B(T)\}] \quad (3.36)$$

where $P_0 = 0.1$ MPa, i.e., ambient pressure to which the data in Tables 3.13, 3.14, 3.15, and 3.16 pertain, and $B(T)/\text{MPa} = 400 + 8.9(\kappa_T/\text{MPa})^{-1}$ and the coefficient 0.1 of the logarithmic term are assumed to be universal [223].

A quantity that is closely related to the volumetric properties of molten salts is their internal pressure:

$$P_{\text{int}} = (\partial U/\partial V)_T = T(\partial P/\partial T)_V - P = T\alpha_P/\kappa_T - P \tag{3.37}$$

For internal pressures at ambient conditions and saturation vapor pressures the last term in the third equality is negligible ($P = 0.1$ MPa and $P_{\text{int}} > 500$ MPa). The internal pressures of the molten salts, P_{int},, are included in Tables 3.15 and 3.16. These internal pressures are manifold larger than those for molecular liquids at ambient conditions. However, whereas for molecular liquids the P_{int} values are commensurate with their cohesive energy densities *ced*, this is not the case for the molten salts, for which $ced > 10P_{\text{int}}$ [155, 227], due to the strong coulombic forces holding the ions together.

3.3.4 Surface Tension

The surface tensions σ of many molten salts have been compiled by Janz and coworkers [3, 214, 236] as functions of the temperature and the data have been supplemented subsequently in several other publications [138, 237–242]. The values diminish linearly with increasing temperatures and are compared in Table 3.17 at a suitable corresponding temperature, $1.1T_m$ according to Reiss et al. [138]. For alkali metal halides, excepting the lithium salts, the surface tension at that temperature was correlated with the cube of the melting point [138].

The surface tensions σ at $1.1T_m$ of a large number of highly ionic molten salts of different types, 1:1, 1:2, 2:1, correlate well with the cohesive energy densities *ced* of the salts, Table 3.12, as shown by Marcus in Fig. 3.4 [156]. One line pertains to 45 salts with univalent anions:

$$\sigma/\text{mJ}\cdot\text{m}^{-2} = (58.0 \pm 3.1) + (2.90 \pm 0.11)(ced/\text{GPa}) \tag{3.38}$$

with a correlation coefficient of 0.9693. A second line pertains to 16 alkali metal salts with divalent anions, with practically the same slope, though with worse scatter, $r_{\text{corr}} = 0.8683$:

$$\sigma/\text{mJ}\cdot\text{m}^{-2} = (95.3 \pm 12.4)) + (2.88 \pm 0.42)(ced/\text{GPa}) \tag{3.39}$$

The outliers from the correlation (3.38) have considerably smaller values of the surface tension than expected from the correlation: MgF_2, $MgCl_2$, $CaCl_2$, CaI_2, and $Ca(NO_3)_2$, and no apparent explanation is found for these discrepancies. The 1:3 molten lanthanide chlorides are excluded from correlation (3.38), Fig. 3.4. They have quite small surface tension values, in view of their moderately large cohesive energy densities. These salts may not be so highly ionic, i.e., they possibly are somewhat associated at the surface. The post-transition-metal halides are also excluded from the correlation (3.38), although AgBr and $PbCl_2$ are rather near this line as is $TlNO_3$ that does conform well. However, $AgNO_3$ does not conform,

Table 3.15 The adiabatic compressibility, κ_S/GPa^{-1}, isothermal compressibility, κ_T/GPa^{-1}, and internal pressure P_{int}/MPa of molten univalent metal salts at the corresponding temperatures $T = 1.1(T_m/K)$

Salt	$1.1T_m$	κ_S	κ_T	P_{int}
LiF	1233		0.093[j]	3690
LiCl	971	0.176[a]	0.216[a]	1290
LiBr	905	0.198[a]	0.235[a]	1020
$LiNO_3$	578	0.178[a]	0.197[a]	690
$LiClO_4$	560		0.230[e]	750
Li_2CO_3	980	0.075[b]	0.091[h]	2190
Li_2SO_4	1245		0.099[i]	2620
Li_2MoO_4	1071		0.105[i]	1850
Li_2WO_4	1114		0.122[i]	1720
NaF	1395		0.133[j]	2960
NaCl	1181	0.247[a]	0.343[a]	1250
NaBr	1122	0.280[a]	0.361[a]	1120
NaI	1026	0.313[a]	0.436[a]	840
$NaNO_2$	613		0.180[e]	1440
$NaNO_3$	637	0.173[a]	0.198[a]	1040
$NaClO_3$	537		0.29[e]	860
Na_2CO_3	1244	0.102[b]	0.118[h]	2430
Na_2SO_4	1273		0.132[i]	2320
Na_2CrO_4	1177		0.138[i]	1890
Na_2MoO_4	1056		0.121[i]	2010
Na_2WO_4	1068		0.133[i]	1900
Na_3AlF_6	1407		0.718[j]	970
KF	1238		0.186[j]	2350
KCl	1147	0.308[a]	0.442[a]	1030
KBr	1108	0.330[a]	0.465[a]	960
KI	1049	0.380[a]	0.572[a]	750
KSCN	495		0.265[e]	960
KNO_3	672	0.193[a]	0.234[a]	1150
KCH_3CO_2	635	0.466[c]		
K_2CO_3	1289	0.146[b]	0.178[h]	1720
K_2SO_4	1484		0.207[i]	1910
K_2WO_4	1323		0.228[i]	2560
$K_2Cr_2O_7$	738		0.187[e]	1240
RbF	1217		0.176[k]	2570
RbCl	1095		0.429[l]	1050
RbBr	1062		0.499[l]	880
$RbNO_3$	641	0.179[d]	0.208[m]	1230
Rb_2CO_3	1221	0.158[b]	0.209[h]	1430
Rb_2SO_4	1481		0.251[i]	1600
CsF	1074		0.228[k]	1710
CsCl	1010	0.306[a]	0.461[a]	870

(continued)

Table 3.15 (continued)

Salt	$1.1T_m$	κ_S	κ_T	P_{int}
CsBr	1000	0.362[a]	0.584[a]	700
CsI	989	0.51[e]	0.69[e]	550
$CsNO_3$	756	0.268[d]	0.308[m]	1230
Cs_2CO_3	1172	0.170[b]	0.229[h]	1300
Cs_2SO_4	1421		0.275[i]	1330
AgCl	801	0.071[f]	0.101[e]	1750
AgBr	778	0.083[f]	0.108[e]	1520
AgI	912	0.139[f]	0.141[h]	1320
$AgNO_3$	534		0.113[d]	1540
Ag_2SO_4	1031		0.072[i]	2280

[a][137], [b][226], [c][235], [d][231], [e][225], [f][233], [g][224], [h][221], [i][88], [j][228], [k][229], [l][230], [m][232], [n][234]

its σ value being considerably larger than expected in view of its modest *ced* value. The σ values of AgCl and of the zinc, cadmium, and bismuth halides, and $SnCl_2$ and $GaCl_3$ are well below the correlation line. In the cases of these salts it is surmised that their partial covalent natures increases the *ced* values appreciably compared with molten salts with alkali metal or alkaline earth metal cations. Therefore, for given σ values, these salts fall below the correlation line in Fig. 3.4.

Although empirical, correlation (3.38) has some theoretical basis as shown for organic liquids and molten metals [242]. For a definite proportionality between σ and *ced* to hold, a dimensionally correct relationship requires the former to be divided by a length, e.g., $V^{1/3}$. Essentially a similar linear dependence of $\sigma/V^{1/3}$ and *ced* was found as (3.38) ($r_{corr} = 0.9443$), due to the limited variability of $V^{1/3}$, which is 3.7 ± 0.5 cm·mol$^{-1/3}$ for the salts under consideration.

Many models have been proposed in the literature, but most of them pertain only to the alkali metal halides. A "simplified corresponding states" correlation for the alkali metal halides (except RbI) was suggested by Harada et al. [141], rewritten as:

$$\sigma/\text{mJ}\cdot\text{m}^{-2} = \left[3.52 - 1.55 \times 1.1T_m/T^*\right]/\left[\left(N_A{}^{-2/3}k_B\right)V^{*2/3}/T^*\right] \qquad (3.40)$$

with averaged values of $V^* \approx 0.957\,V(1.1T_m)$ and $T^* \approx 0.972T_m$ except for the lithium salts, for which $T^* \approx 1.051T_m$. Reasonable values of the calculated σ result from this expression.

A model by Yajima et al. [224] for the molten alkali metal halides that involves charge electroneutrality near the surface was also capable of the prediction of the surface tensions, and these were deemed to be nearer the experimental values than those according to the corresponding states model values in [138] and [141].

A model due to Aqra [243, 246] traceable to the significant structure theory [142, 147] yielded the expression:

Table 3.16 The adiabatic compressibility, κ_S/GPa^{-1}, isothermal compressibility, κ_T/GPa^{-1}, and internal pressure P_{int}/MPa of molten multivalent metal salts at the corresponding temperatures $T = 1.1(T_m/K)$

Salt	T	κ_S	κ_T	P_{int}
MgF_2	1690		0.123[c]	3060
$MgCl_2$	1079	0.550[a]	0.769[d]	270
$MgBr_2$	1086	0.448[a]	0.717[d]	280
MgI_2	1015	0.391[a]	0.839[d]	260
CaF_2	1860		0.064[c]	4650
$CaCl_2$	1161	0.124[a]	0.384[d]	480
$CaBr_2$	1103	0.161[a]	0.528[d]	340
CaI_2	1163	0.263[a]	0.718[d]	360
SrF_2	1840		0.068[c]	5980
$SrCl_2$	1263	0.116[a]	0.434[d]	630
$SrBr_2$	1008	0.116[a]	0.403[d]	510
SrI_2	867	0.151[a]	0.453[d]	420
BaF_2	1752		0.089[c]	4860
$BaCl_2$	1359	0.130[a]	0.547[d]	550
$BaBr_2$	1235	0.163[a]	0.679[d]	410
BaI_2	1114	0.157[a]	0.635[d]	410
$ZnCl_2$	650	0.434[a]	0.547[d]	220
$ZnBr_2$	738	0.513[a]	0.679[d]	300
ZnI_2	791	0.576[a]	0.635[d]	450
$CdCl_2$	925	0.285[a]	0.448[d]	520
$CdBr_2$	924	0.405[a]	0.563[d]	450
CdI_2	726	0.424[a]	0.668[d]	280
$PbCl_2$	851		0.537[e]	970
$PbBr_2$	711		0.623[e]	340
GaI_3	512	0.663[b]	0.793[b]	430
InI_3	531	0.547[b]	0.596[b]	360
ThF_4	1521		0.088[c]	2200
UF_4	1440		0.13[c]	1730

[a][225], [b][221], [c][229], [d][232], [e][234]

$$\sigma = \varphi^{-1}(V_{sd}/V)^2 k_B T_m [0.0481 E_s/RT] \tag{3.41}$$

applicable to molten metal halides at their melting points. The numerical constant 0.0481 pertains to σ /mN m^{-1}, V_{sd} and V are the molar volume of the solid and molten salt at T_m, E_s is the enthalpy of sublimation of the salt, and $\varphi = 1.091 (V_{sd}/N_A)^{2/3}$. A simplified expression: $\sigma = 0.0481 E_s d/V$ was subsequently applied to molten rare earth trihalides [247], where d was the internuclear distance between cation and anion. Agreement for this model was achieved with the not very many experimental data with which these expressions were tested, and values for many divalent metal and rare-earth halides for which no experimental data were available to the author were predicted.

Table 3.17 The surface tensions, σ/mJ·m^{-2}, of molten salts at $T = 1.1(T_m/K)$ from [3, 214, 236] unless otherwise noted

Salt	$1.1T_m$	σ	Salt	$1.1T_m$	σ
LiF	1232	241	MgF_2	1690	223[k]
LiCl	971	131[b]	$MgCl_2$	1057	67
LiBr	902	123[c]	CaF_2	1860	284[k]
LiI	816	91[c]	$CaCl_2$	1151	140
$LiBO_2$	1234	257[d]	$CaBr_2$	1103	115
$LiNO_2$	542	114[e]	CaI_2	1163	83
$LiNO_3$	578	113	$Ca(NO_3)_2$	906	102
$LiClO_3$	441	85	SrF_2	1925	268[k]
Li_2CO_3	1109	245	$SrCl_2$	1261	161
Li_2SO_4	1245	225	$SrBr_2$	1008	146
Li_2MoO_4	1076	214[f]	SrI_2	867	143
Li_2WO_4	1117	217[f]	$Sr(NO_3)_2$	966	128
NaOH	656	151[g]	BaF_2	1805	221[k]
NaF	1395	189	$BaCl_2$	1359	156
NaCl	1181	106	$BaBr_2$	1265	144
NaBr	1122	93	BaI_2	1114	130
NaI	1026	80	$Ba(NO_3)_2$	955	134
NaOH	656	151[h]	$ZnCl_2$	650	54
$NaBO_2$	1363	182[d]	$ZnBr_2$	734	50
$NaNO_2$	599	118	$CdCl_2$	925	84
$NaNO_3$	638	116	$CdBr_2$	924	64
$NaClO_3$	573	87[i]	$SnCl_2$	572	110
$NaBF_4$	749	84	$PbCl_2$	851	138
Na_2CO_3	1244	206	$GaCl_3$	386	24
Na_2SO_4	1273	185	$LaCl_3$	1260	105
Na_2MoO_4	1058	189[f]	$PrCl_3$	1165	104
Na_2WO_4	1064	193[f]	$SmCl_3$	1051	93
Na_3AlF_6	1407	118	$ErCl_3$	1154	80
KOH	696	151[c]	$YbCl_3$	1263	81
KF	1244	135	$BiCl_3$	556	65
KCl	1154	91	$BiBr_3$	540	69[l]
KBr	1108	82	ThF_4	1521	238[m]
KI	1049	71	$ThCl_4$	1440	110[m]
KSCN	495	95	UF_4	1070	195[m]
KBO_2	1342	108[d]	UCl_4	949	75[m]
KNO_2	761	103			
KNO_3	668	106			
$KClO_3$	643[a]	80			
KBF_4	927	73[j]			
K_2CO_3	1289	159			
K_2SO_4	1476	143[c]			
K_2MoO_4	1311	142			

(continued)

Table 3.17 (continued)

Salt	$1.1T_m$	σ	Salt	$1.1T_m$	σ
K_2WO_4	1314	148			
$K_2Cr_2O_7$	738	139			
RbF	1217	118			
RbCl	1087	91			
RbBr	1051	84			
RbI	1012	75			
$RbNO_2$	761	99^e			
$RbNO_3$	641	105			
Rb_2SO_4	1456	125			
CsF	1074	100			
CsCl	1011	85			
CsBr	1000	77			
CsI	989	69			
$CsNO_2$	746	94^e			
$CsNO_3$	756	87			
Cs_2CO_3	1173	128^c			
Cs_2SO_4	1406	112^c			
AgCl	801	176			
AgBr	778	152			
$AgNO_3$	534	146			
$TlNO_2$	505	108^e			
$TlNO_3$	531	90			

[a]At $T_m + 2 = 643$ K, [b][138], [c][237], [d][245], [e][12], [f][239], [g][240], [h][244], [i][238], [j][241], [k][242], estimated at T_m from the scaled particle theory, [l]The temperature coefficient −0.872 given in [236] is obviously a misprint, leading to negative values of σ; instead, −0.072 yields a reasonable value commensurate with that of $BiCl_3$, [m][243] at T_m

The products of the surface tensions of molten salts and their isothermal compressibilities, $\sigma\kappa_T$, show interesting regularities. These products for liquids in general follow the expression:

$$\sigma\kappa_T = a\left(1 - 3y + y^3\right)/4(1 + 2y)^2 \qquad (3.42)$$

according to Mayer [248] and the scaled particle theory (SPT) [249], where d represents the diameter of the hard spheres to which the SPT pertains and $y = (N_A\pi/6)d^3/V$ is their packing fraction (cf. Eqs. 3.29 and 3.35). With appropriate selection of the mean diameter of the ions, d, the compressibility can be obtained with good accuracy from the experimental surface tension and vice versa. The product $\sigma\kappa_T$ has the dimension of a length and is approximately 0.07 L [250], where $\exp(-r/L)$ is the damping factor of the pair correlation function (cf. Eq. 3.1). Only very few molten salts were considered in these studies. Marcus [251] has assembled the products $\sigma\kappa_T$ of many ionic molten salts that present a considerable span of

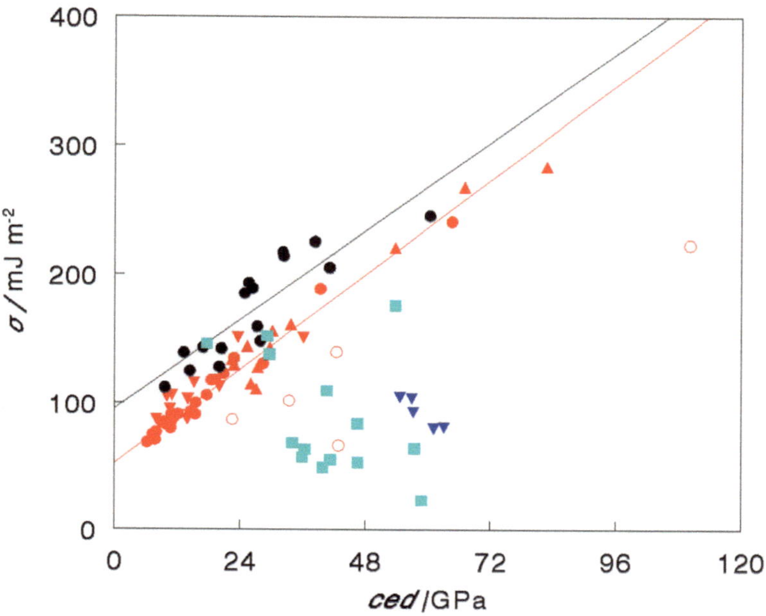

Fig. 3.4 The surface tension, σ, of molten salts plotted against their cohesive energy density, *ced*. Equation (3.3) pertains to the red symbols: alkali metal halides (●), alkaline earth metal halides (▲), other alkali metal salts with univalent anions (▼); Eq. (3.4) pertains to the alkali metal salts with divalent anions (●); outliers from Eq. (3.3) (○), also post-transition metal halides and $AgNO_3$ [■], and lanthanide chlorides (▼) (From Marcus [156] by permission of the publisher (Elsevier))

values, from 16 to 98 pm. They depend solely on the charges of the ions z_+ and z_-, the number n_O of oxygen atoms in oxy-anions, and independent values of the cohesive energy density, *ced*, for a reasonable fit. For 42 alkali metal and alkaline earth metal halides, Fig. 3.5:

$$\sigma\kappa_T/\text{pm} = z_+(z_+ + 1)\left[7 + 138(ced/\text{GPa})^{-1}\right] \tag{3.43}$$

and for 15 alkali metal nitrates, carbonates, sulfates, molybdates, and tungstates:

$$\sigma\kappa_T/\text{pm} = z_+z_-(z_+ + z_-)(0.5z_-n_O)\left[7 + 237(ced/\text{Gpa})^{-1}\right] \tag{3.44}$$

The larger the *ced*, the smaller is this product due to the inverse dependence on it of the compressibility that is not completely compensated by the direct dependence of the surface tension.

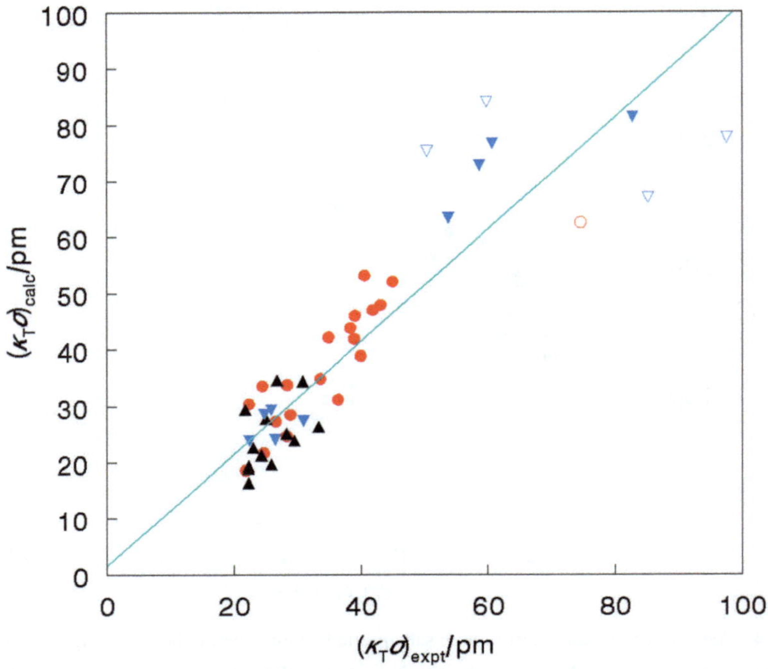

Fig. 3.5 The calculated $\kappa_T\sigma$/pm products according to Eqs. (3.43) and (3.44) plotted against the experimental values: alkali metal halides (●), alkali metal salts with oxy-anions (▲), and alkaline earth halides (▼). Outliers are marked as empty symbols (From Chhabra and Hunter [251] by permission of the publisher (Am. Inst. Phys.))

3.4 Transport Properties

3.4.1 Viscosity

The viscosity of molten salts determines their applications in many areas, and their conductivity depends reciprocally on it. Therefore the viscosity of many molten salts has been measured in the range of their thermal stability, mostly at ambient pressures. Two methods have been mainly employed: the capillary flow method and the oscillation method. The latter comprises an oscillating disk or sphere, and in recent years the oscillating cup has found many supporters [252]. These methods are generally adequate up to 1400 K.

The dynamic viscosity of high temperature molten salts diminishes with increasing temperatures and generally follows the Arrhenius expression, where B_η is the activation energy:

$$\eta = A_\eta \exp(B_\eta/RT) \quad (3.45)$$

A comprehensive list of values of the parameters A_η and E_η from [214], with a few more recent data, is shown in Table 3.18, the numerical data of these parameters pertain to η and A_η in mPa·s and to B_η in J·mol^{-1}.

The fluidity is the reciprocal of the viscosity, $\Phi = \eta^{-1}$, and for molten salts has been related by Marcus [256] to their molar volume V, as both vary with the temperature according to the Hildebrand and Lamoreaux [257] relationship, Fig. 3.6:

$$\Phi = -B + (B/V_0)V \quad (3.46)$$

The parameters B and V_0 resulting from linear plots of Φ of molten salts against V are shown in Table 3.18 too. These parameters are independent of the temperature and the V_0 values are in good agreement with those derived by Chhabra and Hunter [253] as seen in the Table. They correspond to the volume of the virtual molten salt that has no free volume, but in which the ions are free to rotate. Such a volume should be close to that of the crystalline salt at the melting point, but there are no accurate data for this quantity for comparison. Bockris and Richards [137] reported values of $v_0^{1/3}$ for alkali metal halides and nitrates, where v_0 is the incompressible volume per ion. These values for the salts, shown in *italics* in Table 3.18 are comparable with the V_0 values reported there. The V_0 values of molten salts are on the average $87 \pm 6\,\%$ of the molar volumes of the liquid salts at the corresponding temperature of $1.1T_m$ (Tables 3.13 and 3.14).

The meaning of the B parameters is rather obscure. It is expected that the larger the attractive forces between the ions, the less ready would they be to move in an external force gradient, hence the smaller the fluidity. This expectation is borne out only for the alkali metal fluorides, for which B decreases linearly with their cohesive energy densities. For 51 other molten salts of the 1:1, 1:2, and 1:3 types the B parameters *increase* linearly with the cohesive energy (that is, with $ce/z_+^2z_-^2$), but with considerable scatter [256].

Young and O'Connell [140] presented a corresponding states correlation of alkali metal salt viscosities, from which a relationship with the molar volumes may be derived, which is, however, very much more complicated than Eq. (3.46). Janz et al. [258] subsequently extended this treatment to 1:2 and 2:1 molten salts. The resulting expressions are:

$$\begin{aligned} \ln(\Phi/\Phi^*) &= a - b(T^*/T) + c(T/T^*)^8 \\ \Phi^* &= dV^{*2/3}(2\,T^*\mu)^{-1/2} \\ V^* &= V/\left[e + f\,T/T^* + g(T/T^*)^2\right] \end{aligned} \quad (3.47)$$

with 7 substance-independent parameters ($a \ldots g$), and where $T^* = 0.4/\alpha_P$ (proportional to the reciprocal of the isobaric expansibility) and $\mu = M_+M_-/(M_+ + M_-)$ is the reduced mass. The salts are then characterized by the ionic masses and

Table 3.18 The viscosity of molten salts expressed by the A_η and B_η parameters of $\eta = A_\eta \exp(B_\eta/RT)$ [214], and their fluidity expressed by the B and V_0 parameters of $\Phi = \eta^{-1} = -B + (B/V_0)V$ [256]. The V_0 values in parentheses are from Potapov et al. [253], those in *italics* are the sum of the incompressible ionic volumes from Bockris and Richards [137]

Salt	A_η/mPa·s	B_η/J·mol^{-1}	B/s·mPa^{-1}	V_0/cm^3·mol^{-1}
LiF	0.18359	21,832	3.36	12.4
LiCl	0.10852	19,111	6.52	25.4 (26.7) *22.4*
LiBr	0.14030	17,246	6.24	31.8 (32.3) *28.0*
LiI	0.12650	17,386	5.96	40.5 (40.0)
$LiNO_2$	0.00297	34,342		
$LiNO_3$	0.08237	18,575	7.21	38.2 (38.0) *40.3*
$LiClO_3$	0.00198	32,690	2.65	44.2
$LiClO_4$	0.09436	19,523	4.60	51.5
Li_2CO_3	0.08660	36,863	3.65	39.0
Li_2SO_4	0.02412	53,768	3.86	53.0
Li_2MoO_4	0.21989	29,457	2.86	58.2
Li_2WO_4	0.02244	55,019	3.90	60.4
NaOH	0.07211	20,657	7.37	21.9
NaF	0.1197	26,468	3.68	18.2
NaCl	0.08927	21,960	5.00	31.5 (34.5) *34.2*
NaBr	0.10340	20,478	4.73	37.1 (37.4) *36.8*
NaI	0.09940	19,095	5.49	47.2 (49.0) *53.0*
$NaNO_2$	0.04875	19,095	5.83	36.0
$NaNO_3$	0.1041	16,259	6.50	41.7 *54.6*
NaSCN	0.04935	19,397	3.78	42.1
$NaClO_3$	0.02439	25,110	3.48	49.3
$NaBF_4$	0.0832	19,620	4.98	52.2
Na_2CO_3	0.18937	28,834	2.65	49.3
Na_2SO_4	0.148	41,799	1.73	65.7
Na_2MoO_4	0.15264	29,570	3.10	70.0
Na_2WO_4	0.07974	36,683	3.73	73.4
Na_3AlF_6	0.01792	51,780	3.08	87.1
KOH	0.02295	25,845	6.75	30.6
KF	0.1068	23,779	3.96	26.6
KCl	0.06166	25,047	5.56	41.5 (42.6) *46.2*
KBr	0.07370	23,543	4.93	48.0 (47.0) *53.0*
KI	0.10230	20,521	4.22	57.7 (58.6) *66.8*
KNO_2	0.1645	14,326	2.73	42.1
KNO_3	0.07737	18,469	5.22	50.9 (50.9) *69.2*
KSCN	0.00858	27,004	3.95	46.2
KBF_4	0.0946	4530		
K_2CO_3	0.18751	27,038	2.96	65.5
K_2MoO_4	0.28952	21,360	5.94	87.5
K_2WO_4	0.07656	36,065	2.09	95.0
$K_2Cr_2O_7$	0.08051	28,782	2.05	124.3

(continued)

Table 3.18 (continued)

Salt	A_η/mPa·s	B_η/J·mol^{-1}	B/s·mPa^{-1}	V_0/cm^3·mol^{-1}
RbF	0.09711	24,325	3.87	30.7
RbCl	0.06783	24,805	4.91	46.9 (51.4)
RbBr	0.08060	23,550	4.42	53.2 (51.4)
RbI	0.07640	23,081	4.97	65.0 (63.6)
$RbNO_2$	0.08754	18,807		
$RbNO_3$	0.1296	16,636	3.85	55.8
Rb_2CO_3	0.15847	28,318	3.17	74.2
CsF	0.10164	22,177	4.34	35.2
CsCl	0.07148	23,619	5.00	53.5 (57.7) *63.0*
CsBr	0.08470	22,920	4.48	60.3 *68.6*
CsI	0.07720	22,701	5.35	73.5 (73.4)
$CsNO_2$	0.1058	17,628		
$CsNO_3$	0.03936	23,193	4.89	73.6
Cs_2CO_3	0.11478	29,468	3.74	86.6
CuCl	0.1042	21,234	7.51	26.1
AgCl	0.3098	12,197	4.74	30.4
AgBr	0.3806	12,920	3.97	35.0
AgI	0.1418	22,004	5.77	44.6
$AgNO_3$	0.1159	15,145	5.66	48.7
TlCl	0.173	14,226	3.00	40.9
TlBr	0.3157	11,772	3.93	45.7
TlI	0.2284	14,523	4.06	49.3
$TlNO_3$	0.0843	15,301	6.17	52.2
$MgCl_2$	0.17939	20,559	5.64	52.2
$MgBr_2$	0.00341	60,213	14.44[c]	69.9
$CaCl_2$	0.10215	30,351	4.62	50.0
$SrCl_2$	0.09638	34,918	3.92	54.2
$BaCl_2$	0.07993	39,522	3.49	61.0
$Ba(NO_2)_2$	0.00129	47,347		
$ZnCl_2$	2.89×10^{-4}	75,136	2.38	55.1
$ZnBr_2$	7.79×10^{-5}	82,481	3.47	66.1
$CdCl_2$	0.2405	16,368	4.04	49.3
$CdBr_2$	0.1893	19,063	3.87	61.3
CdI_2	0.05208	29,473	3.01	82.1
$PbCl_2$	0.05619	28,293	4.81	53.8
$PbBr_2$	0.08165	24,573	4.03	62.5
$BiCl_3$	0.3787	19,636	0.47[d]	76.0
$LaCl_3$	0.02061	54,598	3.88	73.8
$CeCl_3$[a]	0.09084	34,837	2.83	70.1
$PrCl_3$[a]	0.09421	31,429	4.85	72.3
$NdCl_3$[a]	0.08114	36,123	2.93	70.6
$SmCl_3$[a]	0.04999	40,688	3.43	70.8

(continued)

Table 3.18 (continued)

Salt	A_η/mPa·s	B_η/J·mol^{-1}	B/s·mPa^{-1}	V_0/cm^3·mol^{-1}
$EuCl_3$[b]	0.06752	36,792		
$DyCl_3$[a]	0.05984	42,109	2.90	72.0
$ErCl_3$[a]	0.02503	52,252	2.97	73.5

[a][254], [b][255], [c]This value is apparently too large, [d]This value is apparently too small

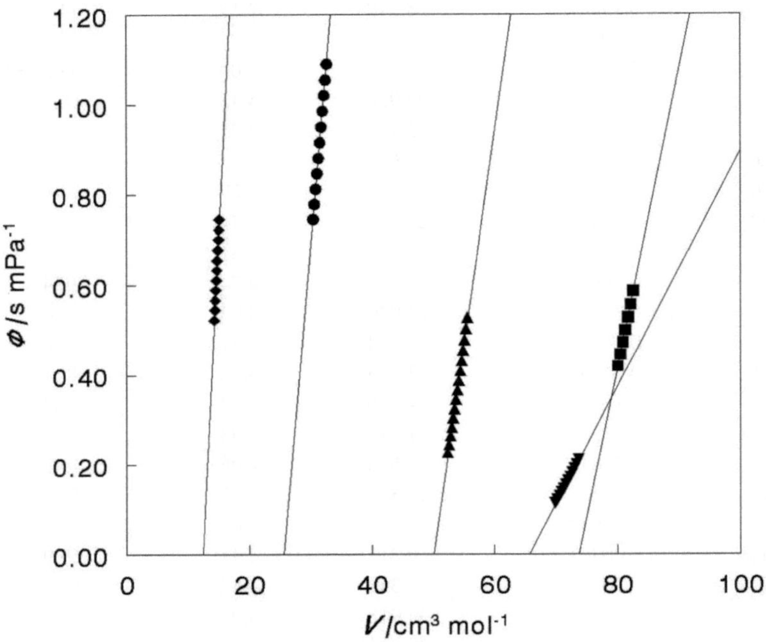

Fig. 3.6 The fluidities Φ of some molten salts plotted against their molar volumes over a suitable temperature range, limited to that where the viscosity data are available: LiF (♦), KF (●), $CaCl_2$ (▲), Na_2SO_4 (▼), and $CsNO_3$ (■) (From Marcus [256] by permission of the publisher (Elsevier))

expansibilities, and the viscosities could be predicted by this set of expressions to within 3 %.

Another application of the corresponding states approach is that of Abe and Nagashima [259], who used the Tosi-Fumi [112] potential function to obtain the two parameters $d = r_+ + r_-$ for the distance and ε for the potential energy. The resulting expression for the fluidity of a molten salts is:

$$\Phi = d(M\varepsilon)^{-1/2}\left[-a + bV/N_A d^3(k_B T/\varepsilon)^{1/2}\right] \tag{3.48}$$

with 2 substance-independent parameters (a and b). A variant of this approach is that of Tada et al. [260], based on listed values of the characteristic temperature T^* and volume V^* of the alkali metal halides and two functions of the ionic masses, m_S and μ_S:

$$\Phi = (V^*/N_A)^{2/3}(k_B T^* m_S)^{-1/2}[a\exp(bT^*/T) + \{c(T^*/T) - d\}\mu_S] \quad (3.49)$$

where $m_S = 2\,M_+^{1/2}M_-^{1/2}/\left(M_+^{1/2} + M_-^{1/2}\right)$ and $\mu_S = \left(M_+^{1/2} - M_-^{1/2}\right)/\left(M_+^{1/2} + M_-^{1/2}\right)$ with four substance-independent parameters ($a \ldots d$). Other theoretical expressions and empirical representations of the viscosities or the fluidities of molten salts have also been reported, but they are less able to predict these values.

3.4.2 Electrical Conductivity

The electrical conductivity of molten salts is again an important feature and has been reported for many salts. The methodology has been described many years ago by Tomlinson [261] and by Sundheim [262] and much more recently by Nunes et al. [263]. The critically compiled conductance data in [214] and many of the subsequently reported data are in terms of the specific conductance κ, which like the viscosity follows an Arrhenius-type expression:

$$\kappa = A_\kappa \exp(-B_\kappa/RT) \quad (3.50)$$

The conductance increases with the temperature, so that the minus sign before the activation energy B_κ for the conductance should be noted, contrary to the positive values of numerator of the exponent for the viscosity. The equivalent conductance of the molten salt, which is the product of the specific conductance with the molar volume of the molten salt, also follows an Arrhenius-type expression:

$$\Lambda = V\kappa = A_\Lambda \exp(-B_\Lambda/RT) \quad (3.51)$$

Values of A_Λ and B_Λ are listed in Table 3.19 as far as they are known or could be derived from reported specific conductance data [214] by means of the molar volumes and their temperature dependence in Tables 3.13 and 3.14. The derived equivalent conductivities at the corresponding temperature of $1.1T_m$ are also recorded in this Table as far as the relevant data are known. Else, the values of A_κ and B_κ are shown in Table 3.19 in parentheses. Some of the values recorded in Table 3.19 appear to be unreliable: for instance the low values of the equivalent conductivities of the lanthanide chlorides from samarium onwards contrast with the much larger values of the lighter salts, and the fluctuations within the series with the atomic number are also suspect.

Table 3.19 The conductivity of molten salts from data in [214] or as noted: the A_Λ and B_Λ parameters of the equivalent conductivity Λ/ohm^{-1} cm^2 equiv.$^{-1}$ = $A_\Lambda \exp(-B_\Lambda/RT)$. Values in parenthesis are the corresponding parameters for the specific conductivity κ/ohm^{-1} cm^{-1} = $A_\kappa \exp(-B_\kappa/RT)$

Salt	A_Λ	B_Λ/J mol^{-1}	Λ at $1.1T_m$
LiF	320.7	2143	260.2
LiCl	508.2	2015	396.0
LiBr	585.3	2117	441.8
LiI	569.5	1810	436.0
$LiNO_2$	(44.82)	(17,426.7)	
$LiNO_3$	967.8	3589	458.6
$LiClO_3$	(186.87)	(24,476.4)	
$LiClO_4$	(14.831)	(13,016.6)	
Li_2CO_3	754.5	4438	473.6
Li_2SO_4	394.8	2932	297.4
Li_2MoO_4	(22.493)	(20,736.3)	405.6
Li_2WO_4	(16.969)	(18,768.4)	409.2
NaOH	668.2	3120	375.1
NaF	344.7	2964	267.0
NaCl	544.6	2990	401.6
NaBr	622.7	3228	440.6
NaI	694.5	3221	476.1
$NaNO_2$	685.7	2949	384.5
$NaNO_3$	705.6	3215	384.5
NaSCN	(43.)	(19,832.5)	
$NaClO_3$	(48.3)	(22,348.8)	
$NaBF_4$	(5.2325)	(6447.7)	214.0
Na_2CO_3	550.2	4199	366.6
Na_2SO_4	550.2	4507	359.4
Na_2MoO_4	779.9	5713	406.9
Na_2WO_4	381.7	4491	260.0
Na_3AlF_6	(9.267)	(12,694.5)	331.2
KOH	520.2	2467	339.6
KF	480.3	3355	346.6
KCl	548.0	3415	383.1
KBr	591.1	3747	393.6
KI	541.2	3442	364.7
KNO_2	777.0	3267	
KNO_3	657.4	3577	346.6
KSCN	787.4	6082	179.6
KBF_4	125.6	4068	74.1
K_2CO_3	544.6	4650	352.9
K_2SO_4	815.0	4722	555.8
K_2MoO_4	(4.72)	(13,549.7)	389.7
K_2WO_4	(8.141)	(19,518.7)	484.5

(continued)

Table 3.19 (continued)

Salt	A_Λ	B_Λ/J mol^{-1}	Λ at $1.1T_m$
$K_2Cr_2O_7$	605.2	8141	160.6
RbF	(9.486)	(10,736)	249.5
RbCl	754.1	4401	465.0
RbBr	611.1	4171	381.0
RbI	568.1	3999	353.2
$RbNO_2$	(10.66)	(14,377.3)	
$RbNO_3$	515.7	3496	267.6
Rb_2CO_3 [a]	(11.31)	(19,630)	
Rb_2SO_4	471.7	4957	315.4
CsF	(12.89)	(13,577)	242.9
CsCl	1102	5110	599.7
CsBr	1169	5533	600.9
CsI	1125	5450	579.8
$CsNO_2$	(6.905)	(12,866.4)	
$CsNO_3$	552.4	3688	307.2
Cs_2CO_3 [a]	(9.38)	(18,800)	
Cs_2SO_4	387.5	4529	264.1
CuCl	153.6	650	138.7
CuBr	(6.342)	(5925)	
AgCl	268.2	1252	222.2
AgBr	210.2	1104	177.2
AgI	239.9	1475	197.5
$AgNO_3$	587.9	2898	306.1
TlCl	546.0	3421	320.6
TlBr	425.8	3511	252.2
TlI	614.5	9163	131.6
$TlNO_2$	393.9	10,057	35.9
$TlNO_3$	633.3	3348	296.7
Tl_2SO_4	760.5	16,979	97.9
$MgCl_2$	263.7	4363	162.1
$MgBr_2$	385.5	5404	211.9
MgI_2	(13.07)	(26,498)	
CaF_2	798.7	21,163	203.3
$CaCl_2$	675.3	5285	390.6
$CaBr_2$	506.7	4901	296.9
CaI_2	440.3	4617	273.1
$SrCl_2$	689.6	5646	402.8
$SrBr_2$	806.5	6183	412.2
SrI_2	610.1	5409	288.1
$BaCl_2$	772.5	6004	454.1
$BaBr_2$	2652	10,296	418.8
BaI_2	831.2	6367	418.0

(continued)

Table 3.19 (continued)

Salt	A_Λ	B_Λ/J mol^{-1}	Λ at 1.1T_m
$Ba(NO_3)_2$	10,217	8104	
MnF_2	(16.903)	(12,561)	
$MnCl_2$	169.5	2694	
$FeCl_2$	490.8	8813	178.0
$CoCl_2$[b]	480.6	6276	
$CoBr_2$[c]	(1790.7)	(30,900)	
$NiCl_2$[b]	1166.8	8861	
CuF_2	(4.031)	(6397)	
$ZnCl_2$	6097	11,785	
$ZnBr_2$	35,684	14,604	
ZnI_2	17,880	12,636	
$CdCl_2$	24.1	2499	174
$CdBr_2$	243.4	3226	159.9
CdI_2	1109.0	6365	386.4
$SnCl_2$	361.8	2727	203.7
PbF_2	(13.041)	(8607)	
$PbCl_2$	588.1	4093	329.8
$PbBr_2$	660.8	4559	305.6
PbI_2	601.8	18,547	30.9
UO_2Cl_2	(15.733)	(41,649)	
$ScCl_3$	(218.4)	(61,225)	
YCl_3	959.2	8827	348.3
$LaCl_3$	469.4	5678	273.0
$LaBr_3$	2652	10,296	
LaI_3	1614.9	29,903	71.4
$CeCl_3$	1030	25,360	81.9
CeI_3	(7.746)	(26,335)	
$PrCl_3$[d]	1430	28,320	77.0
$NdCl_3$[d]	2040	28,050	104.2
$NdBr_3$	3973	11,749	
$SmCl_3$[e]	2799	31,860	73.0
$EuCl_3$[e]	1620	28,734	48.7
$GdCl_3$[e]	575	16,404	75.2
$TbCl_3$[e]	2047	33,032	32.0
$DyCl_3$[e]	2014	34,513	38.4
DyI_3	(7.792)	(31,665)	
$HoCl_3$[e]	652	18,865	81.3
$ErCl_3$[e]	2076	36,425	46.6
$TmCl_3$[e]	748	21,368	89.0
$YbCl_3$[e]	2269	38,464	57.0
$LuCl_3$[e]	2178	38,961	62.2
$BiCl_3$	32.36	981.3	26.1

(continued)

Table 3.19 (continued)

Salt	A_Λ	B_Λ/J mol^{-1}	Λ at 1.1T_m
$BiBr_3$	16.67	535.6	14.8
BiI_3	78.5	7815	22.4
$ThCl_4$	395.0	676.4	194.3
UCl_4 [e]	1162	23,684	57.8

[a][265], [b][219], [c][264], [d][266], [e][220]

The conductivity of molten salts depends on the short-range forces much more than on the long-range coulombic ones, as suggested by Sundheim [262], according to Rice [267], because the relaxation times for the latter are so much longer than those for the former. However, the expected reciprocal dependence of the equivalent conductivity on the viscosity according to the Stokes' law, $\Lambda = \left(F^2/6\pi\right)[z_+/r_+ + z_-/r_-]\eta^{-1}$, is not realized quantitatively. A relationship does exist between the apparent activation energies for viscosity and equivalent conductivity for salts of uni- and divalent cations, on comparison of the entries in Tables 3.18 and 3.19:

$$B_\eta/\text{kJ mol}^{-1} = 5.4 \pm 1.8 + (4.9 \pm 0.4)\left(B_\Lambda/\text{kJ mol}^{-1}\right)^{-1} \quad (3.52)$$

Whereas B_η generally diminishes as the size of the ions increases, B_Λ increases in this direction, Fig. 3.7.

The conductivity of molten salts has been related to the existence of 'free volume' in the melt [268] and it was argued that the Arrhenius activation energy B_Λ should be lower than the corresponding one for ion diffusion in the melt (see below), $B_{D\pm}$ as was in fact found. This would explain why the conductivity does not adhere to the Nernst-Einstein relation $\Lambda = F^2(D_+ + D_-)/RT$ for the diffusion or to Stokes' law as mentioned above. The significant structure theory in this case [160] specifies that only the solid-like particles contribute to the conductivity. Their number per unit volume is $N_A V_{sd} V^{-2}$, where V_{sd} is the molar volume of the solid salt at the melting point. This number of solid-like particles, when multiplied by the sum of cation and anion mobilities that depend reciprocally on the viscosity of the molten salt, yields κ, the specific conductivity. However, the calculated conductivities of the five alkali metal halides tested were some 30 % lower than the experimental values.

The corresponding states approach of Tada et al. [260] uses listed substance-dependent characteristic volumes V^* and temperatures T^* (see above) and two numerical parameters to correlate the specific conductivities of alkali metal halides as:

$$\kappa = 0.129C\exp(-1.03\,T/T^*)V^{*2/3}T^{*1/2}[(M_+ + M_-)/M_+M_-]^{1/2} \quad (3.53)$$

where $C = e^2 N_A{}^{-2/3} k_B{}^{1/2}$.

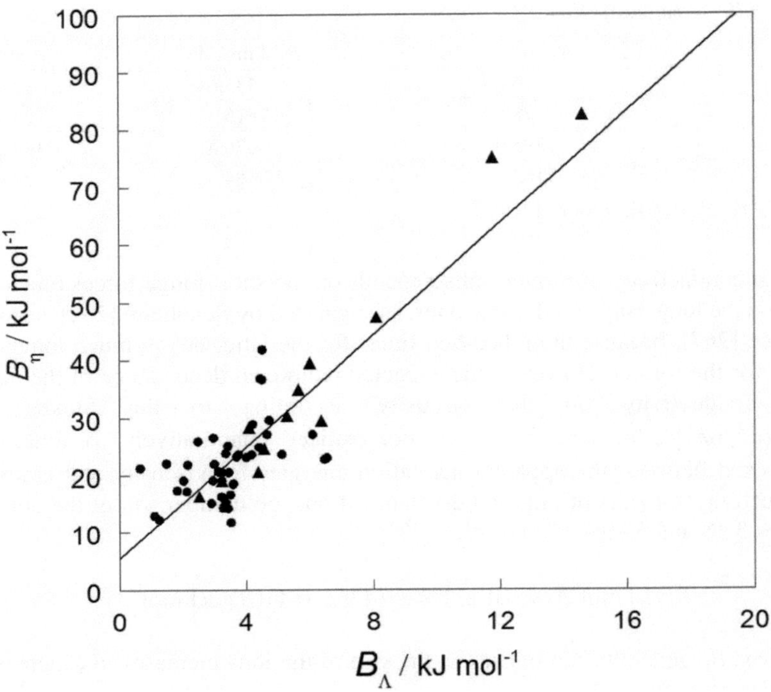

Fig. 3.7 The activation energy for viscous flow, B_η, plotted against the activation energy for their electrical conductance, B_Λ, of uni-univalent molten salts (●) and di-univalent ones (▲)

There has been a long-standing debate regarding the significance of transport numbers in pure single-component molten salts. *External* transport numbers deal with the mobilities of individual ions with regard to an inert wall of an electrochemical cell. These may be measured in a diaphragm cell by means of isotopic tracers as discussed by Klemm [269]. *Internal* transport numbers refer to the difference in the mobilities of the cation and the anion of the molten salt, but in a pure salt with only two ionic species this cannot be measured and has no meaning. (Where there are three ionic species, in common-ion salt mixtures, e.g., MX + MY, such measurements are possible as shown by Haase [270] and are meaningful). A way around this problem is the measurement of the self-diffusion of the ions in a single molten salt by means of isotopic tracers. The ion mobilities $u_\pm$ are then directly proportional to the self-diffusion coefficients $D_\pm$ and internal transport numbers may then be *defined* as: $t_+ = D_+/(D_+ + D_-)$ for the cation. However, these defined internal transport numbers are meaningless, as discussed many years ago and pointed out again recently by Harris [271].

3.4.3 Self Diffusion

Self-diffusion coefficients of ions in pure molten salts are measured by the use of suitable isotopically labeled ions either by diffusion out of capillaries (more rarely into them) into the bulk molten salt (more rarely out of it) that has the same overall chemical nature or, as more recently introduced, by nuclear magnetic resonance (NMR). They can also be determined by molecular dynamics computer simulations, where the velocity self-correlation of an ion is computed. Many of the available data were reported in the compilations by Klemm [269] and by Sjöblom [272] and these and more recent ones are shown in Table 3.20. For each molten salt the self-diffusion coefficient for the cation and the anion are shown separately, and where available its temperature dependence is reported in terms of the parameters A_D and E_D of the Arrhenius-type expression $D = A_D\exp(-B_D/RT)$. Also shown in this Table are the ionic self-diffusion coefficient at the corresponding temperature of $1.1T_m$, or, if the temperature dependence is unknown, at some specified temperature. The values of the ionic self-diffusion coefficients are in the range $1 \leq D/10^{-9}$ $m^2\ s^{-1} \leq 10$, and within a given salt are generally larger for the smaller one of the cation and the anion. Only for the zinc salts that are network forming (Chap. 4) are the self-diffusion coefficients much smaller than this range.

Sjöblom [273] discussed several models for the self-diffusion in molten salts: the hole model, the free volume model, the local density fluctuation model, and the cubic cell model. The hole model of Fürth [275], adopted by Bockris and coworkers (e.g. [137]), considers the diffusion of an ion as its jump into an available neighboring hole of suitable size in the melt. The activation energy B_D is the sum of the energy required for the creation of such a hole and that needed for the ion to jump into it, and is dominated by the former energy. The free volume model concentrates on the low temperature (supercooled) region and modifies the denominator of the Arrhenius-type expression to $R(T - T_0)$ where T_0 is the temperature at which the configurational entropy vanishes (the glass transition temperature) [268, 276]. The local density fluctuation model turned out to be unable to describe properly the temperature dependence of the self-diffusion of ions in molten salts. The cubic cell model stipulates that D/T should be inversely proportional to the kinematic viscosity, $(\eta/\rho)^{-1}$, with the proportionality constant being $Rd^2/24M$, where d is the linear dimension of the cubic cell and M is the molar mass of the ion. This approach does describe the temperature dependence of the self-diffusion adequately, as does the simple Arrhenius-type expression shown in Table 3.20.

Much more recently Harris [271] examined the applicability of the Nernst-Einstein relationship between the molar conductivity Λ and the ionic diffusivities D:

$$\Lambda = \left(F^2/RT\right)\left(\nu_+ z_+{}^2 D_+ + \nu_- z_-{}^2 D_-\right)(1 - \Delta) \tag{3.54}$$

to single binary molten salts in which the electroneutrality $\nu_+ z_+ = -\nu_- z_-$ prevails. Contrary to dilute aqueous electrolyte solutions, where $\Delta = 0$, in molten salts

Table 3.20 The A_D and B_D parameters of the self-diffusion coefficient $D/10^{-9}\ m^2\ s^{-1} = A_D exp(-B_D/RT)$ of molten salts (According to Klemm [265] and Sjöblom [267] or as noted

Salt	Ion	$A_D/10^{-9}\ m^2\ s^{-1}$	B_D/kJ mol^{-1}	D at 1.1T_m or As noted
LiF[a]	F^-			7.2 (at 1125 K)
LiCl[b]	Li^+	125	18.11	13.26
LiCl[b]	Cl^-	36	13.51	6.85
LiBr[c]	Li^+			11.25 (1000 K)
LiBr[c]	Br^-			5.52 (1000 K)
$LiNO_3$	Li^+	247	22.97	2.07
$LiNO_3$	NO_3^-	195	26.53	0.78
Li_2CO_3	CO_3^{2-}	1.4	40.75	0.016
NaF[d]	Na^+			9.80 (1275 K)
NaF[d]	F^-			9.67 (1275 K)
NaCl	Na^+	336	32.89	11.65
NaCl	Cl^-	302	35.10	8.46
NaI	Na^+	63	16.87	8.72
NaI	I^-	43	18.49	4.92
$NaNO_3$	Na^+	129	20.80	2.54
$NaNO_3$	NO_3^-	90	21.25	1.63
Na_3AlF_6[d]	Na^+			9.28 (1325 K)
Na_3AlF_6[d]	F^-			5.81 (1325 K)
Na_2CO_3	Na^+	1000	58.92	3.24
Na_2CO_3	CO_3^{2-}	286	44.43	3.80
KF	F^-			9.5 (at 1125 K)
KCl	K^+	180	28.79	8.95
KCl	Cl^-	180	29.83	8.03
KNO_3	K^+	132	23.14	2.10
KNO_3	NO_3^-	142	24.11	1.90
RbCl	Rb^+	251	33.52	6.32
RbCl	Cl^-	167	31.04	5.52
RbBr[e]	Rb^+			5.0 (955 K)
RbBr[e]	Br^-			5.0 (955 K)
$RbNO_3$[f]	Rb^+	157	23.72	1.84
CsCl	Cs^+	173	30.62	4.51
CsCl	Cl^-	246	32.72	5.00
$CsNO_3$	Cs^+	113	23.48	2.70
$CsNO_3$	NO_3^-	178	26.28	2.72
CuBr[e]	Cu^+			8.4 (810 K)
CuBr[e]	Br^-			4.5 (810 K)
CuI[e]	Cu^+			7.4 (923 K)
CuI[e]	I^-			3.9 (923 K)
AgBr[e]	Ag^+			5.7 (753 K)
AgBr[e]	Br			4.0 (753 K)
$AgNO_3$	Ag^+	49	15.60	1.46
$AgNO_3$	NO_3^-	31	16.07	0.83

(continued)

Table 3.20 (continued)

Salt	Ion	$A_D/10^{-9}$ m^2 s^{-1}	B_D/kJ mol^{-1}	D at $1.1T_m$ or As noted
TlCl	Tl^+	76	18.99	3.81
TlCl	Cl^-	79	19.08	4.06
$MgCl_2$[g]	Mg^{2+}			130 (1573 K)
$MgCl_2$[g]	Cl^-			190 (1573 K)
$CaCl_2$	Ca^{2+}	38	25.65	2.59
$CaCl_2$	Cl^-	190	37.07	3.93
$SrCl_2$	Sr^+	21	22.51	2.45
$SrCl_2$	Cl^-	77	28.79	4.94
$BaCl_2$	Ba^{2+}	64	37.49	2.32
$BaCl_2$	Cl^-	200	38.41	6.69
$ZnCl_2$	Zn^{2+}	5850	64.85	0.017
$ZnCl_2$	Cl^-	13,700	68.20	0.021
$ZnBr_2$	Zn^{2+}	8000	70.71	0.074
$ZnBr_2$	Br^-	11,100	71.34	0.093
$CdCl_2$	Cd^{2+}	110	28.52	2.69
$CdCl_2$	Cl^-	110	28.45	2.72
$PbCl_2$	Pb^{2+}	77	28.36	1.40
$PbCl_2$	Cl^-	90	25.52	2.44
$PbBr_2$	Pb^{2+}	74	27.20	0.74
$PbBr_2$	Br^-	83	25.52	1.10

[a][274], [b][268], [c][272], [d][269], [e][270], [f][271], [g][273]

$\Delta \neq 0$ and is related to the velocity cross-correlation coefficients of like and unlike ions. The Nernst-Einstein expression therefore under-estimates the self-diffusion coefficients because Δ is positive but <1.

3.4.4 *Thermal Conductivity*

Another transport property of molten salts for which there is a considerable amount of data is their thermal conductivity λ_{th}. In earlier years the then available measuring techniques led to the conclusion that λ_{th} increases mildly with increasing temperatures. More modern techniques, such as transient hot wire measurements, yield values of λ_{th} that diminish mildly and linearly with increasing temperatures. The scatter of values reported in the literature is large, however, and they have not been critically compiled so far. Gheribi et al. [277] provided an explicit model expression for λ_{th}, the required inputs for their model being the ionic radii, and the density, velocity of sound, heat capacity, and melting temperature of the salt. Table 3.21 shows the 'recommended' predicted values in terms of the parameters of the linear temperature dependence:

$$\lambda_{\text{th}}(T) = \lambda_{\text{th}}(T_{\text{m}}) - B_{\text{th}}(T - T_{\text{m}}) \quad (3.55)$$

The predicted values of λ_{th} agree as well as may be expected with modern experimental values. The values of λ_{th} at the corresponding temperature of $1.1T_{\text{m}}$ are also shown in Table 3.21, and those for the alkali metal halides depend reciprocally on the molar mass of the salt: $\lambda_{\text{th}}(1.1T_{\text{m}})/\text{W m}^{-1}\text{K}^{-1} \approx 35/\left(M/\text{g mol}^{-1}\right)$, but this cannot be generalized to other salts. A reciprocal dependence on the mass of the molten salts (a fractional power of it) was also noted by Cornwell [278] for 13 salts other than alkali metal halides at temperatures near T_{m}. DiGuilio and Teja [279] reviewed several models for the thermal conductivity of molten salts, and later Hossain et al. [280] employed essentially the same model as in [279] to obtain the reduced thermal conductivity λ_{th}*, but without a clear definition how λ_{th}* is related to the measured thermal conductivity.

3.5 Solvent Properties

The properties of high temperature molten salts as solvents and reaction media need to be discussed in connection with the behavior of the molten salts in mixtures, a wide subject that is outside the scope of the present book. Here only the ability of pure molten salts to dissolve non-reactive gases and some organic compounds, which leads to no profound change of the structure of the melts, is dealt with. Grimes et al. [283] asserted that previous to their work no solubility data of (noble) gases in molten salts appeared to have been published. Tomkins and Bansal compiled the data available at the time [284] and Tomkins later reviewed the experimental methods for the determination of the solubilities in molten salts [285]. Solubilities are available mainly for argon, nitrogen, and carbon dioxide, with few data also for helium, and the salts studied are mainly the alkali metal halides and nitrates. The solubilities of non-reactive gases in molten salts are quite low, so that Henry's law applies. Most of the data in the literature are in terms of K_{P}, the number of moles of gas dissolved per unit volume of the salt divided by the saturation pressure, in units of 10^{-7} mol_{gas} cm^{-3} atm^{-1}. They are transformed into the mole fraction of the gas in the molten salt, x_{gas}, at unit pressure by multiplication with the molar volume of the latter, V_{salt}, because Henry's law constant, $K_{\text{H}} = (K_{\text{P}}V_{\text{salt}})^{-1}$ and $x_{\text{gas}} = P_{\text{gas}}/K_{\text{H}}$. Table 3.22 summarizes the solubilities in terms of $-\log x_{\text{gas}}$ at ambient pressure and $1.1T_{\text{m}}$ (interpolated as necessary) with V_{salt} from Tables 3.13 and 3.14. Also included in the Table are the molar enthalpies of solution, obtained as $\Delta_{\text{solution}}H = R\left[\partial \ln x_{\text{gas}}/\partial(1/T)\right]_P$.

Marcus [155] showed that the gas solubilities in molten salts can be expressed according to a general solubility expression in terms of the cohesive energy density of the salt (the square of its Hildebrand solubility parameter) as:

Table 3.21 The thermal conductivity of molten salts according to Eq. (3.55) from Gheribi et al. [277] or as noted

Salt	$\lambda(T_m)$ W m^{-1} K^{-1}	B 10^{-4} W m^{-1} K^{-2}	λ at $1.1T_m$ W m^{-1} K^{-1}
LiOH	1.2732	6.58	1.217
LiF	1.4360	3.99	1.387
LiCl	0.7265	2.33	0.704
LiBr	0.4748	1.31	0.463
LiI	0.3461	1.52	0.334
$LiNO_3$	0.5856	1.47	0.577
Li_2CO_3	0.9348	1.87	0.916
Li_2SO_4	0.6392	1.26	0.624
NaOH	0.9802	3.68	0.956
NaF	0.9026	2.81	0.863
NaCl	0.4923	2.11	0.467
NaBr	0.3315	1.46	0.315
NaI	0.2392	1.21	0.227
$NaNO_2$	0.5382	2.39	0.524
$NaNO_3$	0.5277	1.99	0.515
$NaHSO_4$	0.46[a]		
Na_2CO_3	0.6176	1.44	0.600
Na_2SO_4	0.4482	1.08	0.434
KOH	0.6866	2.34	0.670
KF	0.5684	2.54	0.537
KCl	0.3974	2.00	0.374
KBr	0.2707	1.49	0.254
KI	0.2105	1.27	0.197
KSCN	0.27[a]		
KNO_2	0.4028	2.39	0.384
KNO_3	0.4047	1.88	0.392
$KHSO_4$	0.34[a]		
K_2CO_3	0.4691	1.06	0.455
K_2SO_4	0.2739	0.856	0.261
RbF	0.4413	3.02	0.405
RbCl	0.2980	1.66	0.280
RbBr	0.2419	1.52	0.226
RbI	0.1526	0.608	0.146
$RbNO_3$	0.3370	1.62	0.327
Rb_2CO_3	0.3482	0.776	0.339
Rb_2SO_4	0.2148	0.527	0.207
CsF	0.3169	2.46	0.290
CsCl	0.2689	1.66	0.252
CsBr	0.2088	1.20	0.197
CsI	0.1304	0.427	0.126
$CsNO_3$	0.2456	1.14	0.237

(continued)

Table 3.21 (continued)

Salt	$\lambda(T_m)$ W m^{-1} K^{-1}	B 10^{-4} W m^{-1} K^{-2}	λ at $1.1T_m$ W m^{-1} K^{-1}
Cs_2CO_3	0.3203	0.454	0.315
Cs_2SO_4	0.1285	0.307	0.124
NH_4HSO_4	0.39[a]		
$AgNO_3$	0.382[b]	1.84[b]	0.472[b]
MgF_2	0.6501	1.41	0.626
$MgCl_2$	0.2040[c]	0.20[c]	
CaF_2	0.5650	0.604	0.554
$CaCl_2$	0.4468	1.15	0.433
$CaBr_2$	0.3880	0.760	0.380
SrF_2	0.3954	0.725	0.382
$SrCl_2$	0.3802	1.18	0.365
$SrBr_2$	0.3550	1.05	0.344
SrI_2	0.2359	0.753	0.229
BaF_2	0.2937	0.642	0.282
$BaCl_2$	0.3279	1.09	0.313
$BaBr_2$	0.2354	0.875	0.225
BaI_2	0.2056	0.792	0.197
$ZnCl_2$	0.30[a]		

[a][281] near T_m, [b][282], [c]These entries in [277] appear to be misprinted, because they are much out of line with those of the other alkaline earth halides

Table 3.22 The (negative logarithm of the) mole fraction solubility $-\log x_{gas}$ of gases in molten salts at $1.1T_m$ and unit pressure from Marcus and Tomkins [155, 285] or as noted and the molar enthalpy of solution, $\Delta_{solution}H$/kJ mol^{-1} in parenthesis

Salt	He	Ar	N_2	CO_2
$LiNO_3$		5.20 (33.5)[b]	5.13[b] (7.6)[b]	3.87[d] (2.0)[d]
NaCl		5.54		4.42[d]
$NaNO_3$	5.32[a] (13.5)[a]	5.49[a] (14.7)[a]	5.00[a] (11.5)[b]	4.08[d] (2.2)[d]
KCl		4.69		4.41[d] (10.3)[d]
KBr		4.31		4.26[e]
KI		4.16		3.89[e]
KNO_3		4.97 (10.6)[b]	4.80[b] (7.2)[b]	4.20[d]
RbCl		4.32		4.51 (12.2)[d]
$RbNO_3$		5.10[c] (20.1)[c]		3.95[d] (2.3)[d]
CsCl		4.66		4.36
$CsNO_3$				4.22[d]
$AgNO_3$		6.09[c]		
$ZnCl_2$				4.04[f]
$ZnBr_2$				3.93[f]
$SnCl_2$				4.65[f]

[a][290], [b][287], [c][286], [d][288], [e][289], [f][291]

$$\log x_{gas} = -(\ln 10)^{-1} V_{gas} (RT)^{-1} ced_{salt} \tag{3.56}$$

within 0.3 units in $\log x_{gas}$, which is the uncertainty within the data reported for a given salt and a given gas by different authors, Fig. 3.8. In this expression the molar volumes, V_{gas}/cm^3 mol^{-1}, of the gases were taken as 2.47, 1.46, and 2.09 for Ar, N_2, and CO_2, as fitting parameters and $ced_{salt} \approx (\delta_{Hsalt} - \delta_{Hgas})^2$, because of the negligible size of δ_{Hgas} compared with $\delta_{Hsalt} = ced_{salt}{}^{1/2}$ (Table 3.12). Gas solubilities in lithium nitrate do not conform to Eq. (3.56) for no apparent reason.

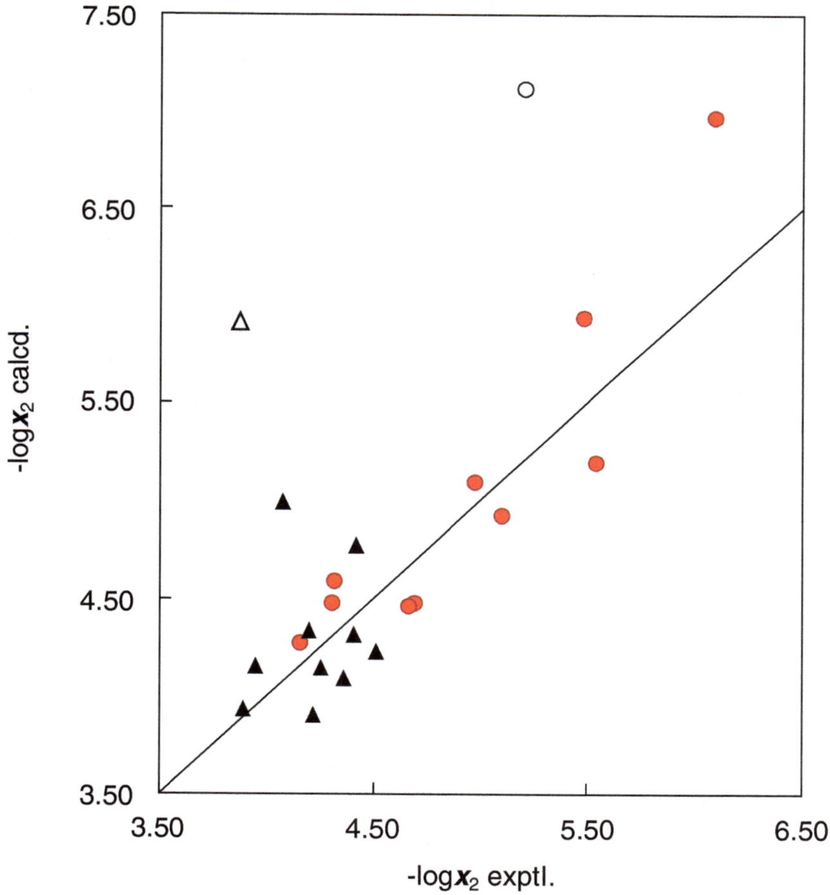

Fig. 3.8 The (negative logarithms of the mole fraction) solubilities of Ar (●) and CO_2 (▲) in molten salts, calculated according to Eq. (3.56), plotted against the experimental values. The *empty symbols* are the outlying $LiNO_3$ data (From Marcus [155] by permission of the publisher (Elsevier))

In earlier attempts of the rationalization of gas solubilities in molten salts Blander et al. [292] used the Uhlig model, according to which the solubility depends on the work required for the production of the cavity in the molten salt in which the gas molecule is to reside. This, in turn, depends on the surface tension of the salt melt and the van der Waals surface area of the gas molecules. This approach, however, under-estimates the actual solubilities [289, 290].

Subsequently Fukase [293] employed a more complicated model to calculate the Henry's law constants K_H, according to which $\ln K_H$ is the sum of (RT/V_{salt}) and a positive hard sphere term (corresponding to the cavity formation work) and substantially two negative terms describing the Lennard-Jones and polarization terms for the interaction of the dissolved gas with the ions of the molten salt. Fair agreement with the experimental $\ln K_H$ values was achieved, but considerably worse agreement with the enthalpies of solution resulted. In a more recent paper Simonin [294] dealt with the effects of polarization on the solubility of gases in molten salts, again including the volume exclusion, dispersion forces, and polarization of the gas molecules and of the ions of the molten salt. He applied the model to the solubilities of noble gases in molten KCl and RbCl at one temperature, 1173 K, with good agreement and no adjustable parameters.

There are hardly any reports of the solubility of organic solutes in molten salts, because these solutes are thermally unstable at the high temperatures involved. Molten KSCN ($T_m = 446$ K) appears to be the only inorganic pure salt in which solutions of certain organic solutes are stable enough to be studied. The indicators methyl red, *m*-cresol purple, and alizarin produce reversible color changes in molten KSCN at 473 K [295]. The solubility of diethylene glycol, triethylene glycol, urea, thiourea, nitrobenzene, anthracene, tribenzylamine, benzophenone, polypropylene glycol, triethanolamine, phthalic anhydride, and glycerol in molten KSCN at 483 K has been reported in [296].

References

1. Lide D (ed) (2001–2002) Handbook of chemistry and physics, 82nd edn. CRC Press, Baton Rouge
2. Barin I, Knacke O (1973) Thermochemical properties of inorganic substances. Springer, Berlin
3. Janz GJ (1967) Molten salts handbook. Academic, New York
4. March NH, Tosi MP (1991) Boiling of alkali halides: an ionic to molecular phase transition. Phys Chem Liq 10:95–98
5. Lely JA, Bijvoet JM (1942) Crystal structure of lithium cyanide. Rec Trav Chim Pays Bas Belg 61:244–252
6. Kondo Y, Schoemaker D, Luty F (1979) Molecular motion and ordering in rubidium cyanide, studied with dielectric and Raman techniques. Phy Rev B 19:4210–4219
7. Poulsen FW (1985) Ionic conductivity of solid and molten lithiumthiocyanate and its hydrate. Acta Chem Scand A 39:290–292

8. Gafurov MM, Aliev AR, Akhmalov IR (2002) Raman and infrared study of the crystals with molecular anions in the region of a solid-liquid phase transition. Spectrochim Acta A 58:2683–2692
9. Farber M, Srivastava RD, Moyer JW, Leeper JD (1985) Electron- impact and thermodynamic studies of potassium metaborate. J Chem Soc Faraday Trans 1(81):913–918
10. Kato Y, Asano M, Harada T, Mizutani Y (1992) Mass- spectrometric study of the thermodynamic properties of rubidium metaborate- cesium metaborate $\{xRbBO_2 + (1 - x) CsBO_2\}$. J Chem Thermodyn 24:1033–1038
11. Honda H, Ishimaru S, Ikeda R (1999) Ionic dynamics in LiNO2 studied by 7Li and 15N solid NMR. Z Naturforsch A 54:519–523
12. Popovskaya NP, Protsenko PI, Eliseeva AF (1968) Surface tension of some univalent metal nitrite and nitrate melts. Russ J Inorg Chem 13:498–501
13. Markowitz MM, Boryta DA, Stewart H Jr (1964) The differential thermal analysis of perchlorates. VI. Transient perchlorate formation during the pyrolysis of the alkali metal chlorates. J Phys Chem 68:2822–2829
14. Simmons JP, Waldeck WF (1931) The system: lithium bromate- water. J Am Chem Soc 53:1725–1727
15. Sahoo MK, Bhatta D (1992) Effect of barium(2+) doping on the thermal decomposition of cesium bromate. Thermochim Acta 197:391–397
16. Karataeva IM, Vinogradov EE (1974) Lithium iodate- rubidium iodate- water and lithium iodate- cesium iodate- water systems at 50.deg. Zh Neorg Khim 19:3156–3160
17. Ram KD, Tripathi R (1990) Study of iodide- iodate isotopic exchange reaction in a few eutectic melts by radiotracer technique. Appl Radiat Isot 41:879–885, extrapolated
18. Bui HL, de Klerk A (2013) Lithium C1- C12 n- alkanoates: thermal behavior from −30 °C to 600 °C. J Chem Eng Data 58:1039–1049
19. Ikeda R, Ishimaru S, Tanabe T, Nakamura D (1995) A noble ionic plastic phase of cesium formate studied by ^{1}H, ^{2}H and ^{133}Cs NMR and electrical conductivity measurements. J Mol Struct 345:151–157
20. Karnowsky MM, Clark RP, Biefeld RM (1978) The phase diagram of the system lithium chromate(VI) – potassium chromate(VI). J Solid State Chem 23:219–223
21. Samuseva RG, Okunev YA, Plyushchev VE (1967) Binary systems from chromates and dichromates of potassium, rubidium, and cesium. Zh Neorg Khim 12:2822–2824
22. Yamawaki M, Oka T, Yasumoto M, Sakurai H (1993) Thermodynamics of vaporization of cesium molybdate by means of mass spectrometry. J Nucl Mater 201:257–260
23. Chang LLY, Sachdev S (1975) Alkali tungstates. Stability relations in the systems alkali monotungstate-tungsten trioxide. J Am Ceram Soc 58:267–270
24. Spitsyn VI, Meerov MA (1952) Pyrosulfates of the alkali elements. Russ J Gen Chem 22:905–912
25. Vesnin YI, Khirpin LA (1966) Polymorphic transformations of alkali metal bichromates. Zh Neorg Khim 11:2216–2221
26. Hatem G, Eriksen KM, Gaune-Escarde M, Fehrmann R (2002) SO_2 oxidation catalyst model systems characterized by thermal methods. Top Catal 19:323–329
27. White KA III, Winnik J (1985) Electrochemical removal of hydrogen sulfide from hot coal gas: electrode kinetics. Electrochim Acta 30:511–519
28. Jensen E (1947) Melting relations of chalcocite. Avhandl Norske Vidensk Akad I 6:1–14
29. Shirokov AV, Kozlov GS (1978) Composition- melting point phase diagram of the calcium nitrate- magnesium nitrate system. Zh Priklad Khim 51:1869
30. Xiao F-S, Xu W, Qiu S, Xu R (1994) New route for dispersion of inorganic salts onto the channel surfaces of microporous crystals: high dispersion of $CuCl_2$ in zeolites using a microwave technique. J Mater Chem 4:735–739
31. Maier CG (1925) Vapor pressure of the common metallic chlorides and a static method at high temperatures. Tech Paper 360:1–54

32. Klement W Jr (1976) Melting temperatures of lead (II) nitrate up to a pressure of 30 kbar. Bull Soc Chim Fr 1656:11–12
33. Ippolitov EG, Makhlachko AG (1970) Phase diagram of the molten and solid barium fluoride- yttrium fluoride system. Izv Akad Nauk SSSR Neorg Mat 6:146–148
34. Dennison DH, Spedding FH, Daane AH (1959) Determination of the melting point, vapor pressure, and decomposition temperature of YI_3. US AEC Rep IS-57:1–16
35. Gong W, Wu Y, Zhang R, Gaune-Escard M (2012) Phase equilibrium in lanthanide halide systems: assessment of $CeBr_3$ and MBr- $CeBr_3$ systems (M=Li, Na, K, Rb, Cs). CALPHAD 35:44–51
36. Chervonnyl AD, Chervonnaya NA (2007) The thermodynamic properties of 4f metal trifluorides. Russ J Phys Chem 81:1543–1559
37. Rycerz L, Gong W, Gaune-Escard M (2013) The $TbBr_3$- LiBr binary system: experimental thermodynamic investigation and assessment of phase diagram. J Chem Thermodyn 56:15–20
38. D'Eye RWM, Martin FS (1957) The barium fluoride- uranium trifluoride system. J Chem Soc 1847–1851
39. Haereid S, Julsrud S, Grande T (1991) On the solubility of noble metals in melts based on zirconium(IV) fluoride. Mater Sci Forum 67–68:291–296
40. Korenev YUM, Sorokin ID, Chirina NA, Novoselova AV (1972) Vapor pressure of hafnium tetrafluoride. Zh Neorg Khim 17:1195–1198
41. Hannebohn O, Klemm W (1936) Measurements on gallium and indium compounds. XI. Fluorides of gallium, indium and thallium. Z Anorg Allg Chem 229:337–351
42. Seifert HJ (2005) Melting points of lanthanide trichlorides. An unsolved problem. J Therm Anal Calorim 82:575–580
43. Cordfunke EHP, Kubaschewski O (1984) The thermochemical properties of the system uranium- oxygen- chlorine. Thermochim Acta 74:235–245
44. Skudlarski K, Miller M (1980) Sublimation and mass spectrometry of copper(I) cyanide. Intl J Mass Spect Ion Phys 36:19–30
45. Goodwin HM, Mailey RD (1908) On the density, electrical conductivity and viscosity of fused salts and their mixtures. V. Phys Rev 25–26:469–489
46. Furakawa Y, Nakamura D (1990) Thallium nuclear magnetic relaxation in solid thallium (I) thiocyanate TlSCN: phase transition and ionic motion. Z Naturforsch 45a:1211–1216
47. Dennis LM, Doan M, Gill AC (1896) Some new compounds of thallium. J Am Chem Soc 18:970–977
48. Moriya K, Matsuo T, Suga H (1988) Thermodynamic properties of alkali and thallium nitrites: the ionic plastically crystalline state. Thermochim Acta 132:133–140
49. Sokolov NM (1979) Study of the properties of thallium salts of fatty acids. Russ J Inorg Chem 24:1938–1940
50. Carlson CM, Eyring H, Ree T (1960) Significant structures in liquids. III. Partition function for fused salts. Proc Natl Acad Sci U S A 46:333–336
51. McQuarrie DA (1962) Theory of fused salts. J Phys Chem 66:1508–1513
52. Kirshenbaum AD, Cahill JA, McGonical PJ, Grosse AV (1962) The density of liquid NaCl and KCl and an estimate of their critical constants together with those of the other alkali halides. J Inorg Nucl Chem 24:1287–1296
53. Pitzer KS (1984) Critical point and vapor pressure of ionic fluids including sodium chloride and potassium chloride. Chem Phys Lett 105:484–489
54. Rebelo LPN, Lopes JNC, Esperança JMSS, Filipe E (2005) On the critical temperature, normal boiling point, and vapor pressure of ionic liquids. J Phys Chem B 109:6040–6043, supporting information Table A
55. Weiss VC (2010) Guggenheim's rule and the enthalpy of vaporization of simple and polar fluids, molten salts, and room temperature ionic liquids. J Phys Chem B 114:9183–9194
56. Guissani Y, Guillot B (1994) Coexisting phases and criticality in NaCl by computer simulation. J Chem Phys 101:490–509

57. Leu A-L, Ma S-M, Eyring H (1975) Properties of molten magnesium oxide. Proc Natl Acad Sci U S A 72:1026–1030
58. Hoch M (1988) The critical point data of refractory metals, aluminum oxide and uranium dioxide using the Hoch- Arpshofen method. J Nucl Mater 152:289–294
59. Zarzycki G (1957) The critical point data of refractory metals, aluminum oxide and uranium dioxide using the Hoch- Arpshofen method. J Phys Radium 18:65A-69A. Study of molten salts by x- ray diffraction. II. Structure in the liquid state of the chlorides LiCl, NaCl, KCl, $BaCl_2$, and of the fluoride CaF_2. General considerations on the structure of molten halides (1958) J Phys Radium 19:13A–19A
60. Levy HA, Danford MD (1964) Diffraction studies of the structure of molten salts. In: Blander M (ed) Molten salt chemistry. Interscience, New York, pp 109–125
61. Ohno H, Igarashi K, Umesaki N, Furukawa K (1994) X-ray diffraction of molten salts. Molten Salt Forum 3:1–230
62. Neilson GW, Adya AK (1997) Neutron diffraction studies on liquids. Annu Rep Program Chem C Phys Chem 93:101–145
63. Neilson GW, Adya AK, Ansel S (2002) Neutron and X- ray diffraction studies on complex liquids. Annu Rep Program Chem C Phys Chem 98:273–322
64. Di Cicco A (1996) Local structure in binary liquids probed by EXAFS. J Phys Condens Matter 8:9341–9345
65. Antonov BD (1975) The x-ray diffraction study of molten alkali metal bromides and iodides. J Struct Chem 16:474–476
66. Ohno H, Furukawa K (1981) X-ray diffraction analysis of molten sodium chloride near its melting point. J Chem Soc Faraday Trans 1(77):1981–1985
67. Levy HA, Agron PA, Bredig M, Danford MD (1960) X- ray and neutron diffraction studies of molten alkali halides. Ann N Y Acad Sci 79:762–780
68. Saito M, Park C, Omote K, Sugiyama K, Waseda Y (1997) Partial structural functions of molten CuBr estimated from the anomalous X- ray scattering measurements. J Phys Soc Jpn 66:633–640
69. Furukawa K (1961) Structure of molten salts near the melting point. Disc Faraday Soc 32:53–62
70. Zarzycki G (1961) High- temperature x- ray diffraction studies of fused salts. Structure of molten alkali carbonates and sulfates. Disc Faraday Soc 32:38–48
71. Okamoto Y, Shiwaku H, Yaita T, Suzuki S, Minato K, Tanida H (2004) Local structure of molten $CdCl_2$ systems. Z Naturforsch 59a:819–824
72. Iwadate Y, Fukushima K, Nakazawa T, Okamoto Y (2002) Local structure of ZnBr2- KBr melts analyzed by X- ray diffraction, Raman spectroscopy, and molecular orbital calculation. J Non Cryst Sol 312–314:424–427
73. Okamoto Y, Madden PA (2005) Structural study of molten lanthanum halides by X- ray diffraction and computer simulation techniques. J Phys Chem Sol 66:448–451
74. Iwadate Y, Suzuki K, Onda N et al (2006) Local structure of molten LaCl3 analyzed by X- ray diffraction and La- LIII absorption- edge XAFS technique. J Alloy Comp 412:248–252
75. Igarashi K, Kosaka M, Ikeda M, Mochinaga J (1990) X- ray diffraction analysis of neodymium trichloride melt. Z Naturforsch 45a:623–626
76. Iwadate Y, Iida T, Fukushima K, Mochinaga J, Gaune-Escarde M (1994) X- ray diffraction study on the local structure of molten $ErCl_3$. Z Naturforsch 49a:811–814
77. Okamoto Y, Madden PA, Minato K (2005) X- ray diffraction and molecular dynamics simulation studies of molten uranium chloride. J Nucl Mater 344:109–114
78. Ohno H, Igarashi K, Iwadate Y, Murofushi M, Mochinaga J, Furukawa K (1986) X- ray diffraction analysis of some molten salts with low melting point. Proc Electrochem Soc 86:296–306
79. Page DI, Mika K (1971) Partial structure factors of molten cuprous chloride from neutron diffraction measurements. J Phys C 4:3034–3044

80. Edwards FG, Enderby JE, Howe RA, Page DI (1975) Structure of molten sodium chloride. J Phys C 8:3483–3490
81. Biggin S, Enderby JE (1982) Comments on the structure of molten salts. J Phys C 15:L305–L309
82. Howe MA, McGreevy RL (1988) A neutron- scattering study of the structure of molten lithium chloride. Philos Mag B 58:485–495
83. McGreevy RL, Howe MA (1989) The structure of molten lithium chloride. J Phys Condens Matter 1:9957–9962
84. Derrien JY, Dupuy J (1975) Structural analysis of the ionic liquids potassium chloride and cesium chloride by neutron diffraction. J Phys 36:191–198
85. Mitchell EWJ, Poncet PFJ, Stewart RJ (1976) The ion pair distribution functions in molten rubidium chloride. Philos Mag B 34:721–732
86. Eisenberg S, Jal J-F, Dupuy J, Chieux P, Knoll W (1982) Neutron diffraction determination of the partial structure factors of molten cuprous chloride. Philos Mag A 46:195–209
87. Allen DA, Howe RA (1992) A measurement of the structure of the superionic conductor copper(I) bromide in its liquid phase by neutron diffraction. J Phys Condens Matter 4:6029–6038
88. Derrien JY, Dupuy J (1976) Structure of molten silver chloride. Phys Chem Liq 5:71–91
89. Locke J, Messoloras S, Stewart RJ, McGreevy RL, Mitchell EWJ (1985) The structure of molten cesium chloride. Philos Mag B 51:301–315
90. Biggin S, Gay M, Enderby JE (1984) The structures of molten magnesium and manganese (II) chlorides. J Phys C 17:977–985
91. Biggin S, Enderby JE (1981) The structure of molten calcium chloride. J Phys C 14:3577–3583
92. McGreevy RL, Mitchell EWJ (1982) The determination of the partial pair distribution functions for molten strontium chloride. J Phys C 15:5537–5550
93. Edwards FG, Howe RA, Enderby JE, Page DI (1978) The structure of molten barium chloride. J Phys C 11:1053–1057
94. Newport RJ, Howe RA, Wood ND (1985) The structure of molten nickel chloride. J Phys C 18:5249–5257
95. Adya AK, Takagi R, Sakurai M, Gaune-Escard M (1998) Structural and thermodynamic properties of molten $DyCl_3$ and $DyCl_3$- NaCl systems. Proc Electrochem Soc 98–11:499–512
96. Hardacre C (2005) Application of EXAFS to molten salts and ionic liquid technology. Annu Rev Mater Res 35:29–40
97. Di Cicco A, Rosolen MJ, Narassi R, Tossici R, Filiponi A, Rybicki J (1996) Short- range order in solid and liquid KBr probed by EXAFS. J Phys Condens Matter 8:10779–10797
98. Di Cicco AJ (1996) Local structure in binary liquids probed by EXAFS. J Phys Condens Matter 8:9341–9345
99. Li H, Lu K, Wu Z, Dong J (1994) EXAFS studies of molten ZnCl2, RbCl and Rb2ZnCl4. J Phys Condens Matter 6:3629–3640
100. Minicucci M, Di Cicco A (1997) Short- range structure in solid and liquid CuBr probed by multiple- edge x- ray- absorption spectroscopy. Phys Rev B 56:11456–11464
101. Di Cicco A, Minicucci M, Filiponi A (1997) New advances in the study of local structure of molten binary salts. Phys Rev Lett 78:460–463
102. Trapananti A, Di Cicco A, Minicucci M (2002) Structural disorder in liquid and solid CuI at high temperature probed by x-ray absorption spectroscopy. Phys Rev B 66:014202-1/11
103. Inui M, Takeda S, Maruyama K, Shirakawa Y, Tamaki S (1995) XAFS measurements on molten silver halides. J Non Cryst Solids 192–193:351–354
104. Okamoto Y, Yaita T, Minato K (2004) High- temperature XAFS study of solid and molten SrCl2. J Non Cryst Solids 333:182–186
105. Okamoto Y, Fukushima K, Iwadate Y (2002) XAFS study of molten zinc dibromide. J Non Cryst Solids 312–314:450–453

106. Watanabe S, Matsuura H, Akatsuka H, Okamoto Y, Madden PA (2005) Structural investigation on lead fluoride- lithium fluoride at various compositions and temperatures. J Nucl Mater 144:104–108
107. Okamoto Y, Shiwaku H, Yaita T, Narita H, Ranida H (2002) Local structure of molten LaCl3 by K- absorption edge XAFS. J Mol Struct 641:71–76
108. Okamoto Y, Shiwaku H, Yaita T, Suzuki S, Gaune-Escarde M (2013) High- energy EXAFS study of molten $GdCl_3$ systems. J Mol Liq 187:94–98
109. Larsen B (1974) Monte- Carlo calculations on a charged hard sphere model. Chem Phys Lett 27:47–51
110. Larsen B (1976) Studies in statistical mechanics of Coulombic systems. I. Equation of state for the restricted primitive model. J Chem Phys 65:3431–3438
111. Adams DJ, McDonald IR (1974) Rigid- ion models of the interionic potential in the alkali halides. J Phys C: Solid State Phys 7:2761–2773
112. Tosi MP, Fumi FG (1964) Ionic sizes and born repulsive parameters in the NaCl- type alkali halides. I. Huggins- Mayer and Pauling forms. J Phys Chem Solids 25:31–43; Ionic sizes and born repulsive parameters in the NaCl- type alkali halides. II. Generalized Huggins- Mayer forms (1964) J Phys Chem Solids 25:45–58
113. Michielsen J, Woerlee P, van de Graaf F, Ketelaar JAA (1975) Pair potential for alkali metal halides with rock salt crystal structure. Molecular dynamics calculations on sodium chloride and lithium iodide. J Chem Soc Faraday Trans 2(71):1730–1740
114. Harada M, Tanigaki M, Yao M, Kinoshita M (1982) Application of simple perturbation theory to uni- univalent molten salts. J Chem Soc Faraday Trans 2(78):1985–1999
115. Margheritis C, Sinistri C (1988) The soft ion model in Monte Carlo simulation of molten salts. Z Naturforsch 43a:129–132
116. Wilson M, Madded PA, Costa-Cabral BJ (1996) Quadrupole polarization in simulations of ionic systems: application to AgCl. J Phys Chem 100:1227–1237
117. Woodcock LV, Singer K (1971) Thermodynamic and structural properties of liquid ionic salts obtained by Monte Carlo computation. 1. Potassium chloride. Trans Faraday Soc 67:12–30
118. Woodcock LV (1971) Isothermal molecular dynamics calculations for liquid salts. Chem Phys Lett 10:257–261
119. Lewis JWE, Singer K, Woodcock LV (1975) Thermodynamic and structural properties of liquid ionic salts obtained by Monte Carlo computation. 2. Eight alkali metal halides. J Chem Soc Faraday Trans 2(71):308–319
120. Margheritis C, Sinistri C (1988) Interionic potentials and Monte Carlo simulations of molten silver (I) chloride. Z Naturforsch 43a:751–754
121. Baranyai A, Ruff I, McGreevy RL (1986) Monte Carlo simulation of the complete set of molten alkali halides. J Phys C 19:453–465
122. Dixon M, Sangster MJL (1976) A comparison of the structure and some dynamical properties of molten rubidium halides. J Phys C 9:3381–3388
123. Dixon M, Sangster MJL (1977) Computer simulation study of the structural properties of molten cesium halides. J Phys C 10:3015–3023
124. Bassen A, Lemke A, Bertagnolli H (2000) Monte Carlo and reverse Monte Carlo simulations on molten zinc chloride. Phys Chem Chem Phys 2:1445–1454
125. Barreto LS, Mort KA, Jackson RA, Alves OL (2002) Molecular dynamics simulation of anhydrous lithium acetate: crystalline and molten phases. J Non Cryst Solids 303:281–290
126. Wang J, Sun Z, Lu G, Yu J (2014) Molecular dynamics simulations of the local structures and transport coefficients of molten alkali chlorides. J Phys Chem B 118:10196–10206
127. Belashchenko DK, Ostrovski OI (2002) Computer modelling of liquid salts RbBr, CuCl, CuBr, CuI and AgBr. CALPHAD 26:523–538
128. Belashchenko DK, Ostrovski OI (2005) Computer modelling of molten halides using diffraction data. Trans Inst Miner Metall C 114:C154–C159

129. Di Ciccon A, Taglienti M, Minicucci M (2000) Short- range structure of solid and liquid AgBr determined by multiple- edge x- ray absorption spectroscopy. Phys Rev B 62:12001–12012
130. Siquera LJA, Urahata SM, Ribeiro MCC (2003) Molecular dynamics simulation of molten sodium chlorate. J Chem Phys 119:8002–8011
131. Stillinger FH Jr (1964) Equilibrium theory of pure fused salts. In: Blander M (ed) Molten salt chemistry. Interscience, New York, pp 1–108
132. Gillan M (1978) Theories of the thermodynamic properties of pure molten salts. Phys Chem Liq 8:121–141
133. Pitzer KS (1987) Thermodynamic properties of ionic fluids over wide ranges of temperature. Pure Appl Chem 59:1–6
134. Blander M (2000) Fundamental theories and concepts for predicting thermodynamic properties of high temperature ionic and metallic liquid solutions and vapor molecules. Metall Mater Trans B 31:579–586
135. Altar W (1937) A study of the liquid state. J Chem Phys 5:577–586
136. Fürth R (1941) The theory of the liquid state. I. The statistical treatment of the thermodynamics of liquids by the theory of holes. Proc Camb Phil Soc 37:252–275
137. Bockris JO'M, Richards NE (1957) The compressibilities, free volumes, and equation of state for molten electrolytes: some alkali halides and nitrates. Proc R Soc Lond A 241:44–66
138. Reiss H, Mayer SW, Katz JL (1961) Law of corresponding states for fused salts. J Chem Phys 35:820–826
139. Luks KD, Davis HT (1967) Recent statistical mechanical theories of the thermodynamic properties of molten salts. Ind Eng Chem Fund 6:194–208
140. Young RE, O'Connell JP (1971) Empirical corresponding states correlation of densities and transport properties of 1–1 alkali metal molten salts. Ind Eng Chem Fund 10:418–423
141. Harada M, Masataka T, Tada Y (1983) Law of corresponding states of uniunivalent molten salts. Ind Eng Chem Fund 22:116–121
142. Lu W-C, Ree T, Gerrard VG, Eyring H (1968) Significant- structure theory applied to molten salts. J Chem Phys 49:797–804
143. Vilcu R, Miscdolea C (1967) Significant- structure theory of liquids. Heat capacities, compressibilities, and thermal- expansion coefficients of some molten alkali halides. J Chem Phys 46:906–909
144. Cheng D-W, Leu A-L, Ma S-M (1986) Properties of some molten alkali halides. Mater Chem Phys 14:85–95
145. Pandey JD, Chaturvedi BR, Pandey RP (1981) Surface tension of molten salts. J Phys Chem 85:1750–1752
146. Bremse F, Alejandre J (2003) Cavities in ionic liquids. J Chem Phys 118:4134–4139
147. Eyring H, Ree T, Hirai N (1958) Significant structures in the liquid state. Proc Natl Acad Sci U S A 44:683–688
148. Clusius K, Goldmann J, Perlick A (1949) Low- temperature research. VII. The specific heat of the alkali halides lithium fluoride, sodium chloride, potassium chloride, potassium bromide, potassium iodide, rubidium bromide, and rubidium iodide between 10° and 273° abs. Z Naturforsch 4a:424–432
149. Nakamura T (1981) Influences of mass and bond length on the Debye temperatures of ionic and covalent substances. Jpn J Appl Phys 20:L653–L656
150. Duraiswamy S, Haridasan TM (1976) Zero- point motion in alkali halides. Indian J Pure Appl Phys 14:337–340
151. Blum H (1967) The chemistry of molten salts. Benjamin, New York
152. Yosim SJ, Owens BB (1964) Calculation of thermodynamic properties of fused salts from a rigid- sphere equation of state. J Chem Phys 41:2032–2036
153. Cantor S, McDermott DP, Gilpatrick LO (1970) Volumetric properties of molten and crystalline alkali fluoroborates. J Chem Phys 52:4600–4604

154. Marcus Y (2009) Heat capacities of molten salts with polyatomic anions. Thermochim Acta 495:81–84
155. Marcus Y (2010) The cohesive energy of molten salts and its density. J Chem Thermodyn 42:60–64
156. Marcus Y (2013) Surface tension and cohesive energy density of molten salts. Thermochim Acta 571:77–81
157. Bauer SH, Porter RF (1964) Metal halide vapors. Structures and thermochemistry. In: Blander M (ed) Molten salt chemistry. Interscience, New York, pp 607–626
158. Riccardi R, Sinistri C (1965) Application of differential thermal analysis to the evaluation of latent heats of transition and fusion. Ricer Sci 2(A 8):1026–1037
159. Dworkin AS (1972) Enthalpy of lithium fluoroborate from 298–700 deg.K. Enthalpy and entropy of fusion. J Chem Eng Data 17:284–285
160. Hatem G (1985) Lithium sulfate: calorimetric determination of the temperatures and enthalpies of high- temperature phase transitions. Thermochim Acta 88:433–441
161. Cordfunke EHP, Konings RJM, Westrum EF Jr (1988) The thermodynamic properties of cesium metaborate (CsBO2) from 5 to 1000 K. Thermochim Acta 128:31–38
162. Kleppa OJ, Meschel SV (1963) Thermochemistry of anion mixtures in simple fused salt systems. II. Solutions of some salts of MO4- and MO42- anions in the corresponding alkali nitrates. J Phys Chem 67:2750–2753
163. Reshetnikov NA, Baranskaya EV (1967) Heats of fusion and polymorphic transformations of alkali metal hydroxides. Izv Vyssh Ucheb Zaved Khim Khim Tekh 10:496–499
164. Faber M, Srivastava RD, Moyer JW, Leeper JD (1985) Electron- impact and thermodynamic studies of potassium metaborate. J Chem Soc Faraday Trans 1(81):913–918; Slough W, Jones J earlier reported 35.6±4.2 kJ mol^{-1}
165. Leonesi D, Piantoni G, Berchiesi G, Franzosini P (1968) Thermodynamic properties of organic acid salts. III. Enthalpy and entropy of fusion of sodium and potassium. formates. Ricer Sci 38:702–705
166. Marchidan DI, Telea C (1969) Heat of melting in the binary mixtures: rubidium nitrate + potassium nitrate and rubidium nitrate + cesium nitrate. Rev Roum Chim 14:1361–1365
167. Zmbov KF, Margrave JL (1967) Mass spectrometric studies at high temperatures. XIV. Vapor pressure and dissociation energy of silver monofluoride. J Phys Chem 71:446–448
168. Ferloni P, Kenesey C, Westrum EF Jr (1994) Thermodynamics of alkali alkanoates X. Heat capacities and thermodynamic properties of lithium methanoate and lithium ethanoate at temperatures from ≈5 K to 580 K. J Chem Thermodyn 26:1349–1363
169. Kobayashi K, Inoue N, Takano T (1992) Specific heat of solid and molten phases, and latent heat of fusion of some carbonates. Jpn J Thermophys Prop 6:2–7
170. Brunetti B, Piacente V, Scardala P (2008) Vapor pressures and sublimation enthalpies of copper difluoride and silver(I, II) fluorides by the torsion- effusion method. J Chem Eng Data 53:687–693
171. Warnquist B (1980) Comments on thermochemical data and fusion temperature for pure sodium sulfide. Thermochim Acta 37:343–345
172. Hatem G, Abdoun F, Gaune-Escard M, Eriksen KM, Fehrmann R (1998) Conductometric, density and thermal measurements of the $M_2S_2O_7$ (M=Na, K, Rb, Cs) salts. Thermochim Acta 319:33–42
173. Aleixo AI, Oliviera PH, Diogo H, Minas de Piedale ME (2005) Enthalpies of formation and lattice enthalpies of alkaline metal acetates. Thermochim Acta 428:131–136
174. Manolatos S, Tillinger M, Post B (1973) Polymorphism in cesium thiocyanate. J Solid State Chem 7:31–35
175. Kubaschewski O, Alcock CB, Spencer PJ (1993) Materials thermochemistry, 6th edn. Pergamon, Oxford
176. Novikov GI, Baev AK (1962) Vapor pressure of chlorides of trivalent lanthanum, cerium, praseodymium, and neodymium. Zh Neorg Khim 7:1340–1352

177. Rycerz L, Ingier-Stocka E, Ziolek B, Gadzuric S, Gaune-Escard M (2004) Heat capacity and thermodynamic properties of $LaBr_3$ at 300–1100 K. Z Naturforsch 59a:825–828
178. Topor L, Moldoveanu T (1974) Vapor pressures of molten cadmium bromide, cadmium iodide, and magnesium bromide and molecular association in the gaseous phase. Rev Roum Chim 19:985–990 (for $MgBr_2$ the T_b is not known)
179. Schäfer H, Bayer L, Breil G, Etzel K, Krehl K (1955) Saturation pressures of manganese chloride, ferrous chloride, cobaltous chloride, and nickel chloride. Z Anorg Allg Chem 278:300–309 (for $CoCl_2$ $\Delta_V H°/kJ\ mol^{--1} = 144.9$ corrects the entry in [31])
180. Chervonnyi AD (2007) Chervonnaya NA. The thermodynamic properties of 4f metal trifluorides. Russ J Phys Chem A 81:1543–1559
181. Topol LE, Ransom LD (1960) The heats of fusion of the cadmium halides, mercuric chloride, and bismuth bromide. J Phys Chem 64:1339–1340
182. Mayer SW, Yosim SJ, Topol LE (1960) Cryoscopic studies in the molten bismuth- bismuth chloride system. J Phys Chem 64:238–240
183. Coughlin JP (1951) High- temperature heat content of nickel chloride. J Am Chem Soc 73:5314–5315
184. Dworkin AS, Bredig MA (1963) Heats of fusion of some rare earth metal halides. J Phys Chem 67:2499–2950
185. Binford JS, Strohmenger JM, Hebert TH (1967) A modified drop calorimeter. The heat content of aluminum carbide and cobalt(II) fluoride above 25°. J Phys Chem 71:2404–2408
186. Cubicciotti D (1968) Thermodynamic properties of bismuth trifluoride. J Electrochem Soc 115:1138–1163
187. Spedding FH, Henderson DC (1971) High- temperature heat contents and related thermodynamic functions of seven trifluorides of the rare earths: yttrium, lanthanum, praseodymium, neodymium, gadolinium, holmium, and lutetium. J Chem Phys 54:2476–2483
188. Kleppa OJ, Wakihara M (1976) Enthalpies of mixing in the liquid mixtures of zinc fluoride with the fluorides of lithium, sodium and potassium. J Inorg Nucl Chem 38:715–719
189. Wakihara M, Kleppa OJ (1977) Enthalpies of mixing of the liquid mixtures of cadmium fluoride with the fluorides of lithium, sodium, and potassium. High Temp Sci 9:35–43
190. Dworkin AS, Bredig MA (1971) Enthalpy of lanthanide chlorides, bromides, and iodides from 298–1300.deg.K: enthalpies of fusion and transition. High Temp Sci 3:81–90
191. Emons H-H, Bräutigam G, Thomas R (1976) Vapor pressure measurements for binary fused mixtures of alkaline earth and alkali metal chlorides. Chem Zvesti 30:773–782
192. Flesch RM, Knacke O, Münstermann E (1986) Sublimation and dissociation of thorium tetraiodide and thorium oxide diiodide from Knudsen cells. Z Anorg Allg Chem 585:123–134
193. Gaune-Escard M, Rycerz L, Szczepaniak W, Bogacz A (1994) Enthalpies of phase transition in the lanthanide chlorides $LaCl_3$, $CeCl_3$, $PrCl_3$, $NdCl_3$, $GdCl_3$, $DyCl_3$, $ErCl_3$ and $TmCl_3$. J Alloy Comp 204:193–196
194. Sun Y, He M, Qiao ZY (1997) An artificial neural network approach to correlation of enthalpies of fusion for rare earth halides. J Alloy Comp 256:9–12
195. Stolyarova VL, Seetharaman S, Svard D, Semenov GA (1998) A high- temperature mass spectrometric study of the vaporization processes of fluxes based on CaO- $CaCl_2$ and CaO-CaF_2 systems. Rapid Comm Mass Spectrom 12:1335–1343
196. Rycerz L, Gaune-Escard M (2002) Thermodynamics of $SmCl_3$ and $TmCl_3$: experimental enthalpy of fusion and heat capacity. Estimation of thermodynamic functions up to 1300 K. Z Naturforsch 57a:79–84; Thermodynamics of $EuCl_3$: experimental enthalpy of fusion and heat capacity and estimation of thermodynamic functions up to 1300 K (2002) Z Naturforsch 57a:215–220
197. Wagman DD, Evans WH, Parker VB, Schumm RH, Halow I, Bailey SM, Churney KL, Nuttal RL (1982) NBS Tables of thermodynamic data. J Phys Chem Ref Data 11(2):1–391
198. Morris DFC (1958) Lattice energies and related properties derived by use of lyotropic numbers. J Inorg Nucl Chem 6:295–302

199. Ladd MFC, Lee WH (1961) Lattice energies of compounds of the transition- type metals. J Inorg Nucl Chem 23:199–205
200. Jenkins HDB (1975) Enthalpy of solvation, $\Delta H^0_{solv}(CrO_4^{2-})$ (g), of gaseous chromate ion as estimated from lattice energy calculations. Chem Phys Lett 35:417–419
201. Jenkins HDB, Roobottom HK, Passmore J, Glasser L (1999) Relationships among ionic lattice energies, molecular (formula unit) volumes, and thermochemical radii. Inorg Chem 38:3609–3620
202. Marcus Y (2010) Heat capacities of molten salts. In: Wilhelm E, Letcher T (eds) Heat capacities. Royal Society of Chemistry, Cambridhe, pp 472–489, Ch. 22
203. Ngeyi SP, Malik I, Westrum EF Jr (1990) Thermodynamics of alkali alkanoates. VII. Heat capacity and thermodynamic functions of potassium acetate from 4 to 585 K. J Chem Thermodyn 22:91–98
204. Zamfirescu C, Dincer I, Naterer GF (2010) Thermophysical properties of copper compounds in copper-chlorine thermochemical water splitting cycles. Int J Hydrogen Energy 35:4839–4853
205. Block-Bolten A (1955) Sulfur in binary liquid solutions. Arch Gornic Hutnic 3:90–98
206. Dworkin AS, Bredig MA (1963) The heats of fusion and transition of alkaline earth and rare earth metal halides. J Phys Chem 67:697–698
207. Gaune-Escard M, Bogacz A, Rycerz L, Szczepaniak W (1996) Heat capacity of LaCl3, CeCl3, PrCl3, NdCl3, GdCl3, DyCl3. J Alloy Comp 235:176–181
208. Rycerz L, Gaune-Escard M (2002) Enthalpies of phase transitions and heat capacity of $TbCl_3$ and compounds formed in $TbCl_3$- MCl systems (M = K, Rb, Cs). J Therm Anal Calorim 68:973–981
209. Rycerz L, Gaune-Escard M (2004) Heat capacity and thermodynamic functions of $TbBr_3$. J Chem Eng Data 49:1078–1081
210. Leibowitz L, Fischer DF, Chasanov MG (1974) Enthalpy of molten uranium- plutonium oxides. US AEC Report ANL-8082, pp 1–19
211. Janz GJ, Tomkins RPT, Allen CB, Downey JR, Garner GL, Krebs U, Singer SK (1975) Molten salts: volume 4, part 2, chlorides and mixtures. Electrical conductance, density, viscosity, and surface tension data. J Phys Chem Ref Data 4:871
212. Janz GJ, Tomkins RPT, Allen CB, Downey JR, Garner GL, Singer SK (1977) Molten salts: volume 4, part 3. Bromides and mixtures; iodides and mixtures. Electrical conductance, density, viscosity, and surface tension data. J Phys Chem Ref Data 6:409
213. Janz GJ, Tomkins RPT (1981) Molten salts: volume 5, part 1. Additional single and multi-component salt systems. Electrical conductance, density, viscosity, and surface tension data. J Phys Chem Ref Data 9:831–1021
214. Janz GJ (1988) Thermodynamic and transport properties for molten salts: correlation equations for critically evaluated density, surface tension, electrical conductance, and viscosity data. J Phys Chem Ref Data 17(Suppl 2):1–325
215. Marcus Y (2015) Volumetric behavior of molten salts and molten salt hydrates: Chapter 20. In: Wilhelm E, Letcher T (eds) Volume properties. Royal Society of Chemistry, Cambridge, pp 526–574
216. Marcus Y (2013) Volumetric behavior of molten salts. Thermochim Acta 559:111–116
217. Campbell AN, Williams DF (1964) The thermodynamics and conductances of molten salts and their mixtures. III. Densities, molar volumes, viscosities, and surface tensions of molten lithium chlorate, with small additions of water, and other substances. Can J Chem 42:1778–1787
218. Denielou L, Petitet JP, Tequi C (1976) Thermodynamic study of molten salts with polyatomic anions. J Phys (Paris) 37:1017–1024
219. Red'kin AA, Salyulev AB, Smirnov MV, Khokhlov VA (1995) Density and electrical conductivity of NaCl- $CoCl_2$ and NaCl- $NiCl_2$ molten mixtures. Z Naturforsch 50a:998–1002
220. Red'kin AA, Nikolaeva EV, Dedyukhin AE, Zaikov YP (2012) The electrical conductivity of chloride melts. Ionics 18:255–265

221. Marcus Y (2013) The compressibility of molten salts. J Chem Thermodyn 61:7–10
222. Stillinger FH (1961) Compressibility of simple fused salts. J Chem Phys 35:1581–1583
223. Goldman G, Tödheide K (1976) Equation of state and thermodynamic properties of molten potassium chloride to 1320 K and 6 kbar. Z Naturforsch 31a:769–776
224. Yajima K, Moriyama H, Oishi J (1984) A model for the surface tension of molten alkali halides. J Phys Chem 88:4390–4394
225. Cleaver B, Spencer PN (1955) Isothermal compressibilities and thermal pressure coefficients of molten salts. High Temp High Press 7:539–547
226. Zhu HM, Saito T, Sato Y, Yamamura T, Shimakage K, Ejima T (1991) Ultrasonic velocity and absorption coefficient in molten alkali metal nitrates and carbonates. J Jpn Inst Met 55:937–944
227. Marcus Y (2013) Internal pressure of liquids and solutions. Chem Rev 113:6536–6551
228. Fernandez R, Østvold T (1989) Surface tension and density of molten fluorides and fluoride mixtures containing cryolite. Acta Chem Scand 43:151–159
229. Hara S, Ogino K (1989) The densities and the surface tensions of fluoride melts. ISIJ Intl 29:477–485
230. Herie L (1965) Compressibility of molten alkali salts. J Phys Chem 69:2785–2787
231. Knape EG, Torell LM (1975) Hypersonic velocities and compressibilities for some molten nitrates. J Chem Phys 62:4111–4115
232. Cleaver B, Zani P (1978) Adiabatic thermal pressure coefficients of molten salts; an indirect method for the measurement of isothermal compressibility. High Temp High Press 10:437
233. Takesawam K-I, Takeda S, Harada S, Tamaki S (1989) Sound wave propagation in molten silver halides. J Phys Soc Jpn 58:538–543
234. Bockris JO'M, Pilla A, Barton JL (1962) Compressibilities of certain molten alkaline earth halides and the volume change on fusion of some of the corresponding solids. Rev Chim Rep Pop Roum 7:59–77
235. Denisovets VP, Prisyazhii VFD (1988) Ultrasound velocity in melts of alkali metal formates and acetates. Ukr Khim Zh 54:1247–1250
236. Janz GJ, Lakshminarayanan GR, Tomkins RPT, Wong J (1969) Molten salts. II. Surface tension data. Nat Stand Ref Data Ser Nat Bur Stand 28:49–111
237. Dutcher CS, Wexler AS, Clegg SL (2010) Surface tensions of inorganic multicomponent aqueous electrolyte solutions and melts. J Phys Chem A 114:12216–12230
238. Campbell AN, van der Kouwe ET (1968) Thermodynamics and conductances of molten salts and their mixtures. V. The density, change of volume on fusion, viscosity, and surface tension of sodium chlorate and of its mixtures with sodium nitrate. Can J Chem 46:1279–1286
239. Gossink RG, Stevels JM (1971) Density and surface tension of molten alkali molybdates and tungstates in connection with structure and glass formation. J Non Cryst Solids 5:217–236
240. Patrov BV, Yurkinskii VP (2004) Surface tension and density of a sodium hydroxide melt. Russ J Appl Chem 77:2029–2930
241. Lubova Z, Danek V, Nguyen DK (1997) Surface tension of melts of the system KF- KCl-KBF4. Chem Papers 51:78–83
242. Gardon JL (1977) Critical review of concepts common to cohesive energy density, surface tension, tensile strength, heat of mixing, interfacial tension, and butt joint strength. Coll Interf Sci 59:582–594
243. Aqra F (2014) Novel estimated surface tension data of actinide halide salts in the molten state. J Nucl Mater 448:230–232
244. BPatrov BV, Yurkinskii VP (2004) Surface tension and density of a sodium hydroxide melt. Zh Priklad Khim 77:2054–2055
245. Shartsis L, Capps W (1952) Surface tension of molten alkali borates. J Am Ceram Soc 35:169–173
246. Aqra F (2014) Surface tension of molten metal halide salts. J Mol Liq 200:120–121
247. Aqra F (2014) Molten rare earth tri- halides: prediction of surface tension. J Mol Liq 200:229–231

248. Mayer SW (1963) A molecular parameter relation between surface tension and liquid compressibility. J Phys Chem 67:2160–2164
249. Reiss H, Frisch HL, Lebowitz JL (1959) Statistical mechanics of rigid spheres. J Chem Phys 31:369–380
250. Eggelstaff PA, Widom B (1970) Liquid surface tension near the triple point. J Chem Phys 53:2667–2669
251. Marcus Y (2013) The compressibility and surface tension product of molten salts. J Chem Phys 139(124509):1–4
252. Nagashima A (1991) Measurement of transport properties of high- temperature fluids. Intl J Thermophys 12:1–15
253. Chhabra RP, Hunter RJ (1981) The fluidity of molten salts. Rheol Acta 20:203–208
254. Potapov A, Khokhlov V, Sato Y (2003) Viscosity of molten rare earth metal trichlorides I. $CeCl_3$, $NdCl_3$, $SmCl_3$, $DyCl_3$ and $ErCl_3$. Z Natorforsch A 58:457–463
255. Potapov A, Sato Y (2011) Viscosity of molten rare earth metal trichlorides. II. Cerium subgroup. Z Natorforsch A 66:649–655
256. Marcus Y (2014) The fluidity of molten salts re- examined. Fluid Phase Equilib 366:57–60
257. Hildebrand JH, Lamoreaux RH (1976) Viscosity of liquid metals: an interpretation. Proc Natl Acad Sci U S A 73:988–989
258. Janz GJ, Yamamura T, Hansen MD (1989) Corresponding- states data correlations and molten salts viscosities. Intl J Thermophys 10:159–171
259. Abe Y, Nagashima A (1981) The principle of corresponding states for alkali halides viscosity. J Chem Phys 75:3977–3985
260. Tada Y, Hiraoka S, Uemura T, Harada M (1958) Corresponding states correlation of transport properties of uniunivalent molten salts. Ind Eng Chem Res 27:1042–1049
261. Tomlinson JW (1959) In: Bockris JO'M (ed) Physicochemical measurements at high temperatures. Butterworths, London, 257 ff
262. Sundheim BR (1964) Transport properties of liquid electrolytes. In: Sundheim BR (ed) Fused salts. McGraw-Hill, New York, pp 208–210
263. Nunes VMB, Lourenco MJV, Santos FJV, Lopes MLSM, Nieto de Castro CA (2010) Accurate measurement of physicochemical properties on ionic liquids and molten salts. In: Gaune-Escard M, Seddon KR (eds) Molten salts and ionic liquids: never the twain? Wiley, New York, pp 229–263
264. Woyakowska A, Plinska S, Josiak J, Krzyzak E (2006) Electrical conductivity of molten cobalt dibromide + potassium bromide mixtures. J Chem Eng Data 51:1256–1260
265. Kojima T, Miyazaki Y, Nomura K, Animoto K (2007) Electrical conductivity of molten Li_2CO_3– X_2CO_3 (X = Na, K, Rb, and Cs) and Na_2CO_3- Z_2CO_3 (Z: K, Rb, and Cs). J Electrochem Soc 154:F222–F230
266. Gaune P, Gaune-Escard M, Rycerz L, Bogacz A (1996) Electrical conductivity of molten $LnCl_3$ and M_3LnCl_6 compounds (Ln = La, Ce, Pr, Nd; M = K, Rb, Cs). J Alloy Comp 233:143–149
267. Rice SA (1962) Kinetic theory of ideal ionic melts. Trans Faraday Soc 58:499–510
268. Angell CA (1965) Diffusion- conductance relations and free volume in molten salts. J Phys Chem 69:399–403
269. Klemm A (1964) Transport properties of molten salts. In: Blander M (ed) Molten salt chemistry. Interscience, New York, pp 538–606
270. Haase R (1991) Internal and external transport numbers in ionic melts. Z Phys Chem (Muenchen) 174:77–87
271. Harris KR (2010) Relations between the fractional stokes- einstein and nernst- einstein equations and velocity correlation coefficients in ionic liquids and molten salts. J Phys Chem B 114:9572–9577
272. Sjoblom CA (1968) Self-diffusion in molten salts. A comparison between diffusion theories and experimental data. Z Naturforsch A 23:933–939

273. Lenke R, Uelenhack W, Klemm A (1973) Self-diffusion in molten lithium chloride. Z Naturforsch 28A:881–884
274. Rollet A-L, Sarou-Kanian V, Bessada C (2010) Self-diffusion coefficient measurements at high temperature by PFG NMR. C R Chim 13:399–404
275. Fürth R (1941) The theory of the liquid state. III. The hole theory of the viscous flow of liquids. Proc Camb Phil Soc 37:281–290
276. Adam G, Gibbs JH (1965) The temperature dependence of cooperative relaxation properties in glass- forming liquids. J Chem Phys 43:139–146
277. Gheribi AE, Torres JA, Chartrand P (2014) Recommended values for the thermal conductivity of molten salts between the melting and boiling points. Solar Energy Mater Solar Cells 126:11–25
278. Cornwell K (1971) Thermal conductivity o molten salts. J Phys D Appl Phys 4:441–445
279. DiGuilio RM, Teja AS (1992) A rough hard-sphere model for the thermal conductivity of molten salts. Intl J Thermophys 13:855–871
280. Hossain MZ, Kassaee MH, Jeter S, Teja AS (2014) A new model for the thermal conductivity of molten salts. Intl J Thermophys 35:246–255
281. Turnbull AG (1961) The thermal conductivity of molten salts. II. Theory and results for pure salts. Aust J Appl Sci 12:324–329
282. Bloom H, Dobroszkowski A, Thicklebank SB (1965) Molten salt mixtures. IX. The thermal conductivities of molten nitrate systems. Aust J Chem 18:1171–1176
283. Grimes WR, Smith NV, Watson GM (1958) Solubility of noble gases in molten fluorides. I. In mixtures of NaF- ZrF_4 (53-47 mole %) and NaF- ZrF_4- UF_4 (50-46-4 mole %). J Phys Chem 62:862–866
284. Tomkins RPT, Bansal N (1991) Solubility data series 45–46. Pergamon Press, Oxford
285. Tomkins RPT (2003) Solubility of gases in molten salts and molten metals. Wiley Ser Solut Chem 6:173–217
286. Cleaver B, Mather DE (1970) Solubilities of helium, argon, and nitrogen in molten nitrates at pressures up to 1 kilobar. Trans Faraday Soc 66:2469–2482
287. Green WG, Field PE (1980) Interaction of gases in ionic liquid. 2. The solubility of argon and nitrogen in molten lithium nitrate and potassium nitrate. J Phys Chem 84:3111–3114
288. Sada E, Katoh S, Yoshii H, Takemoto I, Shiomi N (1981) Solubility of carbon dioxide in molten alkali halides and nitrates and their binary mixtures. J Chem Eng Data 26:279–281
289. Bratland D, Grjotheim K, Krohn C, Motzfeld K (1966) Solubility of carbon dioxide in molten alkali halides. Acta Chem Scand 20:1811–1826
290. Field PE, Green WG (1971) Interactions of gases in ionic liquids. I. Solubility of nonpolar gases in molten sodium nitrate. J Phys Chem 75:821–825
291. Sada E, Katoh S, Beniko H, Yoshii H, Kayano M (1980) Solubility of carbon dioxide in molten salts. J Chem Eng Data 25:45–47
292. Blander M, Grimes WR, Smith NV, Watson GM (1959) Solubility of noble gases in molten fluorides. II. In the LiF- NaF- KF eutectic mixture. J Phys Chem 63:1164–1167
293. Fukase S (1983) Solubility of rare gases in molten salts. J Phys Chem 87:1768–1776
294. Simonin JP (2011) Effect of polarization on the solubility of gases in molten salts. J Chem Phys 134(054508):1–9
295. Brough BJ, Kerridge DH, Mosley M (1966) Indicators in fused salts. J Chem Soc A 1556–1558
296. Burnasheva IA, Gordienko AA, Toropov AP (1971) Solubility of organic substances in some fused low- melting salts and salt eutectics. Teor Rastvorov 248–50

Chapter 4
Network Forming Ionic Liquids

4.1 Binary Network Forming Salts

Some inorganic salts associate in the molten state to large aggregates forming a network of bonds. Such a molten salt on cooling form a glass rather than crystallize. The glass is characterized by infinitely large viscosity and a loss of configurational entropy [1]. Among inorganic salts the foremost examples are BeF_2, $ZnCl_2$, B_2O_3, SiO_2, and GeO_2, the physicochemical data of which are shown in Table 4.1.

There is some information in the literature regarding these molten salts beyond the data in Table 4.1. The structure of the network forming salts has been studied by means of the methods already discussed in Sect. 3.2 for highly ionic high-melting salts. A common feature is the existence in the MX_2 melts of a close-packed anion ($X = F^-$, Cl^-, O^{2-}) structure with the multivalent cations ($M = Be^{2+}$, Zn^{2-}, Si^{4+}, Ge^{4+}) occupying tetrahedral sites in this structure. Thus networks of $MX_4{}^{2\text{ or }4-}$ tetrahedra that share corners are the common structural feature of these salts. This information resulted from the application of a variety of methods to the molten MX_2 salts: x-ray diffraction [22–24], neutron diffraction [24–27], XAFS [28, 29], Raman spectroscopy [30, 31], and computer simulations [24, 32–36]. Most of these reports dealt with $ZnCl_2$ [22, 24–28, 30, 32, 35], some of them with $ZnBr_2$ (a less pronounced network former) [25, 29, 30], and only very few with BeF_2 [33], SiO_2 [31, 32, 36], and GeO_2 [34]. (Studies of the structures of mixtures involving these compounds, e.g. LiF + BeF_2 and various non-stoichiometric silicates, are much more prolific).

The structure of molten B_2O_3 differs, in that three-fold coordination of boron atoms with oxygen atoms prevails in it [37–40]. The structure, obtained by x-ray diffraction at 923 K [37, 39] and up to 1073 K by neutron diffraction [38], thus involves hexagonal boroxol rings and independent BO_3 triangles, the proportion of the latter increasing with the temperature. Raman and luminescence studies of molten B_2O_3 at higher temperatures (up to 1700 K) showed further fragmentation to bent triatomic $BO_2{}^-$ units [40].

Y. Marcus, *Ionic Liquid Properties*, DOI 10.1007/978-3-319-30313-0_4

Table 4.1 Properties of binary glass-forming molten salts

Property/units	BeF_2	$ZnCl_2$	B_2O_3	SiO_2	GeO_2
Molar mass, M/kg mol^{-1}	0.04701	0.13632	0.06962	0.06008	0.10464
Melting point, T_m/K[a]	825	591	723	1996	1389
Normal boiling point, T_b/K[a]	1442	1005	2316	3070[t]	
Glass transition temperature, T_g/K	523[b]	375[c]	545[d]	1261[d]	786[d]
Molar enthalpy of melting, $\Delta_m H$/kJ mol^{-1}[a]	8.4	10.3	22.2	9.6	16.8
Molar heat capacity at 1.1T_m, C_P/J K^{-1} mol^{-1}[a]	87.9	100.8	127.6	85.8	80.1
Surface tension at 1.1T_m, σ/mJ m^{-2}	200[e]	54[f]	67[q]	319[h]	271[h]
Density at 1.1T_m, ρ/g cm^{-3}	1.990[f]	2.492[f]	1.630[g]	2.209[o]	3.488[i]
Molar volume at 1.1T_m, V/cm^3 cm^{3}mol^{-1}	23.6[f]	54.7[f]	42.7[g]	27.2[o]	30.0[i]
Isobaric expansibility at 1.1T_m, $10^6 \alpha_P$/K^{-1}	260[f]	186[f]	82.4[g]	37[o]	8.9[i]
Isothermal compressibility at 1.1T_m, κ_T/GPa^{-1}		0.547[l]	0.05[i]	0.007[i]	0.121[i]
Internal pressure 1.1$T_m \alpha_P/\kappa_T$, P_{int}/MPa		221	1311	10,550	112
Viscosity coefficient, A_η	22.4[f]	8.15[f]	6.3[k]	16.7[j]	6.16[m]
Viscosity coefficient, B_η/kJ	244.1[f]	75.1[f]	89.2[k]	516.0[j]	186.3[m]
Viscosity at 1.1T_m, $10^{-3}\eta$/Pa s	20.62[f]	0.315[f]	1.67[k]	96.7[j]	44.4[m]
Equivalent conductivity coefficient, A_Λ	44.37[n]	11.46[p]			
Equivalent conductivity coefficient, B_Λ/kJ	−197[n]	−0.92[p]			
Equivalent conductivity at 1.1T_m, Λ/S Scm^2 cm^{2}mol^{-1}	0.0002[n]	9.67[p]	9.2×10^{-6}[s]	7.5×10^{-4}[r]	

The viscosity entries pertain to $\ln(\eta/\text{Pa s}) = A_\eta \exp(B_\eta/RT)$ and the conductivity entries to $\ln(\Lambda/\text{S cm}^2\,\text{mol}^{-1}) = A_\Lambda \exp(-B_\Lambda/RT)$

[a][2], [b][3], [c][4], [d][5], [e][6] (assumed value), [f][7], [g][8], [h][9], [i][10], [j][11], [k][12], [l][13], [m][14] used (T/K – 199) instead of T/K in the denominator, [n][15], [o][16], [p][17], [q] [18], [r][19], with κ read from a curve, [s][20], with κ read from a curve, [t][21]

The glass transition temperature T_g (the temperature at which on cooling a liquid the viscosity reaches the value of 10^{12} Pa s) recorded in Table 4.1 can be interpreted in several manners. These include the temperature at which, on super-cooling of the melt, the free volume V_f vanishes or the heat capacity undergoes a deep decline, ΔC_P, to below the crystal values. For glass forming substances the Arrhenius-type dependence of transport properties, $\ln(\eta/\text{Pa s}) = A_\eta \exp(B_\eta/RT)$ for the viscosity, may have to be changed to the Vogel- Fulcher-Tammann- (VFT) form:

$$\eta/\text{Pa s} = A_\eta{}'\exp\left(B_\eta{}'/R(T - T_0)\right) \tag{4.1}$$

where $T_0 < T_g$. At temperatures not far from the melting point the transport phenomena (viscosity, conductivity, self-diffusion of the ions) depend on the temperature according to the Arrhenius expression with the coefficients A_η and B_η for the viscosity (and corresponding ones for the conductivity) shown in Table 4.1. However, deviations may occur at higher temperatures and definitely do so for the super-cooled liquids. The apparent activation energy B_η declines at higher temperatures in the cases of B_2O_3 and SiO_2 and appears to do so also in those of BeF_2 and GeO_2 [41], a behavior attributed to the breakdown of the network to smaller fragments. This finding was challenged in the case of B_2O_3 [12], where true Arrhenius-type behavior of the viscosity was found above 1073 K, and in the case of BeF_2 [42] where such behavior was found all the way up to 1250 K. On the other hand, the VFT expression was found to describe accurately the viscosity of SiO_2 and GeO_2 over at least 12 orders of magnitude [14], with $T_0 = 530$ K and 199 K respectively. The original data were re-examined here and the VFT expression holds very well for $ZnCl_2$ at $598 \leq T/\text{K} \leq 716$ [43] or $600 \leq T/\text{K} \leq 893$ [44] and for $ZnBr_2$ at $673 \leq T/\text{K} \leq 813$ [44] with the same $T_0 = 260$ K (The $T^{1/2}$factor used in [43] is unnecessary, because it varies only by 9 % over the T-range studied). The same VFT expression with $T_0 = 260$ K holds also for the viscosity, specific conductance, and self-diffusion coefficients of zinc and chloride ions in molten $ZnCl_2$, albeit for much fewer data [17].

4.2 Molten Borates and Silicates

Physicochemical data of borate and silicate melts having stoichiometric compositions are shown in Table 4.2. Although a large amount of information regarding mixtures of molten oxides or molten silicates and borates can be found in the literature, data regarding pure, stoichiometric single compounds is relatively scarce.

Some information beyond what is shown in Table 4.2 has been reported. The adiabatic compressibility, obtained from sound velocity and density data (Sect. 3.3.3) was reported [55] as κ_S/GPa^{-1} at 1473 K for Na_2SiO_3 (0.588), $Na_2Si_2O_5$ (0.643), K_2SiO_3 (1.27), $K_2Si_2O_5$ (0.791), and at 1173 K for $LiBO_2$ (1.205) and $Li_2B_4O_7$ (1.29). The isothermal compressibility κ_T/GPa^{-1} (the reciprocal of the bulk

Table 4.2 The molar masses, M, melting points, T_m, molar enthalpies of melting, $\Delta_m H$, constant pressure molar heat capacities, C_p, (from [2]), and the molar volumes, V, isobaric expansivities, α_P, and viscosities, η, at $1.1T_m$ of molten silicates and borates

Salt,	M	T_m	$\Delta_m H$	C_p	V	$10^3\alpha_P$	η
	kg mol^{-1}	K	kJ mol^{-1}	J K^{-1} mol^{-1}	cm^3 mol^{-1}	K^{-1}	Pa s
Li_2SiO_3	0.08997	1474	28.0	167.4	21.8[f]	0.107[f]	
$Li_2Si_2O_5$,	0.15005	1307	53.8	251	23.4[v]		
$LiBO_2$	0.04975	1117	33.9	144.6	26.5[u]	0.247[u]	
$Li_2B_4O_7$	0.16912	1190	120.5	470.6[a]	89.0[i]	0.218[i]	
Na_2SiO_3	0.12206	1362	51.8	177.3	27.2[f]	0.128[f]	132[w]
$Na_2Si_2O_5$	0.18215	1147	35.6	261.2	79.4[u]	0.096[u]	
$NaBO_2$	0.06580	1239	36.2	146	35.5[y]	0.359[y]	
$Na_2B_4O_7$	0.20122	1016	81.2	444.9	99.8[p]	0.427[p]	7.0×10^{-3}[n]
K_2SiO_3	0.15428	1249[b]	50.2[z]	167.4[z]	35.3[f]	0.160[f]	
$K_2Si_2O_5$	0.21436	1318[c]	35.2[z]	275.3[z]	95.9[u]	0.125[u]	
$K_2B_4O_7$	0.23344	1088	104.2	473.2[a]	125[m]		
$MgSiO_3$	0.10039	1850	75.3	146.4[z]	39.3[h]	0.07[h]	0.164[k]
Mg_2SiO_4	0.14069	2171	71.1	205.0[z]			
$CaSiO_3$	0.11616	1813	56.1	151	44.7[h]	0.07[h]	0.120[k]
CaB_2O_4	0.12570	1433	74.1	258			
CaB_4O_7	0.19532	1263	113.4	444.8			
$SrSiO_3$	0.16370	1851[d]	56[e]				0.139[k]
Sr_2SiO_4	0.26732	2598[e]					
SrB_4O_7	0.24286	1267[l]	139.8[l]				
$BaSiO_3$	0.21341	1877[d]	139.8				0.117[k]
BaB_4O_7	0.29257	1182[l]					
$MnSiO_3$	0.13102	1543	66.9	153.8			0.043[k]
Mn_2SiO_4	0.20196	1613	94.6	230.1			180[x]
Fe_2SiO_4	0.20377	1493	92	240.6	18.3[x]	7.0[x]	520[x]
$ZnSiO_3$	0.14149	1710[d]					
Zn_2SiO_4	0.22290	1783[d]					
$CdSiO_3$	0.18849	1516[d]					
Cd_2SiO_4	0.31691	1516[d]					
$PbSiO_3$	0.28328	1037	26.0	130.1			
Pb_2SiO_4	0.50648	1016	51.0	189.1			

[a]At 1.1 T_m, [b][45], [c][46], [d][47], [e][49], [f][48] (according to [53] $V = 69.0$ cm^3 mol^{-1} at $1.1T_m$), [g][8], [h][50], [i][51], [j][11], [k][12], [l][52], [m][53], [n][54], [o][16], [p][18], [u][55], [v][56] at 1673 K, [w][57], [x][58] (at 1873 K for Mn_2SiO_4), [y][59], [z][71]

modulus) is 0.050 for $MgSiO_3$ at 1913 K and 0.040 for $CaSiO_3$ at 1836 K [50, 60]. The viscosity of $Na_2Si_2O_5$ follows the VFT expression (4.1) with $A_\eta' = 0.0611$, $B_\eta' = 31.5$ kJ mol^{-1}, and $T_0 = 454.5$ K [57]. The surface tension of calcium silicates, σ/mN m^{-1}, hardly varies between 1780 and 1870 K: it is 530 for $CaSiO_3$ and 450 for $CaSi_2O_5$ (read from a curve [61]). The thermal conductivity of Na_2SiO_3 follows the

Arrhenius expression $\ln(\lambda_{th}/\mathrm{W\,m^{-1}K^{-1}}) = -14.5 + 20319/T$, i.e., with an activation energy $B_\lambda = 170$ kJ mol^{-1} [62].

Molecular dynamics simulation of Mg_2SiO_4 [63] yielded values for the isothermal compressibility $\kappa_T = 0.025$ GPa^{-1} at 2100 K, the density at 2110 K and at 2 GPa $\rho = 2.75$ g cm^{-3} (read from a figure) and the molar volume $V = 51$ cm^3 mol^{-1}, and an average molar isochoric heat capacity $C_V = 200$ J K^{-1} mol^{-1}. The self-diffusion coefficients for Mg_2SiO_4 at low pressures are $D/\mathrm{m^2\,s^{-1}} = 2.6 \times 10^{-7}$ $\exp(66.4\ \mathrm{kJ\,mol^{-1}}/RT)$ for Mg and 1.7×10^{-7} exp(79.2 kJ mol^{-1}/RT) for Si, and the dynamic viscosity is $\eta/\mathrm{Pa\,s} = 4.5 \times 10^{-4}\exp(41.0\mathrm{kJ\,mol^{-1}}/RT)$.

The structures of only a few of the stoichiometric molten borates and silicates (among those listed in Table 4.2) has been determined. The molecular dynamics simulation of Mg_2SiO_4 [63]. at relatively low pressures showed that >75 % of the silicon atoms are tetrahedrally surrounded by oxygen atoms, the rest mainly in fivefold coordination defining distorted trigonal bipyramidal polyhedra. At low pressure, the average coordination number of the Mg atoms is ≈5.5. As the pressure increases the average coordination number increases (~7.5) near 100 GPa. Octahedrally coordinated Mg attains a maximum at about 20 GPa and decreases systematically as the pressure increases. X-ray absorption spectroscopy of liquid Fe_2SiO_4 at 1575 K and ambient pressure [64] showed ~11 % shortening of the Fe–O distance in the melt compared with the crystal at the melting point, and a similar increase in the volume, indicating a decrease of the average coordination number of the Fe from 6 in the crystal to 4 in the melt, i.e., both Fe(II) and Si(IV) are tetrahedrally coordinated. Neutron diffraction with isotope substitution (see Sect. 3.2) was applied to molten $Li_2B_4O_7$ [65] and to molten $CaSiO_3$ [66]. For the former the Li–O distance decreased somewhat and that of Li–B increased at 1073 K compared to the glassy state, signifying an increase in the non-bridging oxygen content in the lithium coordination in the melt. In the latter salt the melt comprises primarily six- and sevenfold Ca–O coordination. Short chains of edge-shared Ca-octahedra feature in the structure of molten $CaSiO_3$.

A molecular dynamics simulation [67] showed similarities between alkaline earth silicates and alkali metal fluoroberyllates of the stoichiometries $MSiO_3$ with $M'BF_3$ and M_2SiO_4 with M'_2BeF_4 at corresponding temperatures ~$1.1T_m$ when having cations of similar sizes (Li^+ and Mg^{2+}, Na^+ and Ca^{2+}, K^+ and Ba^{2+}). The structures contain monomer, dimer, chain, and sheet units and experimental determinations of properties of fluoroberyllates in the more readily accessible range 620–1070 K could replace those on silicates, pertinent to geological problems, at the higher range of 1700–3000 K.

Another topic that pertains to the molten borate and silicate salts and to oxide melts in general is their acid-base behavior, in which oxygen atoms take the place of the hydrogen ions commonly encountered in aqueous solutions. Flood and Förland [68] established the concept of 'oxoacidity' and introduced the quantity pO = −loga_O, the negative of the logarithm of the oxygen activity in the melt, analogous to the pH. Dron [69] applied this framework to molten silicates, starting with equilibria such as $MSiO_3$ + MO ⇆ M_2SiO_4, involving free O^{2-} ions and bridging and non-bridging oxygen atoms. Konakov [70] reviewed the application of the acid-base concept to oxide melts via the pO scale and discussed the use of the stabilized zirconia electrode (ZrO_2 stabilized by 5 % CaO or Y_2O_3) to its measurement. The electrochemical cells to be used consist of molten silica, SiO_2, as the standard, and the pO of other oxide melts is measured relative to this standard, assigned pO = 7. A secondary standard of Na_2SiO_3 was established, with pO = 5.88. Applications to binary alkali metal borate, silicate, and germanate systems were reviewed.

References

1. Angell CA (1966) The importance of the metastable liquid state and glass transition phenomenon to transport and structure studies in ionic liquids. I. Transport properties. J Phys Chem 70:2793–2803
2. Barin I, Knacke O (1973) Thermochemical properties of inorganic substances. Springer, Berlin
3. Baldwin CM, Mackenzie JD (1979) Preparation and properties of water-free vitreous beryllium fluoride. J Non-Cryst Solids 31:441–445
4. Kartini E, Collins MF, Mezei F, Svensson EC (1998) Neutron scattering studies of glassy and liquid $ZnCl_2$. Physica B 241:909–911
5. Ozhovan MI (2006) Topological characteristics of bonds in SiO_2 and GeO_2 oxide systems upon a glass-liquid transition. J Exp Theor Phys 103:819–829
6. Cantor S, Ward WT, Moynihan CT (1969) Viscosity density in molten beryllium fluoride-lithium fluoride. J Chem Phys 50:2874–2879
7. Janz GJ, Lakshminarayanan GR, Tomkins RPT, Wong J (1969) Molten salts. II. Surface tension data. Nat Stand Ref Data Ser NBS 28:49–111
8. Macedo PB, Capps W, Litovitz TA (1966) Two-state model for the free volume of vitreous B_2O_3. J Chem Phys 44:3357–3363
9. Kingery WD (1959) Surface tension of some liquid oxides and their temperature coefficients. J Am Ceram Soc 42:6–10
10. Dingwell DB, Knoche R, Webb SL (1993) A volume-temperature relationship for liquid GeO_2 and some geophysically relevant derived parameters for network liquids. Phys Chem Miner 19:445–453
11. Urbain G, Bottinga Y, Richet P (1982) Viscosity of liquid silica, silcates, and aluminosilicates. Geochim Cosmochim Acta 46:1061–1072
12. Napolitano A, Macedo PB, Hawkins EG (1965) Viscosity density of borontrioxide. J Am Ceram Soc 48:613–616
13. Bockris JO'M, Pilla A, Barton AL (1962) Densities of solid salts at elevated temperatures and molar-volume change on fusion. Rev Chim Rep Pop Roum 7:59–77
14. Sipp A, Bottinga Y, Richet P (2001) New high viscosity data for 3D network liquids and new correlations between old parameters. J Non-Cryst Solids 288:166–174
15. Kim KB, Sadoway DR (1992) Electrical conductivity measurements of molten alkaline-earth fluorides. J Electrochem Soc 139:1027–1033 (the entry is $10^{-6}A_{\Lambda}$)
16. Ghiorso MS, Kress VC (2004) An equation of state for silicate melts. II. Calibration of volumetric properties at 105 Pa. Am J Sci 304:679–751
17. Bockris JO'M, Richards SR, Nanis L (1965) Self-diffusion and structure in molten Group II chlorides. J Phys Chem 69:1627–1637
18. Shartsis L, Capps W (1952) Surface tension of molten alkali borates. J Am Ceram Soc 35:169–172
19. Panish M (1959) The electrical conductivity of molten silica. J Phys Chem 63:1337–1338

20. Mackenzie JD (1956) Viscosity, molar volume, and electric conductivity of liquid boron trioxide. Trans Faraday Soc 52:1564–1568
21. Schick HL (1960) A thermodynamic analysis of the high-temperature vaporization properties of silica. Chem Rev 60:331–362
22. Neuefeind J, Tödheide K, Lenke A, Bertagnolli H (1998) The structure of molten $ZnCl_2$. J Non-Cryst Solids 224:205–215
23. Heusel G, Bertagnolli H, Neuefeind J (2006) X-ray diffraction studies on molten zinc bromide at high pressure. J Non-Cryst Solids 352:3210–3216
24. Zeidler A, Chirawatkul P, Salmon PS, Usuki T, Kohara S, Fischer H, Howells WS (2014) Structure of the network glass-former $ZnCl_2$: From the boiling point to the glass. J Non-Cryst Solids 407:235–245
25. Allen DA, Howe RA, Wood ND, Howells WS (1991) Tetrahedral coordination of zinc ions in molten zinc halides. J Chem Phys 94:5071–5076
26. Neuefeind J (2001) On the partial structure factors of molten zinc chloride. Phys Chem Chem Phys 3:3987–3993
27. Soper AK (2004) The structure of molten $ZnCl_2$: a new analysis of some old data. Pramana 63:41–50
28. Li H, Lu K, Wu Z, Dong J (1994) EXAFS studies of molten $ZnCl_2$, RbCl and Rb_2ZnCl_4. J Phys Condens Matter 6:3629–3640
29. Okamoto Y, Fukushima K, Iwadate Y (2002) XAFS study of molten zinc dibromide. J Non-Cryst Solids 312–314:450–453
30. Yannopoulos SN, Kalanpounias AG, Crissanthopoulos A, Papatheodorou GN (2003) Temperature induced changes on the structure and the dynamics of the "tetrahedral" glasses and melts of $ZnCl_2$ and $ZnBr_2$. J Chem Phys 118:3197–3214
31. Kalanpounias AG, Yannopoulos SN, Papatheodorou GN (2006) Temperature-induced structural changes in glassy, supercooled, and molten silica from 77 to 2150 K. J Chem Phys 124:014504, 1–15
32. Wilson M, Madden PA (1994) Polarization effects on the structures and dynamics of ionic melts. J Phys Condens Matter 6:A151–A155
33. Heaton RJ, Brookes R, Madden PA, Salanne M, Simon C, Turq P (2006) A first-principles description of liquid BeF_2 and its mixtures with LiF: 1. Potential development and pure BeF_2. J Phys Chem B 110:11454–11460
34. Hawlitzky H, Horbach J, Spas S, Krack M, Binder K (2008) Comparative classical and 'ab initio' molecular dynamics study of molten and glassy germanium dioxide. J Phys Condens Matter 20:285106, 1–15
35. Bassen A, Lemke A, Bertagnolli H (2000) Monte Carlo and reverse Monte Carlo simulations on molten zinc chloride. Phys Chem Chem Phys 2:1445–1454
36. Vashishta P, Kalia RK, Rino JP (1990) Interaction potential for silica: a molecular-dynamics study of structural correlations. Phys Rev B 41:12197–12209
37. Miyake M, Suzuki T (1984) Structural analysis of molten boron oxide (B_2O_3). J Chem Soc Faraday Trans 1(80):1925–1931
38. Misawa M (1990) Structure of vitreous and molten boron oxide (B_2O_3) measured by pulsed neutron total scattering. J Non-Cryst Solids 122:33–40
39. Sakowski J, Herms GJ (2001) The structure of vitreous and molten B_2O_3. J Non-Cryst Solids 293–295:304–311
40. Voron'ko YK, Sobol AA, Shukshin VE (2012) Study of a structure of boron-oxygen complexes in the molten and vapor states by Raman and luminescence spectroscopies. J Mol Struct 1008:69–76
41. Mackenzie JD (1961) Viscosity-temperature relation for network liquids. J Am Ceram Soc 44:598–601
42. Moynihan CT, Cantor S (1968) Viscosity and its temperature dependence in molten beryllium fluoride. J Chem Phys 48:115–119

43. Easteal AJ, Angell CA (1972) Viscosity of molten zinc chloride and supercritical behavior in its binary solutions. J Chem Phys 56:4231–4234
44. Šušić M, Mentus S (1975) Viscosity and structure of molten zinc chloride and zinc bromide. J Chem Phys 62:744–745
45. Kracek FC (1932) The ternary system: K_2SiO_3-Na_2SiO_3-SiO_2. J Phys Chem 36:2529–2542
46. Adams LH, Cohen LH (1966) Enthalpy changes as determined from fusion curves in binary systems. Am J Sci 264:543–561
47. Jaeger FM, Van Klooster HS (1916) Investigations in the field of silicate chemistry. IV. Some data on the meta- and orthosilicates of the bivalent metals: beryllium, magnesium, calcium, strontium, barium, zinc, cadmium and manganese. Proc Kon Ned Akad Wet 18:896–913
48. Bockris JO'M, Tomlinson JW, White JL (1956) Structure of the liquid silicates: partial molar volumes and expansivities. Trans Faraday Soc 52:299–310
49. Huntelaar ME, Cordfunke EHP, Scheele A (1993) Phase relations in the strontium oxide-silica-zirconium dioxide system I. The system SrO-SiO_2. J Alloys Comp 19:187–190
50. Matsui M (1996) Molecular dynamics simulation of structures, bulk moduli, and volume thermal expansivities of silicate liquids in the system CaO-MgO-Al2O_3-SiO_2. Geophys Res Lett 23:395–398
51. Anzai Y, Terashima K, Kimura S (1993) Physical properties of molten lithium tetraborate. J Cryst Growth 134:235–239
52. Slough W, Jones GP (1974) Compilation of thermodynamic data for borate systems. Natl Phys Lab Rep Chem Phys Lab Rep Chem 12:1–20
53. Volarowich MP, Leont'ewa AA (1935) The determination of the specific volumes of melts at temperatures up to 1400°. Z Anorg Allg Chem 225:327–332
54. Volarovich MP (1934) Investigation of the viscosity of the binary system sodium tetraborate-monosodium phosphate in the fused state. J Soc Glas Technol 18:201–208
55. Bockris JO'M, Kojonen E (1960) The compressibilities of certain molten alkali silicates and borates. J Am Chem Soc 82:4493–4497
56. Bottinga Y, Weill DF (1970) Densities of liquid silicate systems calculated from partial molar volumes of oxide components. Am J Sci 269:169–182
57. Richet P (1984) Viscosity and configurational entropy of silicate melts. Geochim Cosmochim Acta 48:471–483
58. Aune RE, Hayashi M, Sridhar S (2005) Thermodynamic approach to physical properties of silicate melts. Ironmak Steelmak 32:141–150
59. Riebling EF (1967) Volume relations in sodium oxide-boron oxide and sodium oxide-silicon dioxide-boron oxide melts at 1300°. J Am Ceram Soc 50:46–53
60. Rivers ML, Carmichael ISE (1987) Ultrasonic studies of silicate melts. J Geophys Res 92:9247–9270
61. Ejima A, Shimoji M (1970) Effect of alkali and alkaline-earth fluorides on surface tension of molten calcium silicates. Trans Faraday Soc 66:99–106
62. Gier EJ, Carmichael ISE (1996) Thermal conductivity of molten Na_2SiO_3 and $CaNa_4Si_3O_9$. Geochim Cosmochim Acta 60:355–357
63. Ben Martin G, Spera SP, Ghiorso MS, Nevins D (2009) Structure, thermodynamic, and transport properties of molten Mg_2SiO_4: molecular dynamics simulations and model EOS. Am Mineral 94:693–703
64. Jackson WE, Mustre de Leon J, Brown GE Jr, Waychunas GA, Conradson SD, Combes J-M (1993) High-temperature XAS study of ferrous silicate liquid: reduced coordination of ferrous iron. Science 262:229–233
65. Majerus O, Cormier L, Calas G, Beuneu B (2003) Structural modifications between lithium-diborate glasses and melts: implications for transport properties and melt fragility. J Phys Chem B 107:13044–13040
66. Skinner LB, Benmore CJ, Weber JKP, Tumber S, Lazareva L, Neuefeind J, Santodonato L, Du J, Parise JB (2012) Structure of molten $CaSiO_3$: neutron diffraction isotope substitution with aerodynamic levitation and molecular dynamics study. J Phys Chem B 116:13439–13447

67. Umesaki N, Ohno H, Igarashi K, Furukawa K (1992) A computer simulation study of the structural similarities between [alkali metal] fluoroberyllate and alkaline earth silicate melts. J Non-Cryst Solids 150:302–306
68. Flood H, Förland T (1947) The acidic and basic properties of oxides. Acta Chem Scand 1:592–604; (1947) The acidic and basic properties of oxides. III. Relative acid-base strengths of some polyacids. Acta Chem Scand 1:790–798
69. Dron R (1982) Acid-base reactions in molten silicates. J Non-Cryst Solids 53:267–278
70. Konakov VG (2011) From the pH scale to the pO scale. The problem of the determination of the oxygen ion O^{2-} activity in oxide melts. J Solid State Electrochem 15:77–86
71. Kubaschewski O, Alcock CB, Spencer PJ (1993) Materials thermochemistry, 6th edn. Pergamon Press, Oxford, Revised

Chapter 5
Low-Melting Ionic Salts

5.1 Salts Melting Between 100 and 250 °C

There are many salts that melt between 100 and 250 °C that are highly ionic and are to be discussed in this section, whereas room temperature ionic liquids (RTILs), melting below 100 °C, are discussed in Chap. 6, although this distinction is artificial. Such highly ionic molten salts are to be contrasted with liquid compounds that are formally classified as salts, but are highly covalent: mercury halides, tetravalent tin and lead halides, etc., which are outside the scope of this book. Some salts that are on the borderline between mainly covalent and mostly ionic are nevertheless included here, e.g., gallium and bismuth trihalides. Some relevant salts that melt below 250 °C (523 K) have already been mentioned among the high-melting salts in Chap. 3, including some of the thiocyanates, nitrates, chlorates, formates, and hydrogensulfates of univalent metal cations and a few other salts. An important class of salts that belongs to this category is the quaternary ammonium salts: mainly the symmetrical ones, the unsymmetrical ones melt mostly at <100 °C and are classified as RTILs.

Attention is here directed at the low-melting ionic salts: Table 5.1 shows the available physicochemical data for the inorganic low-melting salts, gleaned from the relevant tables in Chap. 3 with some additions, and Table 5.2 shows them for the quaternary ammonium salts with a few quaternary phosphonium salts thrown in.

As the tables reveal, especially Table 5.1, there are large amounts of 'unexplored territory' that ought to be filled with experimentally measured data. These would be valuable for further discussion of the relation of the properties of these salts to those of its constituent ions (mainly their sizes), and eventually to the use of such salts. The melting points reported for the quaternary ammonium salts by diverse authors general agree, noteworthy exceptions are the low values reported recently by Bhatt and Gohil [39, 40], based on differential scanning calorimetric endothermic peaks, which may pertain to solid-solid transitions rather than to melting.

Ammonium salts constitute a special category, because they are rather unstable in the molten state. The following temperatures have been recorded [41] for the melting

Y. Marcus, *Ionic Liquid Properties*, DOI 10.1007/978-3-319-30313-0_5

Table 5.1 Temperature of melting, T_m/K, the molar heat of melting, $\Delta_m H$/kJ mol^{-1} and the molar heat capacity C_P/J K^{-1} mol^{-1}, and the cohesive energy density, *ced*/MPa, density, ρ/g cm^{-3}, isobaric expansivity, α_P/K^{-1}, molar volume, V/cm^3 mol^{-1}, surface tension, σ/mN m^{-1}, at 1.1 T_m of highly ionic inorganic salts melting between ~100 and ~250 °C (370–530 K), from tables in Chap. 3 or as annotated

Salt	T_m [a]	$\Delta_m H$ [d]	C_P [m]	*Ced* [n]	ρ [o]	$10^3\alpha_P$ [o]	V [o]	σ [q]
LiCN	433							
$LiNO_2$	473	17.1[d]						115[r]
$LiNO_3$	525	26.7	152.7	20.1	1.7410	0.236	39.6	113
$LiClO_3$	401	10.3[e]	122.2[e]	16.0	2.0545	0.087	44.0	85
$LiClO_4$	509	17.0	161.1	13.1	1.9944	0.307	53.3	
$LiAlCl_4$	421[b]							
$NaClO_3$	488	21.7[f]	133.1	14.0	2.0718	0.466	51.4	87
$NaHCO_2$	531	16.9[g]	142.3[m]					
$NaHSO_4$	455	17.3[h]	246[h]		2.196[p]	0.422[p]	54.6[p]	89[s]
$NaAlCl_4$	426[b]							
KSCN	446	12.8		10.7	1.5623	0.512	54.5	95
KHF_2	512	6.6[c]			1.9213	0.296	40.7	
$KHCO_2$	440	11.9						
$KHSO_4$	488	16.6[h]	287[h]		2.115[p]	0.409[p]	64.3[p]	85[s]
$KAlCl_4$	534[b]							
RbSCN	457							
$RbHF_2$	483[c]	4.7[c]						
$RbHCO_2$	443	12.5[i]		9.9	1.9460	0.494	73.4	
$RbCH_3CO_2$	519	8.9[j]						
$RbHSO_4$	479	12.8[h]						
CsSCN	479	6.0						
$CsHF_2$	453	3.1[c]						
$CsHCO_2$	536							
$CsCH_3CO_2$	464	11.0		8.6	2.3723	0.414	80.5	
$CsHSO_4$	491	9.6[h]						
$AgNO_3$	485	12.6	128.0	16.2	3.3125	0.325	51.2	146[r]
$AgClO_3$	488							
TlSCN	507							
$TlNO_2$	459	6.9			4.8238	0.325	51.9	108[r]
$TlNO_3$	483	8.8[k]	130.0[k]	12.3	4.8158	0.360	55.3	90
$TlHCO_2$	395	10.9[l]						
$TlCH_3CO_2$	401	17.6[l]						
$GaCl_3$	351							24
$GaBr_3$	395							
GaI_3	465			36.6	3.5610	0.668	126.5	
$BiCl_3$	503	23.7	143.5	57.4	3.8453	0.598	82.0	65
$BiBr_3$	491	21.7	157.7	34.1	4.6205	0.523	97.1	69

[a]Tables 3.1 and 3.2, [b][1], [c][2], [d]Table 3.9, [e][3], [f][4], [g][5], [h][6], [i][7], [j]Table 3.8, [k][8], [l][9]., [m]Table 3.11, [n]Table 3.10, [o]Table 3.11, [p][10], [q]Table 3.17, [r][11], [s][12]

Table 5.2 Temperature of melting, T_m, the molar heat of melting, $\Delta_m H$, and the density, ρ, isobaric expansivity, α_P, and molar volume, V, at 1.1 T_m (from [15]) of symmetrical tetraalkylammonium and tetraalkylphosphonium salts melting between ~100 and ~250 °C (370–530 K)

Salt	T_m/K	$\Delta_m H$ [a]/KJ mol^{-1}	ρ/g cm^{-3}	$10^3\alpha_P$/K^{-1}	V/cm^3 mol^{-1}
Me_4N NTF$_2$	403[w]				
Et_4N NO$_3$	553[u] dec., 388[x]		1.029[x]	0.62[x]	
Et_4N SCN	538[v], 535[x]		1.157[x]	0.63[x]	
Et_4N BF$_4$	364[x]		1.132[x]	0.71[x]	
Et_4N PF$_6$	355[x]		1.074[x]	0.57[x]	
Et_4N NTF$_2$	377[g], 378[w]	9.0[g]			
Pr_4N SCN	432[v]				
Pr_4N ClO$_4$	512[o], 514[s]	14.2[s]			
Pr_4N BF$_4$	511[n], 521[t]	19.2[n]	0.8791	0.730	310.7
Pr_4N PF$_6$	510		1.0624	0.303	311.8
Pr_4N BPh$_4$	479	12.1	0.9065	1.592	557.7
Pr_4N Pic	391[f]		1.1093[f]	0.863[f]	373.5[f]
Bu_4N Br	395	15.5	0.9829	0.716	328.0
Bu_4N I	419, 419[s]	9.3	1.0596	0.792	348.7
Bu_4N SCN	399[v], 383[y]				
Bu_4N NO$_3$	392, 388[u]	14.6	0.9036[i]	0.665[i]	284.3[i]
Bu_4N ClO$_4$	483[o], 488[s]	12.3[s]			
Bu_4N BF$_4$	429[n], 344[y]	12.1[n]	0.9122	0.637	361.0
Bu_4N PF$_6$	524[m], 344[y]	16.4[m]	0.9501	0.690	407.8
Bu_4N BBu$_4$	384[h]		0.7734[j]	0.519[j]	499.8[j]
Bu_4N BPh$_4$	510	38.5	0.8576	0.577	654.9
Bu_4N Pic	363[k]		1.0764[l]	0.653[l]	437.5[l]
Pe_4N Br	376	41.4			
Pe_4N I	412, 407[s]	39.3			
Pe_4N NO$_3$	387[a], 384[s]	28.5[a]			402[a]
Pe_4N NO$_2$	370[o]				
Pe_4N SCN	323[a]	19.7[a]			
Pe_4N ClO$_4$	383[a], 389[s]	18.4[a]	0.9545[c]	0.617[c]	417.0[c]
Pe_4N BF$_4$	391[b]		0.9067	0.611	425.0
Pe_4N Pic	347[k]		1.0128[d]	0.567[d]	727.1[d]
Hx_4N Br	377[a]	15.9			
Hx_4N I	378[e], 376[s]	17.7[s]	0.838[e]	0.730[e]	575.0[e]
Hx_4N ClO$_4$	383[a], 377[s]	18.4			539[a]
Hx_4N BF$_4$	367[a], 364[t]	19.2[a]	0.8964[t]	0.644[t]	492.4[t]
Hp_4N I	396[a]	37.2[a]			
Hp_4N ClO$_4$	399[a]	31.8[a]			
Oc_4N I	401[s]	48.7[s]			
Oc_4N NO$_3$	381[s]	41.0[s]			
Oc_4N ClO$_4$	407[s]	46.5[s]			
Pr_4P BF$_4$	509[p]	12.1[p]			
Pr_4P PF$_6$	504[p]	5.8[p]			

(continued)

Table 5.2 (continued)

Salt	T_m/K	$\Delta_m H$ [a]/KJ mol^{-1}	ρ/g cm^{-3}	$10^3\alpha_P$/K^{-1}	V/cm^3 mol^{-1}
Bu_4P Br	376[q]		0.9874[q]	0.826[q]	333[q]
Bu_4P I	383[q]		1.3074[q]	0.795[q]	295[q]
Bu_4P BF_4	368[p]	6.4[p]			
Bu_4P PF_6	499[r], 490[p]	14.7[r], 10.3[p]			

[a][16], [b][17], the density data at 1 MPa), [c][18], the surface tension at $1.1T_m$ is 28.4 mN m^{-1}, [d][19], [e][20], [f][21], [g][22], $NTF_2^- = (CF_3SO_2)_2N^-$, [h][23], [i][24], [j][25], [k][26], [l][27], [m][28], [n][29], [o][30], [p][31], [q][32], [r][33], [s][34], [t][35], [u][36], [v][37], [w][38], [x][39], [y][40]

of ammonium salts without decomposition (although they may sublime before melting), T_m/K: NH_4SCN 422, NH_4N_3 433, NH_4HF_2 398, NH_4NO_3 442, NH_4IO_3 423, NH_4HCO_2 389, $NH_4CH_3CO_2$ 387, NH_4HSO_4 428, and NH_4SO_3F 518. These salts still decompose readily slightly above their melting points. Hardly any additional physicochemical data are available for the melts of these ammonium salts. The following salts melt at higher temperatures but decompose on melting: NH_4Cl, NH_4Br, NH_4I, NH_4BF_4, $(NH_4)_2SO_4$, and $(NH_4)_2CrO_4$ [41].

The transport properties of the inorganic low melting salts can be presented in terms of the Arrhenius-type expressions:

$$\eta/\text{mPa}\cdot\text{s} = A_\eta \exp(B_\eta/RT) \tag{5.1}$$

for the viscosity and

$$\Lambda/\text{S}\cdot\text{cm}^2\text{mol}^{-1} = A_\Lambda \exp(-B_\Lambda/RT) \tag{5.2}$$

for the molar conductivity, with the relevant coefficients A and B, the latter being the activation energy for the transport process. The A and B coefficients of Eqs. (5.1) and (5.2) of the inorganic and tetraalkylammonium salts, where known, as well as the η and Λ at $1.1T_m$ are shown in Table 5.3. The viscosities diminish with increasing temperatures whereas the conductivities augment, hence the different signs of the B-coefficients.

Some structural information is obtained from models that relate bulk properties to the structures. An early attempt to apply this approach was that of Lind et al. [35], who dealt with the Walden product of the molar conductivity Λ and the viscosity η of molten R_4N^+ BF_4^-, -PF_6^-, and -BPh_4^- (R = C_3 to C_6). This product is rather insensitive to the temperature and was applied to the melts at 250 °C. The hard-sphere model predicted the product $\Lambda\eta$ to be proportional to r^2/V, where r is the radius of the cation. However, only for the R = C_3 salts with BF_4^- and PF_6^- were the predicted values commensurate with the experimental ones, for the other salts the prediction overestimated the product. For such melts with small globular ions the structure is determined mainly by the coulomb forces as for high-melting salts, but with alkyl chains longer than propyl (C_3) the thermal movements of these chains clog the interstices in the liquid and obstruct the movement of the ions.

Table 5.3 Corresponding temperature, $1.1T_m$, the viscosity, η, at 1.1 T_m and the coefficients of η/mPa s = $A_\eta \exp(B_\eta/RT)$, and the equivalent conductance, Λ, at 1.1 T_m and the coefficients of Λ/S cm^{2}mol^{-1} = $A_\Lambda \exp(-B_\Lambda/RT)$, of molten salts melting between ~100 and ~250 °C (370–530 K)

Salt	$1.1T_m$/ K	η/ mPa s	A_η mPa s	B_η/ kJ mol^{-1}	Λ/ S cm^{2}mol^{-1}	A_Λ S cm^{2}mol^{-1}	$-B_\Lambda$/ kJ mol^{-1}
$LiClO_3$	441	14.74	0.00198	32690	9.2[g]	7326	24476
$LiClO_4$	560	6.25	0.09436	19523	56.2	921	13017
$LiNO_2$	546	5.73	0.00297	34342	31.1[b]	1492[b]	17545[b]
$LiAlCl_4$	463				40.5	567	10152
$NaClO_3$	537	6.8					
KSCN	501	5.61	0.00858	27004	17.1	6090	24477
$KHCO_2$	486				1.06[a, c]		
$RbHCO_2$	503				*0.206*[c]		
$AgNO_3$	534	0.163	0.1159	15149	37.7	490	11385
$TlNO_2$	505				36.1	319	9153
$TlNO_3$	531	2.70	0.0843	15301	26.5	522	13150
$BiCl_3$	553				31.2	296	9410
$BiBr_3$	540				26.0	209	9356
Pr_4NBF_4	573	1.23	0.01309	22665	40.7	2974	20439
Pr_4NPF_6	561	1.81	0.02322	21436	32.9	7610	25397
Pr_4NBPh_4	527	4.56	0.00095	36735	13.87[f]	1068445[f]	10272[f]
Pr_4NPic	430	7.47[d]			5.16[e]		
Bu_4NBr	432				2.12	817140	46195
Bu_4NI	461	27.0[f]			3.95	136940	39305
Bu_4NBF_4	479	4.36	0.03439	19021	9.91[f]	4025	23834
Bu_4NPF_6	572	1.73	0.00317	29953	25.2	4744	24913
Bu_4NBBu_4	422	7.75[j]					
Bu_4NBPh_4	561	2.57	0.00159	34455	14.1	6841	28840
Bu_4NPic	399	16.74[d]			2.22[e]		
Pe_4NPic	382	26.52[d]			1.55[e]		
Hx_4NBF_4	404	3.61[f]	0.000095[f]	43407[f]	0.706[f]	22952	34535

Inorganic salts from [42], tetraalkylammonium salts from [15] except where noted
[a] [43], [b] [44], [c] [13] the value is the specific, not the molar, conductivity, [d] [19], [e] [26], [f] [35] (for Bu_4NI at 422 K), [g] [14]

5.2 Molten Hydrated Salts

Another category of low melting highly ionic salts are the salt hydrates, $C_pA_q{\cdot}nH_2O$, listed in Table 5.4. It should be noted that not in all the cases do the hydrated salts melt congruently or that the melts with the stoichiometric composition crystallize on cooling to the same composition. It is also a moot question whether these melts, which may be formed at quite low temperatures (e.g., 8 °C for $LiClO_3{\cdot}3H_2O$ or "$ZnCl_2{\cdot}4H_2O$, particularly since this is a liquid at ambient temperature" [58]) may not be better considered as concentrated aqueous solutions of the salt.

Table 5.4 The melting points, T_m, type of melting: congruent *c* or non-congruent *nc*, the molar enthalpies of melting, $\Delta_m H$, and the constant pressure molar heat capacities, C_p, and surface tensions, σ, at the corresponding temperature $T = 1.1T_m$ of molten salt hydrates

	T_m	Type	$\Delta_m H$	C_p	σ
Salt hydrate	K		kJ mol^{-1}	J K^{-1} mol^{-1}	mN m^{-1}
$LiClO_3{\cdot}3H_2O$	281	*c*	36.4[a]	345[h]	
$LiClO_4{\cdot}3H_2O$	368		40.6[a]		
$LiI{\cdot}3H_2O$	348				
$LiNO_3{\cdot}3H_2O$	303	*c*	36.4[a]		
$NaOH{\cdot}H_2O$	337	*c*	15.8[a]		
$NaCH_3CO_2{\cdot}3\,H_2O$	331	*nc*	39.3[a]	453[e]	
$Na_2CO_3{\cdot}10H_2O$	306	*nc*	72.0[a]	~980[i]	
$Na_2SO_4{\cdot}10H_2O$	305	*nc*	78.7[a]	~945[k]	
$Na_2S_2O_3{\cdot}5H_2O$	322	*nc*	49.8[a]	950[e]	
$Na_2HPO_4{\cdot}12H_2O$	309	*nc*	119.9[g]	1249[g]	
$KF{\cdot}4H_2O$	292	*c*	43.1[a]	311[f]	
$K_2HPO_4{\cdot}6H_2O$	286	*c*	30.7[b]		
$MgCl_2{\cdot}6H_2O$	390	*nc*	34.9[a]	410[j]	
$Mg(NO_3)_2{\cdot}6H_2O$	363	*c*	41.0[a]	644[e]	
$MgSO_4{\cdot}7H_2O$	316		49.5[c]		
$CaCl_2{\cdot}6H_2O$	304	*nc*	37.2[a]	494[d]	93.5[m]
$CaBr_2{\cdot}6H_2O$	306		35.6[a]		71.2[m]
$CaI_2{\cdot}6H_2O$	315				60.6[m]
$Ca(NO_3)_2{\cdot}4H_2O$	315	*c*	29.7[a]		92.8[l]
$Ba(OH)_2{\cdot}8H_2O$	351	*c*	93.3[a]		
$Al(NO_3)_3{\cdot}9H_2O$	345	*nc*	66.2[a]		
$Al_2(SO_4)_3{\cdot}18H_2O$	361	*nc*	145.6[a]	1786[k]	
$NH_4Al(SO_4)_2{\cdot}12H_2O$	367	*c*	103.4[a]	1387[k]	
$Cr(NO_3)_3{\cdot}9H_2O$	339		77.8[a]		73.2[l]
$Mn(NO_3)_2{\cdot}6H_2O$	306	*cnc*	40.2[a]		91.6[l]
$FeCl_3{\cdot}6H_2O$	310	*c*	50.2[a]	650[d]	67.3[l]
$Fe(NO_3)_3{\cdot}9H_2O$	316	*nc*	36.0[a]		71.5[l]
$Co(NO_3)_2{\cdot}6H_2O$	328		41.4[a]		87.3[l]
$Ni(NO_3)_2{\cdot}6H_2O$	330		49.0[a]		86.7[l]
$ZnCl_2{\cdot}4H_2O$	<298[n]				
$Zn(NO_3)_2{\cdot}6H_2O$	310	*c*	39.8[a]		88.0[l]
$Cd(NO_3)_2{\cdot}4H_2O$	333	*c*	32.6[a]		88.7[l]

[a][45], [b][46], [c][47], [d][48], [e][49], [f][50], [g][51], [h][52], [i][53], [j][54], [k][55], [l][56], [m][57], [n][58]

Nevertheless, a model of molten salt hydrates that is widely accepted considers the hydrated cations to be single particles having a radius $r = r_I + 2r_{H2O}$, of low electrical field strength [59]. Thus, in molten $Ca(NO_3)_2{\cdot}4H_2O$ the cation $[Ca(H_2O)_4]^{2+}$ has a z/r value of 5.3 nm^{-1}, comparable with that of Cs^+, 5.9 nm^{-1}. Molten salt hydrates may, therefore, be treated in the same class of fluids as anhydrous molten salts, albeit being low-melting, rather than be regarded as concentrated aqueous solutions. This should particularly be valid for congruently melting salt hydrates with hydrated multivalent cations.

Table 5.4 shows the melting points, T_m, and whether the melting/crystallization is congruent or not, and the following thermodynamic data: the molar enthalpy of fusion, $\Delta_m H$, the molar constant-pressure heat capacity, C_p, and the surface tension, σ, as far as they are available.

Note that the molar enthalpies of melting of the salt hydrates are generally rather larger than those of similar anhydrous salts melting at a much higher temperature. Compare, for instance the $\Delta_m H$/kJ mol^{-1} of lithium salt trihydrates with the anhydrous salts: nitrate 36.4 vs. 26.7 and perchlorate 40.6 vs. 17.0, the sodium salt decahydrates with the anhydrous salts: carbonate 72.0 vs. 30.0, sulfate 78.7 vs. 23.0, and the calcium salt hexahydrates with the anhydrous salts: chloride 37.2 vs. 28.5, bromide 35.6 vs. 28.9.

The volumetric properties: density, ρ, isobaric expansibility, α_P, and molar volume, V, of molten hydrated salts at the corresponding temperature $T = 1.1T_m$, taken from [60], are shown in Table 5.5.

It should be noted that for several salts different entries are shown, derived from data of different authors as noted. The discrepancies are <2 %, except in the case of $Al(NO_3)_3{\cdot}9H_2O$, where it is larger, probably due to differences in the water contents of the melts.

Also shown in Table 5.5 are the molar electrostriction volumes, $\Delta_{els}V$, of the hydrated salt melts, which are negative quantities. These are the differences between the molar volume of the salt hydrate melts V (at $1.1T_m$) and the sum of the intrinsic volume of the ions, V_{intr}, weighted by their stoichiometric coefficients in $C_pA_q{\cdot}nH_2O$, plus n times the molar volume of water $V_W{}^*$.

$$\Delta_{els}V = V - \left[pV_{intr}(C) + qV_{intr}(A) + nV_W{}^*\right] \tag{5.3}$$

The intrinsic ionic volumes $V_{i\ intr}$ are assumed to be independent of the temperature. For monatomic ions they are given by the Mukerjee expression [74]:

$$V_{intr}/cm^3 \cdot mol^{-1} = 2522[1.213(r_i/nm)]^3 \tag{5.4}$$

where r_i are the crystal ionic radii, and for some of the polyatomic ions they are given in [75, 76]. The molar volume of water at the corresponding temperature $T = 1.1T_m$ is calculated from its expansibility as:

$$V_W{}^*/cm^3 \cdot mol^{-1} = 22.05 - 0.0318(T/K) + 6.2 \times 10^{-5}(T/K)^2 \tag{5.5}$$

The electrostriction volumes have been found to be proportional to the number of water molecules per formula unit: $\Delta_{els}V = (3.3 \pm 0.3)n$ for the 21 molten salt hydrates for which there are data, Fig. 5.1. Therefore, the molar volumes V and the densities of molten salt hydrates $C_pA_q{\cdot}nH_2O$ at their corresponding

Table 5.5 The volumetric properties of molten salt hydrates at $T = 1.1T_m$

Salt	T K	ρ $g \cdot cm^{-3}$	$10^3\alpha_P$ K^{-1}	V $cm^3 \cdot mol^{-1}$	$-\Delta_{els}V$
$LiClO_3 \cdot 3H_2O$	309	(1.689)[c, d]	(0.654)	(71.1)	(21.9)
$LiClO_4 \cdot 3H_2O$	405	1.4927[e]	0.542	107.5	11.7
$LiI \cdot 3H_2O$	383	1.9903[e]	0.619	94.4	10.5
$LiNO_3 \cdot 3H_2O$	333	1.4012[e]	0.554	87.8	5.2
$NaOH \cdot H_2O$	371	1.6228[f]	0.518	35.7	–0.7
$NaCH_3CO_2 \cdot 3\ H_2O$	364	1.2421[f]	0.715	90.3	
$Na_2S_2O_5 \cdot 5H_2O$	354	1.6546[f]	0.453	150.0	
	354	1.6334[g]	0.472	151.9	
$KF \cdot 4H_2O$	321	1.4346[h, i]	0.088	103.4	–7.0
$Mg(NO_3)_2 \cdot 6H_2O$	399	1.5027[j]	0.638	170.6	14.1
	399	1.4906[k]	0.644	172.0	12.7
$CaCl_2 \cdot 6H_2O$	334	1.4990[l]	0.437	146.1	22.8
$CaBr_2 \cdot 6H_2O$	337[a]	1.9082[l]	0.445	161.2	21.7
	337[a]	1.9218[m, n]	0.459	160.1	22.8
$CaI_2 \cdot 6H_2O$	347[b]	2.2124[l]	0.426	181.7	29.6
$Ca(NO_3)_2 \cdot 4H_2O$	347	1.7073[o]	0.471	138.7	11.8
		1.7200[p]	0.417	137.3	13.3
$Al(NO_3)_3 \cdot 9H_2O$	379	1.5120[j]	0.446	248.1	27.1
	379	1.5613[k]	0.607	240.3	34.9
$Al_2(SO_4)_3 \cdot 16H_2O$	406	1.7153[q]	0.374	367.5	94.3
$NH_4Al(SO_4)_2 \cdot 12H_2O$	404	1.3510[k]	0.631	335.6	29.3
$Cr(NO_3)_3 \cdot 9H_2O$	373	1.6329[j]	0.606	245.1	30.5
	373	1.6341[r]	0.560	244.9	30.7
$Mn(NO_3)_2 \cdot 6H_2O$	337	1.7782[j]	0.569	161.4	24.2
$FeCl_3 \cdot 6H_2O$	341	1.5823[s]	1.117	170.8	21.5
$Fe(NO_3)_3 \cdot 9H_2O$	348	1.6653[p]	0.616	242.6	33.1
	348	1.6384[t]	0.702	246.6	29.1
$Co(NO_3)_2 \cdot 6H_2O$	361	1.7469[j]	0.570	166.6	18.3
$Ni(NO_3)_2 \cdot 6H_2O$	363	1.7416[j]	0.562	167.0	17.5
$Zn(NO_3)_2 \cdot 6H_2O$	341	1.7922[r]	0.644	166.0	18.9
	341	1.8207[p]	0.587	163.4	21.5
$Cd(NO_3)_2 \cdot 4H_2O$	366	2.2475[o]	0.481	137.3	12.3
		2.2449[p]	0.485	137.4	12.2

[a]The melting point, 33.5 °C is from [61], [b]The melting point, 42 °C is from [41], [c]See the discussion, [d][52], [e][62], [f][63], [g][64], [h]Extrapolated from data for very concentrated aqueous solutions, see the text, [i][65], [j][66], [k][67], [l][57], [m]Extrapolated from data for very concentrated aqueous solutions, [n][68], [o][69], [p][70], [q][71], [r][72], [s][56], [t][73]

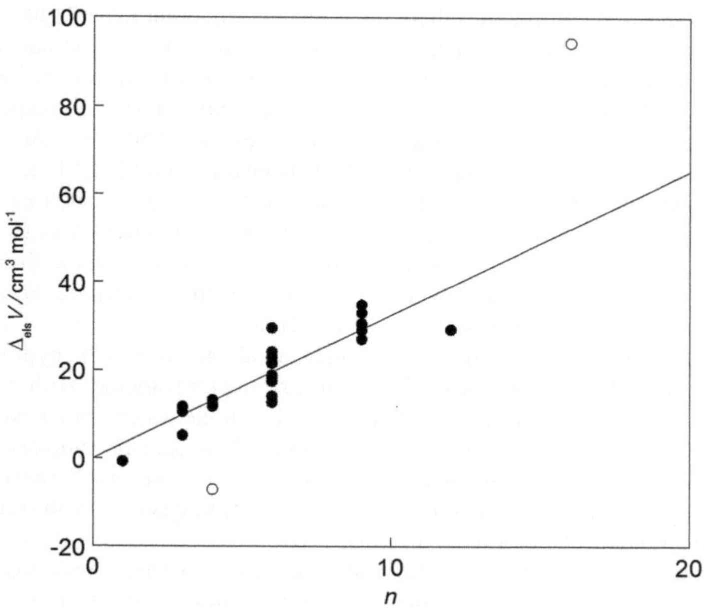

Fig. 5.1 The electrostrictive volume decrease, $\Delta_{els}V$, of hydrated molten salts plotted against n, the number of water molecules in the hydrate, Eq. (5.3).

temperatures, $1.1T_m$, can be estimated from the intrinsic volumes $pV_{intr}(C) + qV_{intr}(A)$ and the molar volume of the water V_W^* according to Eqs. (5.3), (5.4), and (5.5).

The densities of molten salt hydrates at any temperature T can be estimated by means of the expression proposed by Sharma et al. [77], recast in the following form:

$$\rho(T) = \rho(1.1T_m)\{1.0184 - 0.18396[(T/T_m) - 1]\} \tag{5.6}$$

This expression may then be employed for the calculation of the expansibility at any temperature:

$$\alpha_P = 1.0371\left[\rho(1.1T_m)/g \cdot cm^{-3}\right]^2 T_m^{-2}(6.537T_m - T) \tag{5.7}$$

using the densities at the corresponding temperatures listed in Table 5.5.

Hardly any compressibility data for hydrated molten salts have been found. Still, for calcium nitrate hexahydrate (more precisely, $Ca(NO_3)_2{\cdot}5.98H_2O$) the isothermal compressibility κ_T at 0.1 MPa is 0.195 ± 0.016 GPa^{-1} at 298 K diminishing to 0.188 ± 0.015 GPa^{-1} at 423 K spanning the melting point, 315 K and the corresponding temperature, $1.1T_m$ [78] (but note the wrong units in the publication).

For magnesium chloride hexahydrate at the melting point (390 K) the adiabatic compressibility may be estimated as follows. The ultrasound speed above that of water for concentrated aqueous solutions, up to 5.5 mol salt (kg water)$^{-1}$ at 55, 75, and 95 °C [79] can be extrapolated to 9.26 mol salt (kg water)$^{-1}$, corresponding to 6 mol water per mole $MgCl_2$, and to 117 °C to yield $\Delta u = 550$ m s^{-1}. Added to the speed of sound in water (calculated for 117 °C from data in [80]) 1,517 m s^{-1} yields the (extrapolated) speed of sound in the molten hydrate at its melting point of $u = 2{,}067$ m s^{-1}. The molar volume V of this salt is estimated from Eqs. (5.3), (5.4), and (5.5) and $\Delta_{el}V = (3.3 \pm 0.5)n$ for $n = 6$ as 150 cm^3 mol^{-1}, and with the molar mass $M = 203.3$ g mol^{-1} the density is $\rho = 1357$ kg mol^{-1}. Hence, the adiabatic compressibility of $MgCl_2{\cdot}6H_2O$ is $\kappa_S = 0.173$ GPa^{-1}.

The adiabatic compressibilities of molten alkali metal acetate hydrates were measured at 60 °C; for $NaCH_3CO_2{\cdot}3H_2O$ this practically coincides with its melting point: $\kappa_S = 0.250$ GPa^{-1}. Minima in κ_S as a function of the water content were found near this composition also for lithium and potassium acetate hydrates, with κ_S slightly decreasing in the sequence Li > Na > K for the alkali metal cations [81]. All these values of the compressibilities are commensurate with those of the anhydrous molten salts in Tables 3.15 and 3.16.

The transport properties of molten salt hydrates are rather scarce too, and the values that could be obtained for the dynamic viscosity, η, and the molar conductivity, Λ, at the corresponding temperature of $T = 1.1T_m$ are shown in Table. 5.6.

Table 5.6 Transport properties of molten salt hydrates at $T = 1.1T_m$

Salt hydrate	T	η	Λ
	K	mPa s	S cm^2 mol^{-1}
$NaOH{\cdot}H_2O$	371	4.36[a]	
$NaCH_3CO_2{\cdot}3\,H_2O$	364	2.06[b]	
$Na_2S_2O_3{\cdot}5H_2O$	354	5.22[b]	
$Na_2CO_3{\cdot}10H_2O$	338	3.30[b]	
$Na_2HPO_4{\cdot}12H_2O$	340	1.79[b]	
$Mg(NO_3)_2{\cdot}6H_2O$	399	14.53[c]	7.30[e]
$CaCl_2{\cdot}6H_2O$	334	12.34[d]	7.79[d]
$CaBr_2{\cdot}6H_2O$	337	8.70[d]	14.28[d]
$CaI_2{\cdot}6H_2O$	347	8.31[d]	12.15[d]
$Ca(NO_3)_2{\cdot}4H_2O$	347	6.55[d]	8.15[d]
$Cr(NO_3)_3{\cdot}9H_2O$	373	20.93[e]	
$Mn(NO_3)_2{\cdot}6H_2O$	337	44.93[e]	
$FeCl_3{\cdot}6H_2O$	341	12.45[e]	
$Fe(NO_3)_3{\cdot}9H_2O$	348	21.39[e]	
$Co(NO_3)_2{\cdot}6H_2O$	361	23.82[e]	
$Ni(NO_3)_2{}^\circ 6H_2O$	363	23.11[a]	
$Zn(NO_3)_2{\cdot}6H_2O$	341	21.63[e]	
$Cd(NO_3)_2{\cdot}4H_2O$	366	14.10[a]	23.75[e]

[a][82] [b][77], [c][67], [d][57], [e][83]

The relevant temperatures being much lower than for the anhydrous molten salts, the viscosities of the molten salt hydrates are larger and the conductivities considerably smaller than for the anhydrous salts. No details of the temperature dependencies of η and Λ could be shown, however.

References

1. Mohandas KS, Sanil N, Rodriguez P (2006) Development of a high temperature conductance cell and electrical conductivity measurements of MAlCl4 (M = Li, Na and K) melts. Transactions of the Institutions of Mining and Metallurgy, Section C. Mineral Proc Extract Metall 115:25–30
2. Moriya K, Matsuo T, Suga H (1988) Thermodynamic properties of alkali and thallium nitrites: the ionic plastically crystalline state. Thermochim Acta 132:133–140
3. Campbell AN, Nagarajan MK (1964) The thermodynamics and conductances of molten salts and their mixtures. II. The viscosities, heats of fusion, and heat capacities of lithium chlorate and lithium chlorate-lithium nitrate mixtures. Can J Chem 42:1616–1626
4. Brooker MH, Shapter JG, Drover K (1990) Raman study of sodium chlorate as a function of temperature into the melt and the novel high temperature phase. J Phys Condens Matter 2:2259–2272
5. Leonesi D, Piantoni G, Berchiesi G, Franzosini P (1968) Thermodynamic properties of organic acid salts. III. Enthalpy and entropy of fusion of sodium and potassium formats. Ric Sci 38:702–705
6. Hatem G, Eriksen KM, Gaune-Escard M, Fehmann R (2002) SO_2 oxidation catalyst model systems characterized by thermal methods. Top Catal 19:323–330
7. Masuda Y, Morita W, Wang X, Yukawa Y (2000) Comparative study on the thermal phase transitions of rubidium and cesium formats. Thermochim Acta 152–153:61–67
8. Rolla M, Franzosini P, Riccardi R (1961) Cryoscopy of dilute solutions in fused thallous nitrate. Disc Faraday Soc 32:84–89
9. Domalski ES, Hearing ED (1990) Heat capacities and entropies of organic compounds in the condensed phase. Volume II. J Phys Chem Ref Data 19:881–902
10. Fukushima K, Murofushi M, Oki M, Igarashi K, Mochinaga J, Iwadate Y (1994) Intraionic structure of HSO_4- and alkali cation configuration in molten $NaHSO_4$ and $KHSO_4$. Z Naturforsch A 49:785–789
11. Popovskaya NP, Protsenko PI, Eliseeva AF (1968) Surface tension of some univalent metal nitrite and nitrate melts. Russ J Inorg Chem 13:498–501
12. Dutcher CS, Wexler AS, Clegg SL (2010) Surface tensions of inorganic multicom-ponent aqueous electrolyte solutions and melts. J Phys Chem A 114:12216–12230
13. Leonesi D, Berchiesi G, Cingolani A (1975) Electric conductivity in molten binaries of alkali formates and acetates. J Chem Eng Data 20:31–32
14. Wang S-CS, Bennion DN (1983) The electrochemistry of molten lithium chlorate and its possible use with lithium in a battery. J Electrochem Soc 130:741–747
15. Janz GJ, Dampier FW, Lakshminarayanan GR, Lorenz PK, Tomkins RPT (1968) Molten salts. I. Electrical conductance, density, and viscosity. NSRDS-NBS Rep 15:1–143
16. Coker TG, Ambrose J, Janz GJ (1970) Fusion properties of some ionic quaternary ammonium compounds. J Am Chem Soc 92:5293–5297
17. Barton AFM, Spedy RJ (1974) Simultaneous conductance and volume measurements on molten salts at high pressure. J Chem Soc Faraday Trans 1(70):506–527 (the density data at 1MPa)

18. Gordon JE, SubbaRao GN (1978) Fused organic salts. 8. Properties of molten straight-chain isomers of tetra-n-pentylammonium salts. J Am Chem Soc 100:7445–7452 (the surface tension at Tm is 14.91 N m^{-1})
19. Lampreia MI, Barreira F (1976) Transport properties of molten tetra-alkylammonium picrates. I. Viscosity. Electrochim Acta 21:485–489
20. Griffiths TR (1963) Densities of some molten organic quaternary halides. J Chem Eng Data 8:568–569
21. Sugden S, Wilkins H (1929) Parachor and chemical constitution. XII. Fused metals and salts. J Chem Soc:1291–1298
22. Herstedt M, Henderson WA, Smirnov M, Ducasse L, Servant L, Talaga D, Lassegues JC (2006) Conformational isomerism and phase transitions in tetraethylammonium bis(trifluoromethanesulfonyl)imide Et4NTFSI. J Mol Struct 783:145–156 [NTF2– = (CF3SO2)2N–]
23. Hoff RH, Hengge AC (1998) A facile high-yield synthesis and purification of tetrabutylammonium tetrabutylborate. J Org Chem 63:195
24. Furton KG, Poole CF (1987) Thermodynamic characteristics of solute-solvent interactions in liquid organic salt solvents, studied by gas chromatography. J Chromatogr 399:47–67
25. Morrison G, Lind JE Jr (1968) Effect of the internal Coulomb field on the viscosity of a fused salt. J Chem Phys 49:5310–5314
26. Barreira ML, Barreira F (1976) Transport properties of molten tetra-alkylammonium picrates. II. Conductivity. Electrochim Acta 21:491–495
27. Kumar A (1993) Surface tension, viscosity, vapor pressure, density, and sound velocity for a system miscible continuously from a pure fused electrolyte to a nonaqueous liquid with a low dielectric constant: anisole with tetra-n-butyl-ammonium picrate. J Am Chem Soc 115:9243–9248
28. Carvalho PJ, Ventura SPM, Batista MLS, Schröder B, Goncalves F, Esperanca J, Mutelet F, Coutinho (2014) Understanding the impact of the central atom on the ionic liquid behavior: phosphonium vs. ammonium cations. J Chem Phys 140:064505/1–11
29. Zabinska G, Perloni P, Sanesi M (1987) On the thermal behavior of some tetraalkylammonium tetrafluoroborates. Thermochim Acta 122:87–94
30. Gordon JE (1965) Fused organic salts. IV. Characterization of low-melting quaternary ammonium salts. Phase equilibrium for salt-salt and salt-nonelectrolyte systems. Properties of the liquid salt medium. J Am Chem Soc 87:4347–4358
31. Matsumoto K, Harinaga U, Tanaka R, Koyama A, Hagiwara R, Tsunashima K (2014) The structural classification of the highly disordered crystal phases of [Nn][BF4], [Nn][PF6], [Pn][BF4], and [Pn][PF6] salts (Nn(+) = tetraalkylammonium and Pn(+) = tetraalkylphosphonium). Phys Chem Chem Phys 16:23616–23626
32. Pomaville RM, Poole SK, Davis LJ, Poole CF (1988) Solute-solvent interactions in tetra-n-butylphosphonium salts studied by gas chromatography. J Chromatogr A 438:1–14
33. Neves CMSS, Rodriguez AR, Kurina KA, Esperança JMSS, Freire MG, Coutinho JAP (2013) Solubility of non-aromatic hexafluorophosphate-based salts and ionic liquids in water determined by electrical conductivity. Fluid Phase Equilib 358:50–55
34. Nakayama H, Kuwata H, Yamamoto N, Akagi Y, Matsui H (1989) Solubilities and dissolution states of a series of symmetrical tetraalkylammonium salts in water. Bull Chem Soc Jpn 62:985–992
35. Lind JE Jr, Abdel-Rehim HAA, Rudich SW (1966) Structure of organic melts. J Phys Chem 70:3610–3619
36. Verevkin SP, Emel'yanko VN, Krossing I, Kalb R (2012) Thermochemistry of ammonium based ionic liquids: tetra-alkyl ammonium nitrates – experiments and computations. J Chem Thermodyn 51:107–113
37. Coddens ME, Furton KG, Poole CF (1886) Synthesis and gas chromatographic stationary phase properties of alkylammonium thiocyanates. J Chromatogr 356:59–77
38. Matsumoto H., Kageyama H, Miyazaki Y (2002) Room temperature ionic liquids based on small aliphatic ammonium cations and asymmetric amide anions. Chem Comm 1726–1727

39. Bhatt VD, Gohil K (2013) Ion exchange synthesis and thermal characteristics of some [N+2222]-based ionic liquids. Bull Mater Sci 36:1121–1125
40. Bhatt VD, Gohil K (2013) Ion exchange synthesis and thermal characteristics of some [N+4444] based ionic liquids. Thermochim Acta 556:23–29
41. Lide D (ed) (2001–2002) Handbook of Chemistry and Physics, 82nd edn. CRC Press, Baton Rouge
42. Janz GJ (1988) Thermodynamic and transport properties for molten salts: correlation equations for critically evaluated density, surface tension, electrical conductance, and viscosity data. J Phys Chem Ref Data 17(Suppl 2):1–325
43. Leonesi D, Cingolani A, Franzosini P (1973) Electric conductivity of sodium-potassium acetates molten system. J Chem Eng Data 18:391–393
44. Protsenko AV, Protsenko PI (1964) Physical-chemical properties of the melt $LiNO_2$-$LiNO_3$. Izv Vyshch Zaved Ucebn Zaved Metall 7:35–38
45. Guion J, Sauzade JD, Laugt M (1983) Critical examination and experimental determination of melting enthalpies and entropies of salt hydrates. Thermochim Acta 67:167–179
46. Voigt W, Zheng D (2002) Solid-liquid equilibria in mixtures of molten salt hydrates for the design of heat storage materials. Pure Appl Chem 74:1909–1920
47. Levitskij EA, Aristov YI, Tokarev MM, Parmon VN (1996) "Chemical Heat Accumulators": a new approach to accumulating low potential heat. Sol Energy Solar Cells 44:219–235
48. Meisingset KK, Grønvold F (1986) Thermodynamic properties and phase transitions of salt hydrates between 270 and 400 K. IV. Calcium chloride hexahydrate, calcium chloride tetrahydrate, calcium chloride dihydrate, and iron chloride (FeCl3) hexahydrate. J Chem Thermodyn 18:159–173
49. Yinping Z, Yi J, Yi J (1999) A simple method, the T-history method, of determining the heat of fusion, specific heat and thermal conductivity of phase-change materials. Meas Sci Technol 10:201–205
50. Günther E, Mehling H, Werner M (2007) Melting and nucleation temperatures of three salt hydrate phase change materials under static pressures up to 800 MPa. J Phys D Appl Phys 40:4636–4641
51. Sandnes B, Rekstad J (2006) Supercooling salt hydrates: stored enthalpy as a function of temperature. Sol Energy 80:616–625
52. Gawron K, Schröder J (1977) Properties of some salt hydrates for latent heat storage. Energy Res 1:351–363
53. Grønvold F, Meisingset KK (1983) Thermodynamic properties and phase transitions of salt hydrates between 270 and 400 K. II. Sodium carbonate monohydrate and sodium carbonate decahydrate. J Chem Thermodyn 15:881–889
54. Pilar R, Svoboda L, Honcova P, Oravova L (2012) Study of magnesium chloride hexahydrate as heat storage material. Thermochim Acta 546:81–86
55. Grønvold F, Meisingset KK (1982) Thermodynamic properties and phase transitions of salt hydrates between 270 and 400 K. I. Ammonium aluminum sulfate, potassium aluminum sulfate, aluminum sulfate, zinc sulfate, sodium sulfate, and sodium thiosulfate hydrates. J Chem Thermodyn 14:1083–1098
56. Jain SK (1978) Density, viscosity, and surface tension of some single molten hydrated salts. J Chem Eng Data 23:170–173
57. Jain SK, Prashar S, Jain SK (1999) Physical properties of some molten hydrated calcium salts. Indian J Chem 38A:778–782
58. Duffy JA, Ingram MD (1977) Metal aquo ions in molten salt hydrates. A new class of mineral acid? Inorg Chem 16:2988
59. Angell CA (1965) A new class of molten salt mixtures. The hydrated dipositive ion as an independent cation species. J Electrochem Soc 112:1224–1227
60. Marcus Y (2015) Volumetric behavior of molten salts and molten salt hydrates. In: Wilhelm E, Letcher T (eds) Volume properties. Royal Society for Chemistry, Cambridge, Ch. 20, pp 526–541

61. Feilchenfeld H, Fuchs J, Sarig S (1985) The melting point adjustment of calcium chloride hexahydrate by addition of potassium chloride or calcium bromide hexahydrate. Sol Energy 34:199–201
62. Mashovets VP, Baron NM, Zavodnaya GE (1969) Density of congruently melting crystal hydrates in solid and molten states. Russ J Phys Chem 43:971–974
63. Sharma SK, Jotshi CK, Singh A (1987) Density of molten salt hydrates – experimental data and an empirical correlation. Can J Chem Eng 65:171–174
64. Bhattacharjee C, Ismail S, Ismail K (1986) Density and viscosity of sodium thiosulfate pentahydrate ($Na_2S_2O_3{\cdot}5H_2O$) + potassium nitrate melt. J Chem Eng Data 31:117–118
65. Novotny P, Söhnel O (1988) Densities of binary aqueous solutions of 306 inorganic substances. J Chem Eng Data 33:49–55
66. Jain SK (1977) Volumetric properties of some single molten hydrated salts. J Chem Eng Data 22:383–385
67. Minevich A, Marcus Y, Ben-Dor L (2004) Densities of solid and molten salt hydrates and their mixtures and viscosities of some of them. J Chem Eng Data 49:1451–1455
68. Scatchard G (1926) International critical tables of numerical data, physics, chemistry and technology, 3rd edn. McGraw-Hill, New York, p 73
69. Jain SK (1973) Density and partial equivalent volumes of hydrated melts. Tetrahydrates of calcium nitrate, cadmium nitrate, and their mixtures with lithium, sodium, and potassium nitrate. J Chem Eng Data 18:397–399
70. Ramana KV, Sharma RC, Gaur HC (1986) Volumetric properties of molten hydrated salts. 7. Mixtures of ferric nitrate nonahydrate with hydrates of calcium, cadmium, zinc, and magnesium nitrates. J Chem Eng Data 31:288–291
71. Ornek D, Gurkan T, Oztin C (1998) Physical and chemical properties of a highly viscous aluminum sulfate melt. Ind Eng Chem Res 37:2687–2690
72. Jain SK (1978) Refractive index of molten Lewis acid salt hydrates: mixtures of chromium(III) nitrate nonahydrate + calcium nitrate tetrahydrate. J Chem Eng Data 23:216–218
73. Gupta S, Sharma RC, Gaur HC (1981) Volumetric properties of molten hydrated salts. 5. Fe $(NO_3)3{\cdot}8.75H_2O$ + MNO_3 system. J Chem Eng Data 26:187–191
74. Mukerjee P (1961) Ion-solvent interactions. I. Partial molal volumes of ions in aqueous solutions. II. Internal pressure and electrostriction of aqueous solutions of electrolytes. J Phys Chem 65:740–744
75. Marcus Y (2012) The standard partial molar volumes of ions in solution. Part 5. Ionic volumes in water at 125 to 200°C. J Phys Chem B 116:7232–7239
76. Marcus Y (2013) Volumetric properties of molten salt hydrates. J Chem Eng Data 58:488–491
77. Sharma SK, Jotshi CK, Singh A (1984) Viscosity of molten sodium salt hydrates. J Chem Eng Data 29:245–246
78. Pickston L, Smedley SI, Wooddall G (1977) The compressibility and electrical conductivity of concentrated aqueous calcium nitrate solutions to 6 kbar and 150 °C. J Phys Chem 81:581–585
79. Millero FJ, Vinokurova F, Fernandez M, Hershey JP (1987) PVT properties of concentrated electrolytes. VI. The speed of sound and apparent molal compressibilities of sodium chloride, disodium sulfate, magnesium chloride and magnesium sulfate solutions from 0 to 100 °C. J Solut Chem 16:269–281
80. Okazaki N (2000) Temperature rule for the speed of sound in water: a chemical kinetics model. Chem Eur J 6:3339–3345
81. Deki S, Iwabuki H, Kajinami A, Kanaji Y (1993) Adiabatic compressibility of hydrated alkali-metal acetate melts. Proc Electrochem Soc 93:121–130
82. Sharma SK, Jotshi CK, Singh A (1984) An empirical correlation for viscosity of molten salt hydrates. Can J Chem Eng 62:431–433
83. Bhatia K, Sharma RC, Gaur HC (1978) Conductivity of molten hydrated salts: manganese nitrate hexahydrate + ammonium nitrate system. Electrochim Acta 23:1367–1369

Chapter 6
Room Temperature Ionic Liquids

Room temperature ionic liquids (RTILs) are salts that melt below 100 °C, generally but not necessarily above 0 °C. They typically comprise a large organic cation, having a basic cyclic structure (e.g., imidazolium, pyrrolidinium, or pyridinium) or quaternized nitrogen or phosphorus atoms, all with attached alkyl chains, and any of a varied list of anions. Before discussing such ionic liquids in the rest of this chapter it is appropriate to mentions some other ionic liquids that do not belong to these categories. Among these are protic ionic liquids that were reviewed by Greaves and Drummond [1]. Some examples of such ionic liquids are the following.

The first known room temperature ionic liquid, ethylammonium nitrate, $C_2H_5NH_3^+$ NO_3^-, was described by Walden a hundred years ago [2] and has since been widely studied. It is a representative of a good number of protic alkylammonium RTILs [3–5]. Ethylammonium nitrate melts at 14 °C [6] (lower melting points, 12 °C [7] and 9 °C [3] are probably due to impurities, such as water). The molar enthalpy of melting is 10.64 kJ mol^{-1} [8] (but a ten times larger value was reported in [3] that appears to be unreasonable). In the molten state at 25 °C it has a very low vapor pressure, ~5.3 Pa [9]. The molar enthalpy of vaporization (at 25 °C) is 105.3 kJ mol^{-1}, the vaporization being to ion pairs below 146 °C and to dissociated $EtNH_2 + HNO_3$ above this temperature [10]. Physicochemical properties of this RTIL at several temperatures are shown in Table 6.1. Evans et al. [7] were among the first to point out the similarity of this molten salt to water, in terms of the thermodynamics of solubilities of inert gases (Kr, CH_4, C_2H_6, and C_4H_{10}) and the ability of surfactants to form micelles in them. These authors suggested that molten ethylammonium nitrate forms a three-dimensional hydrogen bonded network, since the number of donor hydrogen atoms and acceptor sites on the oxygen atoms are well matched.

Another unconventional RTIL, trifluoromethanesulfonic acid monohydrate H_3O^+ $CF_3SO_3^-$ (hydronium trufluoromethane sulfonate), was first characterized by Gramstad and Haszeldine [16] as a white stable solid with a melting point of 34 °C. Some of its properties as an ionic liquid were subsequently reported by Corkum and Milne [17], who refined its freezing point to 36 °C (whereas the

Y. Marcus, *Ionic Liquid Properties*, DOI 10.1007/978-3-319-30313-0_6

Table 6.1 Properties of two protic RTILs

Property	$C_2H_5NH_3^+$ NO_3^-			H_3O^+ $CF_3SO_3^-$
Melting point, t_m/°C		14[c]		36[k]
Heat of melting, molar, $\Delta_m H°$/kJ mol^{-1}		10.64[d]		
Boiling point, t_b/°C		240[e]		220[l]
Heat of vaporization, molar, $\Delta_v H°$/kJ mol^{-1}		105.3[e]		
Temperature	15 °C	25 °C	45 °C	25 °C
Density, ρ/g cm^{-3}	1.2156[a]	1.2091[a]	1.1966[a]	1.713[k]
Expansibility, $10^3\alpha_P$/K^{-1}		0.522[a]		0.715[m]
Molar volume, V/cm^3 mol^{-1}	88.9[a]	89.4[a]	90.3[a]	98.1[k]
Vapor pressure, p/Pa		5.3[f]		
Heat capacity, C_P/J K^{-1} mol^{-1}		206[a], 209[g]		
Surface tension, σ/mN m^{-1}		47.3[h], 49.2[i]	46.6[j]	
Refractive index, n_D	1.453[b]	1.4524[c]	1.444[b]	
Molar refractivity, R_D/cm^3 mol^{-1}		31.72[h]		
Permittivity, static, ε_s	27.1[b]	26.2[e]	23.6[b]	15.1[n]
Viscosity, dynamic, η/mPa s	50.0[b]	34.9[b]	19.3[b]	17.4[m]
Conductivity, specific, κ/S cm^{-1}	0.024[b]	0.0269[b]	0.039[h]	0.64[k]
Conductivity, molar, Λ/S cm^2 mol^{-1}	2.12[b]	2.52[b]	3.51[b]	62.8[m]
Coefficients $X = A_X \exp[B_X/R/(T - T_0)]$				
X = viscosity, A_η/mPa s		0.0018[h]		0.0102[m]
B_η/kJ mol^{-1}		24.5[h]		16.18[m]
T_0/K		0		79[m]
X = electrical conductivity, A_Λ/S cm^2 mol^{-1}		366[h]		1126[m]
B_Λ/kJ mol^{-1}		−12.4[h]		−13.47[c]
T_0/K		0		79[c]

[a][12], [b][11], [c][6], [d][8], [e][1, 10], [f][9], [g][13], [h][3], the σ at 27 °C, [i][15], [j][14], the σ at 50 °C, [k][17], [l][18], [m][20] at 65 °C = 1.1(T_m/K), [n][19]

anhydrous acid, $H^+F_3CSO_3^-$ melts already at −45.5 °C). The density of the undercooled molten salt at 25 °C, read from a curve, is 1.713 g cm^{-3}, slightly above that of the anhydrous acid, $\rho = 1.696$ g cm^{-3}. The viscosity of the molten salt was not measured, that of the undercooled melt is >40 mPa s^{-1}. The vapor pressure of the molten salt was compared isopiestically by Sarada et al. [18] with that of phosphoric acid, and the results were reported in a figure as a function of the temperature. The static permittivity of the molten salt was measured by Barthel et al. [19], yielding $\varepsilon_s = 15.1$ for the undercooled salt at 25 °C, diminishing to 11.8 at the corresponding temperature of $1.1T_m$, 338 K. These and other properties of trifluoromethanesulfonic acid monohydrate are shown in Table 6.1, placing it as a (rather unconventional) protic RTIL.

The RTILs dealt with in the rest of this chapter have cations from a large list, but the most popular cations include 1-alkyl-3-methylimidazolium (abbreviated as $C_n$mim, where n refers to the number of carbon atoms in the 1-alkyl substituent), N-alkylpyridinium (abbreviated as C_nPy), and quaternary ammonium or – phosphonium. Among the latter cations with open alkyl chains the most widely studied are of the form RR'_3N^+ or RR'_3P^+, and those with rings are mainly 1-methyl-1-alkylpyrrolidinium (abbreviated as C_nMPyrr). Such RTILs are dealt with here in some detail, whereas those based on other cations are mentioned only cursorily. Similarly, the anions of RTILs dealt with here in detail include tetrafluoroborate, hexafluorophosphate, bis(trifluorosulphonyl)imide (abbreviated as NTF_2^-), trifluoromethylsulfonate, dicyanamide, alkylsulfate, small inorganic anions, and carboxylate anions among a few others.

6.1 Structural Aspects of RTILs

6.1.1 Diffraction Studies and Computer Simulations

The structures of RTILs have been determined by x-ray and neutron diffraction or EXAFS and supported by computer simulations. Such structures may have, but need not have, a bearing on the structures of their liquid melts. Certain RTILs, such as 1-alkyl-3-methylimidazolium salts with a range of anions: chloride, bromide, trifluoromethanesulfonate, and bis(trifluoromethanesulfonyl)imide were studied by Hardacre et al. using SAXS and x-ray reflectivity, but they form liquid crystals [21]. Relatively few structures of isotropic liquid RTILs have been reported, as reviewed by Adya [22], who pointed out that the main problem encountered in such structural studies is that the multi-atomic cations and anions are not rigid and have internal vibrations. The assumption of rigid structures, based on ab initio computations on single molecules, does not solve this problem for the larger ions. Furthermore, the largest intramolecular distances within the ions of the RTILs are larger than the shortest intermolecular distances, so that it is impossible to distinguish between them. Therefore, with existing techniques the information on molten

RTILs from neutron and x-ray scattering and EXAFS is at best semi-quantitative. On the whole, what distinguishes RTILs from high-melting salts is that the asymmetry of the ions of the former precludes the coulombic forces between them from ordering the salt in a manner leading to stable crystalline forms at room temperature characterizing the latter. On the other hand, hydrogen bonding, π-stacking, and dispersive forces between alkyl chains dominate in RTILs over the coulombic forces between the large ions, giving rise to special structures of the melts.

In two cases, at least, no abrupt change in structure was observed on heating beyond the melting point: the EXAFS structure of bis(tetrabutylammonium) tetrabromomanganate showed the same Mn–Br distances and angles at 362 K for the solid and at 388 K for the melt according to Crozier et al. [23]. Another EXAFS study by Carmichael et al. dealt with bis(C_2mim) and bis(C_{14}mim) tetrachloronickolates at 404 and 408 K respectively, compared with the solids at 298 K [24]. No significant changes in the Ni–Cl distance were observed on going from the crystal to the melt. The structures of mixtures of molten 1-ethyl-3-methylimidazolium chloride with $AlCl_3$ in the concentration range of 46–67 mol% $AlCl_3$ (where the anions are $AlCl_4^-$ and $Al_2Cl_7^-$) were measured by Takahashi et al. using neutron diffraction techniques [25]. Unlike the tetrahedral $AlCl_4^-$ anion, the geometry of the $Al_2Cl_7^-$ one changes considerably in the melt compared with the isolated anion, implying cation-anion direct association. Neutron diffraction was also applied by Hardacre et al. to molten 1,3-dimethylimidazolium chloride and hexafluorophosphate, and significant charge ordering was found, as in the crystalline salt [26, 27]. Similar structures were observed, but that of the PF_6^- salt was expanded relative to that of the salt with the smaller Cl^- anion.

High energy synchrotron x-ray diffraction was applied by Hagiwara et al. to the highly conducting RTILs C_2mim^+-, C_4mim^+-, and $C_6mim^+H_{2.3}F_{3.3}^-$ with the unusual fluoride anion coordinated with 2.3 HF molecules [28]. Most of the information obtained pertained to internal structures of the cations, but an indication of a layered intermolecular structure of the melt was obtained. Small angle x-ray scattering was applied by Bradley et al. to molten $C_n mim^+NTF_2^-$ ($n = 12$–18) salts, but no layered structures were found, contrary to salts of these cations with anions such as Cl^-, Br^-, and $CF_3SO_3^-$, which formed layered liquid-crytal mesophases [29]. Layered structures of melts of NTF_2^- salts with a variety of cations were, however, found by Mizuhata et al. in small angle as well as wide angle x-ray diffraction [30]. The cations included $C_nMe_3N^+$ ($n = 5$–16), $(C_n)_4N^+$ ($n = 5$ or 6), C_3MPyrr^+ and the corresponding piperidinium one. The layer spacing of the $C_nMe_3N^+NTF_2^-$ salts increased linearly with the alkyl chain length (at 30 °C) the slope being <2. A more diffuse structural term: 'nanoscale structural heterogeneities', rather than layering, was applied by Triolo et al. to molten 1-alkyl-1-methylpiperidinium NTF_2 salts (alkyl C_n, $n = 3$–7) studied by small- and wide-angle x-ray scattering [31].

Computer simulation studies of the structure of molten RTILs require suitable force fields of the constituent ions. Such a force field was supplied in the case of 1-ethyl-3-methylimidazolium tetrachloroaluminate by Andrade et al. [32] and the molecular dynamics simulation was tested against the structure determined by neutron diffraction. One result from this study is that for this RTIL the coulombic interaction energy, -211 kJ mol^{-1}, dominates over the dispersive interactions, -42 kJ mol^{-1}. A subsequent molecular dynamics study by Salane et al. of this RTIL, with polarized chlorine atoms, showed the similarity in the structure of the melt with tetrachloroaluminate salts with alkali metal cations, determined almost wholly by the electrostatic interactions (see Sect. 3.2) [33]. Molecular dynamics simulations on series of 1-alkyl-3-methylimidazolium PF_6^- and NTF_2^- RTILs by Lopes and Padua [34] showed that with alkyl chains of butyl or larger aggregation of these chains into non-polar domains occurs. These nanostructural domains permeate a tridimensional network of ionic channels formed by the anions and by the imidazolium rings of the cations. The length scale of these polar/non-polar domains increase about linearly for $C_n mim^+PF_6^-$ with the alkyl chain lengths: $0.84 + 0.091n$ nm. Such nanostructures were described by Dupont as consisting of $[C_n mim_p X_{p\_q}]^{p+}[C_n mim_{p\_q} X_r]^{p-}$ species, where X^- is the anion and $p \geq 3$ [35].

6.1.2 *Modeling of RTIL Properties*

Here only some examples of the modeling of the properties of RTILs are described. Further models are dealt with in the following sections, pertaining to the properties themselves.

Some structural information on molten RTILs is obtained from models that relate bulk properties to the structures. An early attempt to apply this approach was that of Lind et al. [36], who dealt with the Walden product of the molar conductivity Λ and the viscosity η of low-melting quaternary ammonium salts. The Walden product was more recently used by Ueno et al. to probe the ionicity of RTILs [37]: plots of $\log\Lambda$ against $\log(\eta^{-1})$ yield a straight line with unit slope for a completely ionic electrolyte, such as 0.01 mol dm^{-3} aqueous KCl. Small deviations from such a line occur for highly ionic aprotic RTILs, increasing to somewhat larger deviations for protic RTILs in which more hydrogen bonding prevails, to still increasing deviations for RTILs in which either cation or anion are strong Lewis bases (electron pair donors) or Lewis acids (electron pair acceptors). Typical cations of RTILs have increased Lewis acid characters in the order $C_n MPyrr < RR'_3N^+ < C_n Py < C_n mim$ and typical anions have increasing Lewis base character in the order $PF_6^- < BF_4^- < NTF_2^- < CF_3SO_3^- < CF_3CO_2^- <$ halide. The deviations also increase (ionicity diminishes) for increasing alkyl chain lengths in a given family of RTILs.

A quantitative measure of the iconicity of RTILs is the effective concentration of ions according to Ueno et al. [37]:

$$c_{\text{eff}} = c\Lambda/\Lambda_{\text{NMR}} = c\Lambda\left(N_A e^2/k_B T\right)^{-1}(D_+ + D_-)^{-1} \tag{6.1}$$

where $c = V^{-1}$ is the molar concentration, i.e., the reciprocal of the molar volume, and the Λ_{NMR} is obtained from the NMR-derived diffusion coefficients of the cation

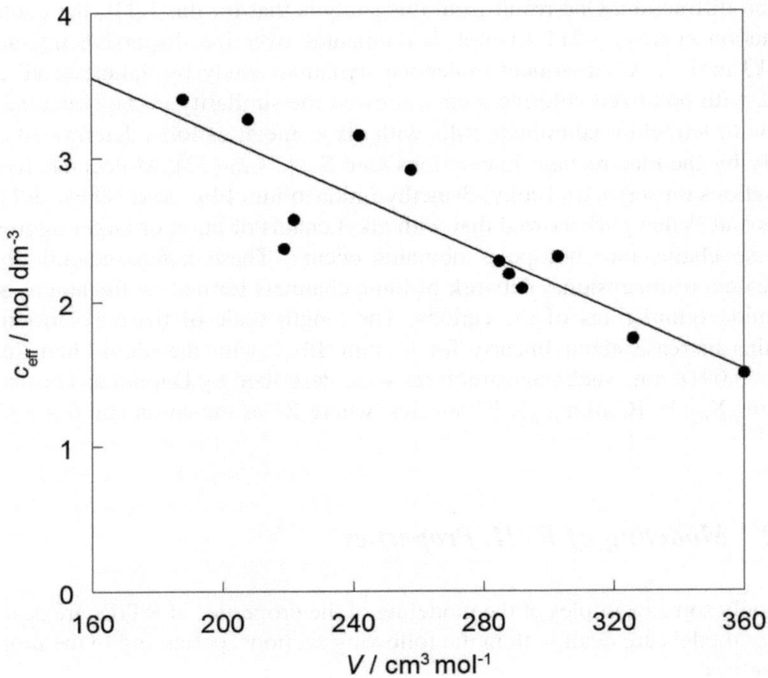

Fig. 6.1 The effective ionic concentration of molten RTILs, c_{eff}, Eq. (6.1), plotted against their molar volumes, V, from data at 30 °C of Ueno et al. [37]

and anion, D_+ and D_-, via the Nernst-Einstein relationship. The iconicity is c_{eff}/c and it takes into account not only the interactions between cations and anions but also their sizes, Fig. 6.1. For C_4mim salts $0.52 \leq c_{eff}/c \leq 0.68$, diminishing in the anion order $BF_4^- > PF_6^- > CF_3SO_3^- > CF_3CO_2^- > NTF_2^-$ and for $C_n$mim$^+$NTF$^-$ salts the ionicity is $0.54 \geq c_{eff}/c \geq 0.76$ diminishing with increasing n from 1 to 8.

The estimation of the melting points of RTILs (in conjunction with the permittivities of the melts) was reported by Krossing et al. [38] as an answer to the question "why are ionic liquids liquid?". The melting point T_m is that temperature, where for an RTIL:

$$\Delta_m G(T_m) = \Delta_m H(T_m) - T\Delta_m S(T_m) = 0 \qquad (6.2)$$

The Gibbs energy at any temperature may be calculated from the Born-Fajans-Haber cycle as the sum of the lattice and the (negative) solvation Gibbs energies: $\Delta G(T) = \Delta_{latt}G(T) + \Delta_{solv}G(T)$.

The lattice Gibbs energy, in turn, is $\Delta_{latt}G(T) = U_{pot}(T) + 2RT - T\Delta_{latt}S(T)$, where the lattice potential energy $U_{pot}(T)$ is transformed to the enthalpy by addition of $2RT$ (for the common uni-univalent RTILs). The lattice potential energy is obtained from the volume-based thermodynamics expression of Jenkins et al. [39]:

$$U_{pot}/\mathrm{kJ\,mol^{-1}} = 1038 + (234.6/\mathrm{nm})/\left(v_m^{1/3}/\mathrm{nm}\right) \tag{6.3}$$

where v_m is the microscopic volume of the RTIL, the sum of its cation and anion volumes (Tables 2.4 and 6.9). No temperature dependence of U_{pot} was considered. The lattice entropy is the difference between the sum of the entropies of the constituting ions in the isolated gas phase (calculated quantum-mechanically) and that of the solid according to Jenkins and Glasser [40]:

$$S(\mathrm{RTIL, cr, 298K})/\mathrm{J\,K^{-1}mol^{-1}} = 15 + 1360\left(v_m/\mathrm{nm}^3\right) \tag{6.4}$$

The solvation Gibbs energy $\Delta_{solv}G(T)$ is calculated for the cation and anion separately as being the quantity pertaining to the condensation of the gaseous ions into the liquid RTIL, which is their solvent, with a Born-type expression according to the COSMO model of Klamt and Schürmann [41]. This depends on the sizes of the ions and on the static permittivity of the liquid RTIL. Very few experimental values of the latter quantity are available, obtainable from dielectric spectroscopy, because of the high conductivity of the RTILs, but the values for 14 RTILs of various compositions are in the narrow range $10.0 \leq \varepsilon_s \leq 15.1$. The resulting $\Delta_{solv}G$(298 K) values are negative and numerically larger than the $\Delta_{latt}G$ (298 K) values, i.e., $\Delta_{melt}G$(298 K) <0, so that the melting point T_m <298 K, as observed by Krossing et al. [38] for 11 out of the 14 RTILs for which melting points are available. Because of the approximations in the method, the predicted melting points are within 10 K of the experimental values for 9 RTILs of those studied, but in two cases ($C_4mim^+CF_3SO_3^-$ and $C_4Py^+NTF_2^-$) they are up to 30 K too low. The method may be inverted for the estimation of the static permittivity ε_s of the RTILs, given their melting points, the results being within five units of the experimental values.

The statistical associated fluid theory (SAFT) combined with the mean spherical approximation (MSA), the latter for the calculation of the electrostatic interactions, permitted the estimation of liquid-vapor equilibria of RTILs as well as some other thermodynamic properties according to Guzman et al. [42]. For an amount of RTIL involving N particles the Helmholz energy per particle, $a = A/N$, is obtained as follows (with $\beta = 1/k_BT$):

$$\beta a = \beta a_{Id} + \beta a_{ref} + \beta a_{chain} + \beta a_{MSA} \tag{6.5}$$

The first three terms pertain to the ideal-, reference-, and chain-contributions of the SAFT model and the fourth to the electrostatic interactions from the MSA model. The cations have m_+ segments and the anions have only one, $m_- = 1$ segment. The ideal term is:

$$\beta a_{Id} = \ln\left(\rho_{seg}T^{-3/2}\right) - 1 \tag{6.6}$$

where $\rho_{seg} = 0.5(m_+ + m_-)\rho$ and $\rho = N/V = (N_+ + N_-)/V$ is the number density of the RTIL. The reference term is proportional to the sum of the Lennard-Jones interactions, $a_{LJ}(T,\rho_{seg})$, of the segments:

$$\beta a_{\text{ref}} = 0.5T^{-1}(m_+ + m_-)a_{\text{LJ}}(T, \rho_{\text{seg}}) \tag{6.7}$$

All segments are assumed to be Lennard-Jones particles with assigned length parameter σ and energy parameter ε. The chain term is:

$$\beta a_{\text{chain}} = [1 - 0.5(m_+ + m_-)]\ln g_{\text{LJ}}{}^{\sigma}(T, \rho_{\text{seg}}) \tag{6.8}$$

where the contact value $g_{\text{LJ}}{}^{\sigma}$ is calculated according to Johnson et al. [43]. The electrostatic term from the MSA theory is:

$$\beta a_{\text{MSA}} = \left[\Gamma(\Gamma + 2/3)(\Gamma + 1)^{-2}\right]T^{-1}\varepsilon_{\text{ion}} \tag{6.9}$$

where $\Gamma = 0.5\left[(1 + 2\kappa)^{1/2} - 1\right], \kappa = (4\pi\rho\varepsilon_{\text{ion}}/T)^{1/2}$ is the reciprocal of the Debye length, and ε_{ion} is the electrostatic energy of two like-charged ions separated by a distance σ.

The SAFT-MSA method was applied by Guzman et al. [42] to $C_n\text{mim}^+X^-$ RTILs, where $n = 1, 2, 4, 6$ and 8 for $X^- = BF_4^-$, $n = 4, 6$ and 8 for $X^- = PF_6^-$, and $n = 4$ and 6 for $X^- = NTF_2^-$, with the Lennard-Jones parameters σ and ε and the number m_+ of cation segments supplied for each RTIL. From the Helmholz energy per particle and its temperature dependence, βa, Eq. (6.5), the vapor pressure was calculated as $p = \rho^2 T(\partial\beta a/\partial\rho)_T$ and the entropy per particle as $s = -\beta a - T(\partial\beta a/\partial T)_\rho$ and the heat of vaporization at the boiling point as $\Delta h(T_b) = T_b\left[s(T_b, \rho_{\text{vap}}) - s(T_b, \rho_{\text{liq}})\right]$ as well as the critical parameters, T_c, ρ_c, and P_c.

The soft SAFT model and EoS were applied by Oliveira et al. to a series of pyridinium RTILs with the NTF_2 anion [44], with the assumption of a constant cation–anion interaction term and alkyl-chain-dependent interactions transferred from the analogous imidazolium series of RTILs. Once fitting parameters for the volumes of the RTILs were established, other quantities, such as the compressibilities and surface tensions were predicted. A soft SAFT model was also applied by MacDowell et al. [45] to imidazolium RTILs with Cl^-, $MeSO_4^-$, and $Me_2PO_4^-$ anions, fitting their temperature dependent densities and then applying the parameters to calculate the viscosity and surface tension. Another modified SAFT model (electrolyte perturbed chain statistical associated fluid theory, ePC-SAFT) was applied by Ji et al. [46] to imidazolium-based RTILs, fitting the densities up to 100 MPa. The solubility of CO_2 in the RTILs could be well predicted by this model, as well as the much lower solubility of CH_4.

The density of RTILs, and more generally their *P-V-T* properties have been described by several models. The COSMO-RS model was applied by Palomar et al. [47] for the prediction of the densities and molar volumes of 40 - imidazolium-based RTILs. The predicted densities had a small positive bias (2.5 %) relative to the experimental values. The Sanchez-Lascombe equation of state (SL-EoS) was applied by Machida et al. [48] to the *P-V-T* properties of $C_4\text{mim}^+$

RTILs with BF_4^-, PF_6^-, and $OcSO_4^-$ anions up to 373 K and 200 MPa. The resulting isothermal compressibilities κ_T increased in this order of salts from 0.4 to 0.5 GPa^{-1} at ambient pressure and diminished at increasing pressures. The free, compressible, volumes at 313 K and ambient pressure were 7.4 %, 6.7 %, and 7.6 % respectively.

Microscopic ionic volumes of RTILs, the sums of the constituent 20 cations and 20 anions: $v_m = v_+ + v_-$, were calculated by several theoretical methods according to Preiss et al. [49] and compared with the crystal volumes obtained from x-ray diffraction (Table 2.4). It was then shown that several physical properties were linear with these microscopic volumes: the molar volume V, the isobaric expansibility α_P, the molar heat capacity C_P, the (logarithm of) the viscosity $\ln\eta$, and the (logarithm of) the molar conductivity $\ln\Lambda$: all these obey the relation $a + bv_m$.

A perturbed hard sphere equation of state (PHS-EoS) was used by Hosseini et al. [50, 51] to model the volumetric properties of RTILs. This EoS has the form:

$$P/\rho k_B T = \left(1 + \eta + \eta^2 - \eta^3\right)(1 - \eta)^{-3} - a(T)\rho/k_B T \tag{6.10}$$

where $\eta = b(T)\rho/4$ is the packing fraction, and involves the two substance-dependent parameters $a(T)$ and $b(T)$. These, in turn, depend on the critical properties of the RTILs [50] (not shown! the latter being problematic as discussed below, Sect. 6.2), or on the reduced temperature $k_B T/\varepsilon$, the ε being the listed [51] non-bonded interaction energy between two hard spheres. The Peng-Robinson cubic EoS plus association (PR-EoS-SSM) method was used to model the P-V-T properties of 25 RTILs by Ma et al. [52], employing four substance-dependent parameters as well as the critical temperature. A perturbed hard-dimer-chain EoS was applied by Hosseini et al. [53] to obtain the densities and isothermal compressibilities of RTILs. Hard dimers associate with hard spheres to form tetramers as the basis for this model.

High pressure densities of imidazolium-based RTILs were modeled by Machida et al. [54] up to 200 MPa by means of the Sanchez-Lascombe SL-EoS (modified by allowance for the temperature dependence of the interaction energy) with six substance-dependent parameters. High pressure densities of phosphonium RTILs were modeled by Tome et al. [55] up to 45 MPa.

6.2 Thermochemical Data

6.2.1 Melting and Decomposition Temperatures

In the following the molar masses and the fundamental thermochemical data for the more commonly applied RTILs are presented. The data pertain, as far as could be ascertained, to carefully dried RTILs, with residual water contents below 0.01 mass %. The thermochemical data include the melting point, T_m, the onset of decomposition temperature, T_{dec}, the molar enthalpy of melting, $\Delta_m H$, the molar heat capacity, C_P, and the surface tension, σ, the latter two quantities at 25 °C. Table 6.2 shows these data for the 1-alkyl-3-methylimidazolium salts, Table 6.3 does it for the 1-alkylpyridinium salts, and Table 6.4 does it for the quaternary ammonium (aliphatic and cyclic) and phosphonium salts. Many such data are available in the recently published IL thermo database [56].

It should be realized that not all RTILs have definite melting points: some form liquid crystals that may transform gradually into isotropic liquids on heating whereas others on cooling do not crystallize but form a glass. In this book only isotropic liquid phases are dealt with so that the lack of a definite T_m should not deter from dealing with the RTIL when it is indeed liquid. Many crystalline RTILs undergo phase transitions somewhat below their melting points, so that the molar enthalpy of melting may be quite small, most of the endothermic enthalpy change having already occurred in the phase transition.

The melting points of RTILs are readily determined experimentally by standard methods, including differential scanning calorimetry (DSC), if they occur at all, i.e., melting and freezing take place at the same temperature within 1 K. Although values are reported in the literature to 0.1 K, these accuracies are not generally realistic. An attempt at the prediction of the melting points of crystalline RTILs from other properties, mainly the microscopic ionic volumes of the cation and anion, has already been described in Sect. 6.1.2. Quantitative structure/activity relationships (QSAR) were applied to the 1-alkyl-3-methylimidazolium tetrafluoroborates and hexafluoro-phosphates by Sun et al. [172], their chlorides and bromides by Yan et al. [108], and to pyridinium bromides by Bini et al. [173] in order to estimate their melting points. Structural parameters were generated by molecular mechanics computations and multiple linear regressions were employed in order to select the statistically best training set that was then employed to obtain the melting points of a test set. Absolute differences between calculated and experimental melting points were <20 K in most cases. The melting points and enthalpies of fusion of 1-alkyl-3-methylimidazolium hexafluorophosphates were estimated by molecular dynamics computer simulations by Zhang and Maginn [85] who confirmed the trends that T_m diminishes with increasing alkyl chain lengths for short chains (up to C_4) and increases again for long chains ($\geq C_{10}$), but generally no clear melting points were obtained for RTILs with chain lengths between these values.

The onset of decomposition of RTILs on heating is generally determined by DSC measurements, and is an indication of the liquid range expected for the RTIL between its melting and decomposition. RTILs do not generally have a normal boiling point, their vapor pressures remain low, and on heating the RTILs generally decompose much before the vapor pressure attains atmospheric pressure. The question whether we understand the volatility of RTILs was raised by Ludwig and Kragl [174] and the relevant data could be obtained mainly for the $C_n mim^+$ NTF_2^- series of RTILs that remain stable to decomposition up to ~600 K. At this temperature or somewhat below it they do have readily measurable vapor pressures and the vaporization takes place to the gaseous ion pairs.

Table 6.2 The molar mass, M, the melting point, T_m, and the temperature of start of decomposition, T_{dec}, the molar enthalpy of melting, and the heat capacity and surface tension at 25 °C of 1-alkyl-3-methylimidazolium ($C_n$mim) salts

Salt	M/g mol^{-1}	T_m/K	T_{dec}/K	$\Delta_m H$/kJ mol^{-1}	C_p/J K^{-1} mol^{-1}	σ/mN m^{-1}
C_1mim BF_4	184.0	373[k]				
C_1mim PF_6	243.1	403[k]				
C_1mim NTF2	377.3	295[c]	717[ff]	24.5[ff]	474[nn]	36.3[bbb]
C_1mim Cl	146.6	398[k]		21.5[qqq]		
C_1mim Br	177.1	449[k]		20.7[qqq]	215[dd]	
C_1mim NO_3	159.2	367[k]				
C_1mim CH_3SO_4	208.2	309[nnn]		16.6[qqq]		65.1[xx]
C_1mim $C_2H_5SO_4$	222.3					58.3[xx]
C_2mim BF_4	198.0	288[a]	556[a]	9.5[gg]	305[ee]	54.4[ccc]
C_2mim PF_6	256.1	333[i]	648[n]	18.0[gg]	258[ee]	54.0[qqq]
C_2mim NTF2	391.3	270[a]	664[a]	24.8[ff]	505[ee]	36.1[aaa]
C_2mim CH_3SO_4	222.4					62.9[xx]
C_2mim $C_2H_5SO_4$	236.4	213[p]		21.4[ii]	394[x]	48.8[pp]
C_2mim Cl	160.6	358[i]	558[n]	21.4[ii]	253[dd]	
C_2mim Br	191.1	352[k]	560[w]	15.7[gg]	247[dd]	
C_2mim I	238.1	352[k]	576[n]	16.5[gg]		
C_2mim SCN	169.2	267[c]			281[qqq]	53.1[ddd]
C_2mim NO_3	173.2	304[a]	557[a]	19.6[hh]		
C_2mim $AlCl_4$	283.0	282[c]		13.8[qqq]		52.4[uu]
C_2mim $FeCl_4$	311.8	291[l]	550[l]			
C_2mim $C_2F_5BF_3$	254.0	272[m]	571[m]			
C_2mim $C_3F_7BF_3$	304.0	281[m]	565[m]			
C_2mim $C_4F_9BF_3$	354.1	253[m]	553[m]			
C_2mim $N(CN)_2$	177.3	252[c]		21.4[ii]	315[bb]	44.1[ss]
C_2mim CH_3SO_3	206.3	278[c]			345[bb]	45.1[hhh]
C_2mim $CH_3C_6H_4SO_3$	282.4	323[nnn]		27.8[nnn]		
C_2mim CF_3SO_3	260.3	264[c]	~713		363[bb]	40.4[ww]
C_2mim CH_3CO_2	170.2	253[c]	413[jjj]	21.4[ii]	322[qqq]	38.1[eee]
C_3mim BF_4	212.0	275[a]	577[a]			51.1[ccc]
C_3mim PF_6	270.1	313[i]	608[n]	21.4[ii]		47.7[vv]
C_3mim NTF2	405.3		672[a]		536[nn]	35.9[ccc]
C_3mim Cl	160.6	325.2[qqq]		10.1[qqq]		
C_3mim Br	205.1	314.7[qqq]		13.5[qqq]	281[qqq]	
C_3mim NO_3	187.2		539[a]	21.4[ii]		54.6[ccc]
C_3mim CH_3SO_4	236.2					52.3[xx]
C_3mim $C_2H_5SO_4$	250.2					48.6[xx]
C_3mim CH_3CO_2	184.1					36.8[eee]
C_4mim BF_4	226.0	191[c]	566[a]	21.4[ii]	366[ee]	46.9[ccc]
C_4mim PF_6	284.2	284[b]	696[j]	19.9[s]	410[ee]	47.5[tt]

(continued)

Table 6.2 (continued)

Salt	M/g mol^{-1}	T_m/K	T_{dec}/K	$\Delta_m H$/kJ mol^{-1}	C_p/J K^{-1} mol^{-1}	σ/mN m^{-1}
C_4mim NTF2	419.3	269[a]	676[a]	20.9[ff]	565[ee]	34.9[ccc]
C_4mim CH_3SO_4	250.3	253[p]			416[x]	43.3[rr]
C_4mim $C_2H_5SO_4$	264.3					41.7[xx]
C_4mim $C_8H_{17}SO_4$	348.6	307[o]	614[o]	12.7[qqq]	635[qqq]	25.2[xx]
C_4mim Cl	174.6	314[b]	423[b]	10.3[gg]	317[dd]	48.2[tt]
C_4mim Br	219.2	350[k]	552[w]	16.3[gg]	311[dd]	
C_4mim I	266.2	201[n]	538[n]	19[hh]	310[z]	54.7[pp]
C_4mim SCN	197.3	244[kkk]		0.28[kkk]	385[qqq]	47.5[zz]
C_4mim NO_3	201.2	296[a]	553[a]	18.0[gg]	354[jj]	50.5[ccc]
C_4mim $AlCl_4$	311.1	263[c]				
C_4mim $FeCl_4$	339.9		550[l]			46.5[pp]
C_4mim $N(CN)_2$	205.3	267[c]		21.4[ii]	375[jj]	45.8[ff]
C_4mim CH_3SO_3	234.3	253[c]	625[e]			28.1[ggg]
C_4mim $CH_3C_6H_4SO_3$	310.4	330[ppp]		16.4[ppp]		
C_4mim CH_3SO_3	250.3	253[lll]				43.7[iii]
C_4mim CF_3SO_3	288.3	286[b]	613[b]	19.4[qqq]	424[x]	35.1[ww]
C_4mim CH_3CO_2	198.3	253[c]	415[d]	19 [hh]	322[bb]	35.2[eee]
C_4mim CF_3CO_2	252.2			19.1[qqq]	408[qqq]	
C_5mim BF_4	240.1	185[k]	601[a]			42.8[ccc]
C_5mim NTF_2	433.3	267[a]	670[a]		598[nn]	35.9[qq]
C_5mim Cl	188.7	350[qqq]		18.9[qqq]		
C_5mim NO_3	201.2					42.8[ccc]
C_5mim CH_3CO_2	212.3					34.1[eee]
C_6mim BF_4	254.1	191[c]	577[a]		429[ee]	39.2[tt]
C_6mim PF_6	312.2	212[c]	634[j]		420[ee]	43.4[tt]
C_6mim $PF_3(C_2F_5)_3$	612.3				730[qqq]	31.6[qqq]
C_6mim NTF2	447.4	266[a]	635[a]	28.3[gg]	645[ee]	33.8[ccc]
C_6mim Cl	202.6	198[c]	526[n]	10.2[qqq]	381[dd]	41.8[tt]
C_6mim Br	247.3	221[k]	540[w]		376[dd]	
C_6mim I	294.2					43.5[tt]
C_6mim NO_3	229.3	196[k]	556[a]	16.6[hh]		39.9[ccc]
C_6mim $N(CN)_2$	233.3					
C_6mim $AlCl_4$	336.1				534[qqq]	39.6[uu]
C_6mim $FeCl_4$	367.9		590[l]			
C_6mim CF_3SO_3	316.4				498[qqq]	32.4[ww]
C_6mim CH_3CO_2	226.3		409[iij]			33.0[eee]
C_7mim Cl	216.8	190[qqq]		10.2[qqq]		
C_8mim BF_4	282.1	194[c]	555[d]	21.4[ii]	498[ee]	30.7[tt]
C_8mim PF_6	340.3	233[c]	572[j]	21.4[ii]		33.9[vv]
C_8mim NTF2	475.5	187[ll]	618[d]	25.2[gg]	692[nn]	27.8[ss]
C_8mim Cl	230.6	273[c]	516[n]	10.3[qqq]	445[dd]	31.9[tt]

(continued)

Table 6.2 (continued)

Salt	M/g mol^{-1}	T_m/K	T_{dec}/K	$\Delta_m H$/kJ mol^{-1}	C_p/J K^{-1} mol^{-1}	σ/mN m^{-1}
C_8mim Br	275.2	285[mmm]			440[qqq]	31.6[bbb]
C_8mim I	322.2					32.7[bbb]
C_8mim NO_3	243.3	221[i]				33.3[bbb]
C_8mim $N(CN)_2$	261.4	268.1[qqq]				36.9[qqq]
C_8mim CF_3SO_3	344.4	278.1[qqq]				30.1[bbb]
C_8mim $FeCl_4$	396.0		550[l]			28.5[bbb]
C_{10}mim BF_4	296.1	269[h]	617[jjj]			
C_{10}mim PF_6	354.3	305[h]				30.7[pp]
C_{10}mim NTF_2	503.5	244[qqq]			754[qqq]	29.5[bbb]
C_{10}mim Cl	244.6	311[h]		30.9[gg]		
C_{12}mim BF_4	310.2	292[k]				25.2[oo]
C_{12}mim PF_6	382.3	328[k]		27.3[mm]		23.6[oo]
C_{12}mim NTF_2	532.4		596[jjj]		820[qqq]	32.3[bbb]

[a][57], [b][58], [c][59], [d][60], [e][61], [f][62], [g][63], [h][64], [i][65], [j][66], [k][67], [l][68], [m] [69], [n][70], [o][71], [p][72], [q][58], [s][73], [w][66], [x][74], [z][75], [bb][76], [cc][77], [dd][78], [ee][79], [ff][80], [gg][81], [hh][75], [ii][82], [jj][83], [ll][84], [mm][85], [nn][86], [oo][87] values at 63 °C, [pp][88] values at 25–30 °C from various sources, [qq][89], [rr][90], [ss][91] values at 20 °C, [tt][92], [uu][93], [vv][94], [ww][95], [xx][96], [zz][97], [aaa][98], [bbb][99], [ccc][57], [ddd][100], [eee][101], [fff][102], [ggg][103], [hhh][104], [iii][105], [jjj][106], [kkk][107], [lll] [72], [mmm][108], [nnn][109], [ppp][110], [qqq][56]

6.2.2 *Vaporization*

Experimental values of the vapor pressures (indeed very small: mPa to fractions of a Pa) of $C_n$mim$^+$ NTF_2^- ($n = 2, 4, 6, 8$) RTILs were obtained by Zaitsau et al. from Knudsen effusion cells [175]. The resulting molar enthalpies of vaporization (at 298 K) are respectively 135, 136, 140, and 150 kJ mol^{-1} for these four RTILs. The enthalpies of vaporization were then estimated for a variety of 1-alkyl-3-methylimidazolium RTILs with Cl^-, BF_4^-, PF_6^-, and NTF_2^- anions from the surface tensions σ and molar volumes V by the expression $\Delta_V H = AN_A^{1/3}\sigma V^{2/3} + B$, Fig. 6.2, yielding values $112 \leq \Delta_V H(298\,\text{K})/\text{kJ}\,\text{mol}^{-1} \leq 169$. Further estimates along these lines were reported for other imidazolium RTILs by Esperança et al. [176] with similar results. For the series of $C_n$mim$^+$ NTF_2^- ($n = 2, 3, 4, 5, 6, 7, 8$) RTILs the values of $\Delta_V H$(298 K)/kJ mol^{-1} were found by Santos et al. [177] from high-accuracy vapor pressure measurements to increase linearly with increasing alkyl chain length as $118 \pm 6 + 9n$. For the same series with $n = 2$, 3, 4, 5, 6, 7, 8, 10 and 12 the values of $\Delta_V H$(298 K)/kJ mol^{-1} were found by Rocha et al. [178] to increase systematically but non-linearly with increasing alkyl chain lengths from 133 to 171, there being a break in the $\Delta_V H(n)$ curve at $n = 6$, attributed to structural modifications of the liquid [178] (this non-linearity was later challenged by Verevkin et al. [86]).

Table 6.3 The molar mass, M, the melting point, T_m, and the temperature of start of decomposition, T_{dec}, the molar enthalpy of melting, and the heat capacity and surface tension at 25 °C of 1-alkylpyridinium salts

Salt	M/g mol^{-1}	T_m/K	T_{dec}/K	$\Delta_m H$/kJ mol^{-1}	C_p/J K^{-1} mol^{-1}	σ/mN m^{-1}
C_2Py Br	188.2	383[b]		12.8[o]		
C_2Py NTF_2	388.3	304[g]		18.9[dd]		35.5[y]
C_2Py $EtSO_4$	233.4	289[t]	483[d]		389[w]	
C_2(3 M)Py NTF_2	402.3	288[c]			491[w]	
C_2(4 M)Py NTF_2	402.3	291[i]		11.0[o]		
C_3Py Cl	156.8	365.1[dd]				
C_3Py Br	202.2	342[o]		11[dd]		
C_3Py BF_4	208.8				363[p]	50.9[z]
C_3Py NTF_2	402.3	318[f]				
C_3(3 M)Py Cl	171.7	365[v]	473[v]			
C_3(3 M)Py NTF_2	416.4	296[c]			517[t]	34.7[aa]
C_4Py Br	216.2	378[d]	480[bb]	20.4[dd]		
C_4Py NO_3	198.4		508[bb]			51.5[cc]
C_4Py BF_4	223.0	280[a]	562[bb]		383[q]	50.9[z]
C_4Py NTF_2	416.4	299[g]	609[bb]	27.9[dd]	395[p]	34.8[y]
C_4Py CF_3SO_3	285.4	301[k]	575[bb]		470[p]	36.5[p]
C_4(2 M)Py BF_4	237.1	301[p]				47.1[y]
C_4(3 M)Py Cl	185.3	385[n]	515[n]	28[n]		
C_4(3 M)Py BF_4	223.0		590[d]		388[s]	44.8[z]
C_4(3 M)Py CF_3SO_3	299.4	307[e]			516[e]	38.2[y]
C_4(3 M)Py NTF_2	430.4		506[d]		548[s]	35.5[y]
C_4(3 M)Py N $(CN)_2$	216.4	278[k]				
C_4(4 M)Py BF_4	223.0					45.8[z]
C_4(4 M)Py NTF_2	430.3	292[i]				35.0[y]
C_4(4 M)Py $MePhSO_3$	321.4	325[l]				
C_5Py NTF_2	430.4	273[g]		22.8[dd]		32.5[dd]
C_5Py Cl	174.8	365.1[dd]				
C_6Py Br	244.2	309[q]				
C_6Py NTF_2	444.4	273[f]	605[d]		612[w]	
C_6(3 M)Py Cl	213.9	355[n]	511[n]	20[n]		
C_6(3 M)Py Br	258.2				343[u]	
C_6(3 M)Py CF_3SO_3	327.4	338[x]		42.0[x]		
C_6(3 M)Py NTF_2	458.5	276[dd]	603[d]		624[u]	
C_8Py Br	272.3	292[q]				
C_8Py CF_3SO_3	341.5	322[m]	553[m]			
C_8Py $C_4F_9SO_3$	491.5	346[m]	567[m]			

(continued)

Table 6.3 (continued)

Salt	M/g mol^{-1}	T_m/K	T_{dec}/K	$\Delta_m H$/kJ mol^{-1}	C_p/J K^{-1} mol^{-1}	σ/mN m^{-1}
C_8Py NTF_2	472.5	261[m]	593[m]			
C_8(2 M)Py CF_3SO_3	355.5	313[m]	448[m]			
C_8(2 M)Py $C_4F_9SO_3$	505.5	351[m]	565[m]			
C_8(3 M)Py Cl	241.9	353[n]	505[n]			
C_8(3 M)Py BF_4	293.2				446[w]	
C_8(3 M)Py NTF_2	486.5				669[u]	
C_8(4 M)Py $C_4F_9SO_3$	505.5	34[m]	571[m]			
C_8(4 M)Py NTF_2	486.6	280[m]	573[m]			
C_{10}Py NTF_2	500.5		617[j]			
C_{12}Py Br	328.4	329[q]				
C_{12}Py NTF_2	528.6		610[j]			
C_{12}(3 M)Py Cl	262.5	361[n]	500[n]	37[n]		

[a][111], [b][112], [c][113], [d][60], [e][114], [f][62], [g][63], [h][115], [i][116], [j][117], [k][118], [l][119], [m][120], [n][121], [o][122], [p][123], [q][124], [r][125], [s][126], [t][127], [u][128], [v][129], [w][130], [x][131], [y][132], [z][133], [aa][134], [bb][106], [cc][135], [dd][56]

On the whole, then, the molar enthalpy of vaporization of the RTILs for which data are available are above 100 kJ mol^{-1} and even above 200 kJ mol^{-1} for NTF_2^- salts with the longer alkyl chain cations, which is a large endothermic value, considerably larger than that of common molecular solvents. This is, of course, directly related to the small vapor pressures of the RTILs.

Theoretical calculations and computer simulations of the enthalpy of vaporization and the normal boiling points again dealt with the 1-alkyl-3-methylimidazolium salts, yielding similar results pertaining to the measured enthalpies [78, 82, 179, 180]. Normal boiling points $T_b > 530$ °C were obtained according to Rane and Errington [179] from three different simulation methods for the $C_n$mim$^+$ NTF_2^- RTILs, but the more likely values were >700 °C, well above the decomposition temperatures.

6.2.3 *Cohesive Energies and Solubility Parameters*

Once reliable estimates of the molar enthalpies of vaporization of RTILs are available, their cohesive energies, $ce = \Delta_V H - RT$, are readily obtained. Note that no ionic dissociation of the vapor is used in this derivation (otherwise, $-2RT$ had to be employed). Hence, the cohesive energy densities, $ced = (\Delta_V H - RT)/V$, shown in Table 6.5, and the Hildebrand solubility parameters, $\delta_H = ced^{1/2}$ are obtained on

Table 6.4 The molar mass, M, the melting point, T_m, and the temperature of decomposition, T_{dec}, and the surface tension and heat capacity at 25 °C of RTIL quaternary ammonium and phosphonium salts

Salt	M/g mol^{-1}	T_m/K	T_{dec}/K	C_P/J K^{-1} mol^{-1}	σ/mN m^{-1}
$MeEt_3N$ CH_3SO_4	213.7	306[hhh]			
$MeEt_3N$ $ClCH_2CO_2$	195.7	403[w]			
$MeBu_3N$ CH_3SO_4	297.9	335[hhh]			
$MeBu_3N$ $ClCH_2CO_2$	279.8	328[w]			
$MeBu_3N$ CF_3SO_3	335.4	356[v]			
$MeOc_3N$ Br	434.4	330[v]			
$MeOc_3N$ NTF_2	648.9				27.8[vv]
EtMe3N $CF_3SO_2NCOCF_3$	334.2	288[nn]			
$PrMe_3N$ NTF_2	368.3	292[mm]			
$BuMe_3N$ NTF_2	382.3	292[q]	673[ff]	559[hhh]	38.1[rr]
$BuEt_3N$ BF_4	231.0	435[l]	645[l]		
$BuEt_3N$ PF_6	289.0	452[l]	627[l]		
$HxMe_3N$ NTF_2	426.8	298[tt]			36.0[rr]
$HxEt_3N$ BF_4	259.1	361[l]	640[l]		
$HxEt_3N$ PF_6	317.1	435[l]	603[l]		
$HxEt_3N$ $N(CN)_2$	238.3	503[kk]			
$HxEt_3$ NTF_2	468.9	219[uu]			35.2[bbb]
$HxBu_3$ NTF_2	536.5	297[ll]			
$HxBu_3$ $N(CN)_2$	322.4	230[kk]			
$OcEt_3N$ BF_4	287.1	375[l]	635[l]		
$OcEt_3N$ NTF_2	495.0				33.1[ss]
$OcEt_3N$ PF_6	345.1	362[l]	523[l]		
$DcMe_3N$ NTF_2	480.9				35.5[bbb]
$DcEt_3N$ NTF_2	523.0				32.4[ss]
$DcEt_3N$ Br	308.3	388[m]			
$DcBu_3N$ Br	420.4	295[m]			
$DcHx_3N$ BF_4	511.4	332[m]			
$DcHx_3N$ $C_{12}H_{25}SO_4$	690.0	292[m]			
$DoMe_3N$ Br	308.7		519[u]		
$DoEt_3N$ Br	336.3	405[u]			
$DoEt_3N$ NTF_2	551.0	405[u]			31.8[ss]
$TdEt_3N$ Br	364.3	371[m]			
$TdBu_3N$ Br	448.5	294[m]			
$TdHx_3N$ BF_4	539.6	330[m]			
$TdHx_3N$ $C_{12}H_{25}SO_4$	718.1	295[m]			
Me2PyrrCF3SO3NCOCF2	360.5	297[nn]			
Me_2Pyrr $C_4F_9SO_3$	399.5	471[jj]			
EtMePyrr CF_3SO_3	263.5	379[ii]			
EtMePyrr NTF_2	394.4	364[bb]			
EtMePyrr $C_4F_9SO_3$	413.5	449[jj]			
EtMePyrr $N(CN)_2$	180.5	263[kk]			

(continued)

Table 6.4 (continued)

Salt	M/g mol^{-1}	T_m/K	T_{dec}/K	C_P/J K^{-1} mol^{-1}	σ/mN m^{-1}
PrMePyrr BF_4	214.5	337[nn]			
PrMePyrr PF_6	273.2	386[z]			
PrMePyrr CF_3SO_3	277.5	351[ii]		435[hhh]	
PrMePyrr $C_4F_9SO_3$	427.5	382[ij]			
PrMePyrr $N(CN)_2$	194.5	238[kk]		502[hhh]	
PrMePyrr $N(FSO_3)_2$	340.6	263[hh]			
PrMePyrr NTF_2	408.6	285[mm]		501[hhh]	35.4[zz]
PrMePyrr TFESI	468.6	281[ee]	576[ee]		
BuMePyrr I	269.2	426[gg]	529[gg]		
BuMePyrr SCN	200.4	295[qq]			49.7[rr]
BuMePyrr BF_4	229.3	411[nn]			
BuMePyrr $C_4F_9SO_3$	441.6	367[ij]			
BuMePyrr $N(FSO_3)_2$	354.6	255[hh]			
BuMePyrr NTF_2	422.6	255[dd]		590[qq]	33.1[xx]
BuMePyrr $N(CN)_2$	208.5	218[kk]		502[oo]	56.1[rr]
BuMePyrr $FeCl_4$	343.1	354[aa]			
PeMePyrr NTF_2	426.7	281[hhh]			
PeMePyrr $C_4F_9SO_3$	455.6	391[ij]			
HxMePyrr NTF_2	440.7	276[hhh]			
HxMePyrr $N(CN)_2$	237.0	262[kk]			
HxMePyrr $C_4F_9SO_3$	470.1	393[ij]			
EtMePip NTF_2	408.6	358[cc]			
PrMePip NTF_2	422.6	288[ff]	691[ff]		35.3[ww]
PrMePip TFESY	482.6	299[ee]	575[ee]		
PrEtPip I	283.2	340[gg]	565[gg]		
BuMePip I	283.2	457[gg]	534[gg]		
BuMePip NTF_2	436.6				34.2[ww]
BuEtPip I	297.2	401[gg]	533[gg]		
HxMePip NTF_2	464.7	278[hh]			
PrMeMor I	285.2	360[gg]	545[gg]		
BuMeMor I	285.2	398[gg]	514[gg]		
BuEtMor I	299.4	376[gg]	511[gg]		
$MeBu_3P$ NTF_2	497.5	289[aaa]	652[aaa]		
$MeBu_3P$ $N(CN)_2$	283.4	279[ddd]	660[ddd]		
$MeBu_3P$ $MeSO_4$	328.5			618[fff]	
$BuEt_3P$ $N(CN)_2$	241.3	278[ddd]	667[ddd]		
$BuEt_3P$ $N(SO_2F)_2$	355.4	257[eee]	586[eee]		
$BuEt_3P$ NTF_2	455.4	328[eee]			
$PeEt_3P$ $N(SO_2F)_2$	369.4	238[eee]	582[eee]		
$PeEt_3P$ NTF_2	469.4	290[eee]	653[eee]		
$PeEt_3P$ $N(CN)_2$	255.3	261[ddd]	666[ddd]		
$HxBu_3P$ PF_6	432.4	414[ggg]			

(continued)

Table 6.4 (continued)

Salt	M/g mol^{-1}	T_m/K	T_{dec}/K	C_P/J K^{-1} mol^{-1}	σ/mN m^{-1}
$OcBu_3P\ BF_4$	402.3		672[aaa]		
$OcBu_3P\ PF_6$	460.5	293[aaa]	636[aaa]		
$OcBu_3P\ NTF_2$	595.7		646[aaa]		
$OcBu_3P$ SCN	373.6		651[aaa]		
$OcBu_3P\ CF_3SO_3$	464.7		681[aaa]		
$OcBu_3P\ MePhSO_3$	486.7		617[aaa]		
$DoBu_3P\ BF_4$	458.5	298[aaa]	664[aaa]		
$DoBu_3P\ NTF_2$	651.8	289[aaa]	656[aaa]		
$TdHx_3P$ Cl	519.3	223[fff]		772[fff]	33.4[bbb]
$TdHx_3P\ NTF_2$	764.0	201[fff]		1293[ccc]	30.1[yy]
$TdHx_3P\ N(CN)_2$	550.0			1024[fff]	31.7[yy]

[l][136], [m][137], [q][138], [u][139], [v][140], [w][141], [z][142], [aa][143], [bb][144], [cc][145], [dd][146], [ee][147], [ff][148], [gg][149], [hh][150], [ii][151], [jj][152], [kk][153], [ll][154], [mm][155], [nn] [156], [oo][157], [pp][158] also $\Delta_m H = 21.9$ kJ mol^{-1}, [qq][159] also $\Delta_m H = 2.1$ kJ mol^{-1}, [rr][160], [ss][161], [tt][162], [uu][163], [vv][91], [ww][134], [xx][164], [yy][165], [zz][166], [aaa][167], [bbb][89], [ccc][168], [ddd][169], [eee][170], [fff][171], [ggg][106], [hhh][56]

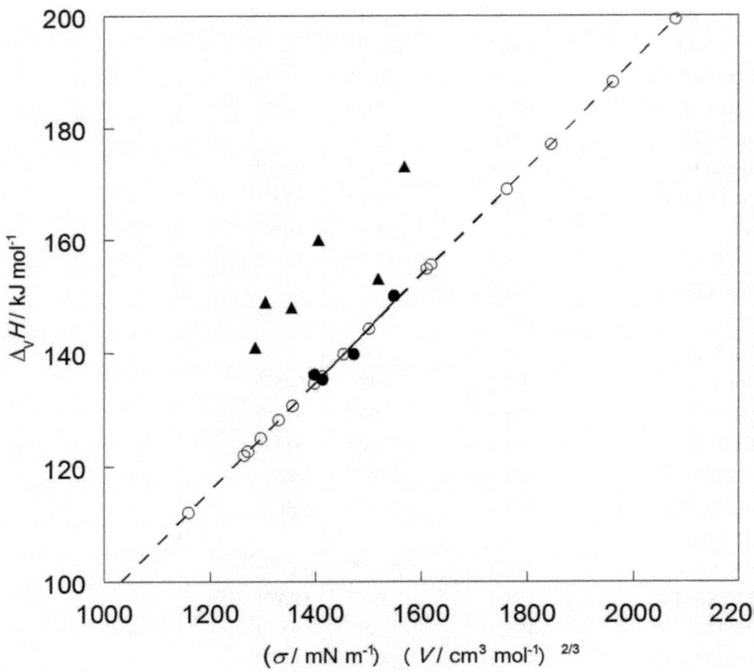

Fig. 6.2 The molar enthalpy of vaporization, $\Delta_v H$, of imidazolium RTILs as a function of the factor $\sigma V^{-2/3}$ of these salts: ● are the data from which Zaitsu et al. [175] constructed their predictive line (- - - -) and their predicted $\Delta_v H$ values (o). However, for some imidazolium RTILs the predictions are lower than the experimental values (▲)

Table 6.5 The cohesive energies, *ce*, their densities, *ced*, and the Gordon parameters, *GP* [181], of RTILs at 25 °C. Values calculated with the use of molar volume data from Tables 6.6 and 6.8 and surface tension data from Tables 6.2 and 6.4 (without references) are in *italics*

Salt	*ce*/kJ mol^{-1}	Refs.	*ced*/MPa	Refs.	*GP*/J m^{-3}
C_1mim PF_6	140	[182]			
C_1mim NTF_2	128, 131	[178, 183]	*530*		0.583
C_1mim CH_3SO_4	104	[184]	1,016,695	[184, 185]	*1.207*
C_1mim $(CH_3)_2PO_4$	142	[186]	801	[186]	0.862[a]
C_2mim BF_4	171	[187]	782,682	[184, 185]	1.004
C_2mim PF_6	139	[188]	1,038,774	[185, 189]	
C_2mim NTF2	121,157	[177, 182]	503	[185]	0.613
C_2mim CH_3SO_4	145	[188]	*786*		*1.105*
C_2mim $C_2H_5SO_4$	151,161	[188, 190]	728	[185]	0.846
C_2mim $(CH_3)_2PO_4$	137	[186]	708	[186]	0.764[a]
C_2mim CH_3SO_3	95	[184]	633	[184]	*0.820*
C_2mim CF_3SO_3	158,146	[188, 191]	656	[185]	*0.705*
C_2mim Cl	137	[192]	1457	[193]	
C_2mim Br	151	[182]			
C_2mim SCN	149	[188]	635	[194]	0.996
C_2mim NO_3	161	[182]			
C_2mim $N(CN)_2$	151	[188]	*938*		*0.811*
C_2mim $B(CN)_4$	147	[195]			
C_2mim CH_3CO_2	165	[82]			
C_2mim $EtCO_2$	116	[196]			
C_2mim CF_3CO_2			653, 1331	[194, 197]	
C_3mim BF_4					0.920
C_3mim PF_6	152	[82]			
C_3mim NTF2	169,129	[177, 178]	811	[197]	0.552
C_3mim Cl	163	[82]	1248	[193]	
C_3mim NO_3					1.008
C_3mim $EtCO_2$	121	[196]			
C_4mim BF_4	125,174	[175, 187]	655	[185]	0.818
C_4mim PF_6	146,184	[182, 198]	667	[185]	0.824
C_4mim NTF2	128,208	[182, 198]	472	[185]	0.463
C_4mim CH_3SO_4	151	[182]	674	[185]	*0.731*
C_4mim $C_8H_{17}SO_4$	176	[188]	357	[185]	*0.366*
C_4mim $(CH_3)_2PO_4$	131	[186]	576	[186]	0.624[a]
C_4mim CF_3SO_3	139,149	[198, 199]	598	[185]	*0.580*
C_4mim Cl	148,91	[184, 192]	748,582	[184, 185]	0.885
C_4mim Br	149	[182]			
C_4mim I	153	[200]			
C_4mim SCN	*112*		607	[194]	*0.835*
C_4mim NO_3	163	[82]			0.904
C_4mim CH_3CO_2	183,119	[201, 202]	645	[202]	*0.598*
C_4mim $EtCO_2$	124	[196]			

(continued)

Table 6.5 (continued)

Salt	*ce*/kJ mol^{-1}	Refs.	*ced*/MPa	Refs.	*GP*/J m^{-3}
C_4mim CF_3CO_2	*230*		1109	[197]	
C_4mim $N(CN)_2$	152	[188]	*784*		*0.791*
C_5mim BF_4					0.726
C_5mim NTF_2	133,140	[78, 178]	427	[185]	0.503
C_5mim NO_3					0.743
C_5mim $EtCO_2$	127	[196]			
C_6mim BF_4	*122*		548	[185]	0.654
C_6mim PF_6	166,194	[187, 190]	554	[185]	0.626
C_6mim NTF2	137,214	[182, 198]	410	[185]	0.438
C_6mim $MeSO_4$	137	[182]			
C_6mim Cl	158	[192]			0.722
C_6mim SCN	112	[194]	559	[194]	
C_6mim NO_3					0.674
C_6mim $EtCO_2$	134	[196]			
C_6mim CF_3CO_2			980	[197]	
C_6mim CF_3SO_3	148	[192]			
C_7mim NTF2	139,149	[78, 178]	*435*		
C_8mim BF_4	159,194	[187, 190]	456	[185]	0.480
C_8mim PF_6	166,213	[190, 198]	499	[185]	0.499
C_8mim NTF2	143,226	[78, 198]	400	[185]	0.414
C_8mim $MeSO_4$					0.462
C_8mim CF_3SO_3	146	[188]	*484*		0.425
C_8mim Cl	163	[192]	509	[185]	0.505
C_8mim Br					0.512
C_8mim I	164	[192]			0.521
C_8mim NO_3					0.535
C_8mim $N(CN)_2$	157	[188]	*609*		
C_8mim CF_3CO_2			882	[197]	
C_{10}mim PF_6	133	[175]			
C_{10}mim NTF_2	149,147	[78, 178]	*372*		0.402
C_{10}mim $B(CN)_4$	203	[195]			
C_{12}mim NTF_2	155,169	[78, 178]	*451*		0.396
C_2Py NTF_2	137,139	[188, 200]	546	[188]	0.592
C_2(2Et)Py NTF_2	143,138	[188, 203]	485	[188]	
C_2Py $C_2H_5SO_4$	168	[188]	918	[188]	
C_2(3 M)Py $C_2H_5SO_4$	161	[188]	807	[188]	
C_3(3Me)Py NTF_2	134	[188]	464	[188]	
C_3(4Me)Py NTF_2	138	[188]	483	[188]	
C_3Py NTF_2	142,138	[188, 200]	514	[188]	0.507
C_4Py NTF_2	140,141	[188, 200]	494	[188]	0.507
C_4Py CH_3CO_2	175	[201]	*928*		
C_4Py CH_3SO_4	168	[188]	827	[188]	

(continued)

Table 6.5 (continued)

Salt	*ce*/kJ mol^{-1}	Refs.	*ced*/MPa	Refs.	*GP*/J m^{-3}
C_4(3 M)Py BF_4	*142*		706	[185]	*0.765*
C_4(3 M)Py CF_3SO_3	*117*		505	[194]	*0.623*
C_4(4 M)Py BF_4	154	[188]	716,759	[185, 188]	*0.784*
C_4(4 M)Py NTF_2	*129*		425	[194]	*0.521*
C_4(4 M)Py SCN	*118*		602	[194]	
C_4(4 M)Py $MePhSO_3$			532	[194]	
C_5Py NTF_2					0.484
C_6Py NTF_2	143	[182]			
$MeOc_3N$ NTF_2	186	[188]	454,319	[188, 191]	*0.330*
$BuMe_3N$ NTF_2	*209*		734	[191]	*0.579*
$BuPrMe_2N$ NTF_2	*225*		713	[191]	
$HxMe_3N$ NTF_2	*222*		686	[191]	*0.524*
$HxEt_3N$ NTF_2	*241*		660	[191]	*0.493*
$HxPrMe_2N$ NTF_2	*228*		666	[191]	
$DcMe_3N$ NTF_2	*258*		605	[191]	*0.472*
$DcPrMe_2N$ NTF_2	*253*		581	[191]	
EtMePyrr NTF_2	136	[188]	488	[188]	
EtMePyrr $N(CN)_2$	149	[188]	857	[188]	
PrMePyrr NTF_2	137	[188]	475,496	[185, 204]	
PrMePyrr $N(CN)_2$	152	[188]	801	[188]	
BuMePyrr CF_3SO_3	*121*		521	[194]	
BuMePyrr NTF_2	141,147	[188, 205]	445,388	[185, 204]	
BuMePyrr $N(CN)_2$	154	[188]	742	[188]	*0.963*
BuMePyrr SCN	*121*		623	[194]	*0.858*
PeMePyrr NTF_2	142	[205]			
HxMePyrr NTF_2	146	[205]	416	[204]	
HpMePyrr NTF_2	151	[205]			
OcMePyrr NTF_2	155	[205]			
DcMePyrr NTF_2	162	[205]	392	[204]	
EtMePip NTF_2	135	[188]	476	[188]	
PrMePip NTF_2	134	[188]	448	[188]	
BuMePip NTF_2	135	[188]	432	[188]	
BuMePip SCN			942	[195]	
PeMePip NTF_2			529	[206]	
HxMePip NTF_2			515	[206]	
Me*i*Bu_3P $MePhSO_3$	215	[194]			
$MeOc_3P$ NTF_2	187	[188]	315	[188]	
$TdHx_3P$ Cl	201	[200]	396	[191]	*0.400*
$TdHx_3P$ NTF_2	214	[188]	350,303	[188, 191]	*0.337*
$TdHx_3P$ $N(CN)_2$	*209*		342	[191]	*0.374*

[a]From Ref. [186]

using the molar volumes V dealt with in Sect. 6.3. These solubility parameters are in the range $18 \leq \delta_H/\mathrm{MPa}^{1/2} \leq 38$,

An indirect method was applied to obtain Hildebrand solubility parameters δ_H of 6 imidazolium RTILs from a linear solvation energy relationship (LSER) correlation of the solvent dependency of a rate constant for a certain reaction by Swiderski et al. [199]. Another indirect method was based by Lee and Lee [198] on the comparison of the viscosity of eight imidazolium RTILs with those of organic solvents. The resulting values of $\delta_H/\mathrm{MPa}^{1/2}$ were in the range 24–32. These values are less reliable than those obtained from $ced = (\Delta_V H - RT)/V$.

Molecular simulation was applied by Derecskei and Derecskei-Kovacs [208] to nine imidazolium RTILs with resulting $\delta_H/\mathrm{MPa}^{1/2}$ values in the range 25–35. Molecular dynamic simulations of four mixtures of imidazolium RTILs with methanol also yielded $ced = \delta_H^{\,2}$ values for the RTILs according to Jahangiri et al. [189]. Molecular dynamic simulations were applied by Sistla et al. [197] to yield solubility parameters for use in the evaluation of their capabilities to absorb CO_2. The simulations were first applied to 16 RTILs involving 4 key cations (C_2mim^+, C_4mim^+, C_6mim^+, and C_8mim^+) and 4 key anions (NTF_2^-, PF_6^-, $CH_3CO_2^-$, and $(C_2F_5)_3PF_3^-$), and the 16 expressions were solved simultaneously to yield the following relationship:

$$\left(\delta_H/\mathrm{MPa}^{1/2}\right)^2 = 0.6084(\delta_{H+}\delta_{H-}) - 0.1958\left(\delta_{H+}^{\,2}\right) - 0.1958\left(\delta_{H-}^{\,2}\right) \qquad (6.11)$$

Individual ionic values of δ_{H+} and δ_{H-} were thus obtained for the key ions, and Eq. (6.11) were then used to obtain δ_H values for altogether 210 RTILs involving also 17 other cations of various types and 6 other anions. The δ_H values diminish as the chain lengths of the alkyl substituents or the size of the anion increases.

An indirect method for arriving at the *ced* was from the corresponding enthalpies of vaporization $\Delta_V H$ and surface tensions σ of 6 imidazolium RTILs, according to Jin et al. [204], noting the proportionality between *ced* and $\sigma V^{-1/3}$, and then calculating *ced* for 12 pyrrolidinium and other quaternary ammonium RTILs. The thus estimated $\delta_H/\mathrm{MPa}^{1/2}$ values are on the low side, ranging from 20 to 27. The Hildebrand solubility parameters of seven imidazolium, eight quaternary ammonium, and nine quaternary phosphonium RTILs were estimated by Kilaru and Scovazzo [191] from their dynamic viscosities η (Sect. 6.5) as $\delta_H = K(T/V)^{1/2}(\ln\eta V)^{1/2}$. The resulting $\delta_H/\mathrm{MPa}^{1/2}$ values increased with the alkyl chain lengths from 25 to 30 for the imidazolium salts to 21–27 for the ammonium salts to 18–23 for the phosphonium salts. The viscosity was also used by Weerachanchai et al. [184] for obtaining the Hildebrand solubility parameters of 11 RTILs with various cations and anions.

The Hildebrand solubility parameters of 18 RTILs were calculated from the inverse gas chromatography derived infinite dilution activity coefficients by Marciniak [194] who later added values for eight additional RTILs [195], and compared the values with those obtained by other methods. Gas chromatography was also used by Paduszynski and Domanska [206] to obtain infinite dilution activity coefficients that were then used for the estimation of the δ_H values of four piperidinium-based RTILs.

The enthalpies of vaporization $\Delta_V H$ of aprotic RTILs were estimated by Schröder and Coutinho [188] by means of the COSMO-RS method and compared with results from other methods. The uncertainty from using this method was estimated as ± 10 kJ mol^{-1}. Values of the *ced* and of δ_H were then derived for 102 RTILs with a variety of cations and anions.

The enthalpy of vaporization, estimated according to Zaitsau et al. [175] as shown above yielded the $ce = \Delta_V H - RT$ and the *ced*, which were then used for other purposes by Singh and Kumar [185]. This approach was also used by Xu et al. [202] to obtain the $\Delta_V H$ and δ_H values of four imidazolium carboxylate RTILs and by Ren et al. [207] to obtain the $\Delta_V H$ and δ_H values of seven imidazolium dialkylphosphate RTILs.

The linear dependence of $\Delta_V H$ on $\sigma V^{2/3}$ noted above gives rise to another characterization of liquids used as solvents, namely the Gordon Parameter $GP = \sigma V^{-1/3}$, which is valuable for the use of RTILs for the self-assembly of amphiphiles according to Greaves and Drummond [209, 210]. These authors reported values for protic RTILs, but later Greaves and Drummond [181] added the *GP* values of aprotic RTILs shown in Table 6.5. Some additional *GP* values were calculated by the author of this book from the surface tensions and molar volumes and are also shown there.

6.2.4 Critical Properties

Critical temperatures T_c have been reported for many RTILs, but have been based on unreliable long extrapolations of experimental data. Although $C_{10/12}mim^+NTF_2^-$ could be distilled without decomposition at reduced pressures, the estimated normal boiling points, $T_b \approx 0.6\ T_c$ of 1-alkyl-3-methylimidazolium salts (with the anions Cl^-, BF_4^-, PF_6^-, and NTF_2^-) according to Rebelo et al. [211] are much higher than the onsets of the decomposition temperatures, T_{dec} (Table 6.2). The critical temperatures, T_c, in turn, were estimated from the Eötvos expression:

$$\sigma V^{2/3} = a + bT \tag{6.12}$$

yielding $T_c = {}^{-}a/b$ as the temperature at which the surface tension, σ, vanishes, and from the Guggenheim expression:

$$\sigma = \sigma^0 (1 - T/T_c)^{11/9} \tag{6.13}$$

yielding T_c again for the temperature at which the surface tension, σ, vanishes. These two approaches were used also in several publications for the estimation of the critical temperature of RTILs. Application to $C_n mim^+NTF_2^-$ ($n = 1, 2, 4, 6, 8$) RTILs [212–214], $C_4mim^+PF_6^-$ [215] and $C_2mim^+An^-$ (An^- being one of eight anions) [216] and $Cat^+CH_3CO_2^-$ [Cat^+ being one of four cations) [217], and $R_3R'N^+An^-$ or $R_3R'P^+An^-$ (An^- being one of several anions) [218, 219] generally yielded values of $T_c \geq 1000$ K and $T_b \geq 900$ K from surface tension and density data at $\leq$350 K. The values of T_c obtained from these two approaches in many cases differed by >100 K, even up to 200 K. The experimental $\sigma(T)$ data do vary linearly within the range of measurement (see below), but very long extrapolations of them are thus required to obtain $\sigma = 0$. Therefore, the reliability of these critical temperature estimates is questionable, and estimation methods of physical quantities of RTILs based on reduced temperatures with respect to the critical temperature should be dealt with cautiously.

Critical temperatures were also obtained on the basis computer simulations (Monte Carlo) of the coexistence of liquid and vapor (assumes to be a single ion pair) as applied by Rai and Maginn [213] to $C_n mim^+NTF_2^-$ ($n = 1, 2, 4, 6$) RTILs. These T_c values were larger than those estimated from the surface tensions and in better accord with estimates from group contributions by Valderrama and Robles [220], available for 50 diverse RTILs and by Valderrama et al. [221] available for 200 RTILs. Estimated normal boiling points, T_b, and critical pressures, volumes, and acentric factors were also provided in the latter report. The ratios of T_b/T_c ranged between 0.634 and 0.794. The estimates by Valderrama and Robles from group contributions were updated by Valderrama et al. [222], including the critical compressibility facto r and extended to nearly 300 RTILs and further updated by them [223] and extended by them [224], and appear to yield reliable values.

6.2.5 *Heat Capacities*

The isobaric molar heat capacities, C_P, of RTILs (shown for 25 °C in Tables 6.2, 6.3, and 6.4), have been recently reviewed and critically compiled by Paulechka [83], dealing with values available to that time, and a few more recent C_P data of RTILs are included in these tables. The temperature dependence of the values is described in [83] by empirical third degree power series, although linear dependencies are adequate for many RTILs [58, 225]. The C_P of the RTILs studied increases slowly with the temperature, up to 0.2 % per K, the more, the longer the alkyl chain(s) of the RTILs. This contrasts with the substantially temperature-invariance of the C_P values of the high-melting salts shown in Tables 3.3.3 and 3.3.4. The pressure (up to 60 MPa) and temperature (up to 50 °C) dependencies of the C_P of four imidazolium tetrafluoroborates was reported by Sanmamed et al. [226]. Within these ranges the pressure dependencies were very small, ~0.3 % at the maximal pressure studied at ambient temperatures.

Volume-based thermodynamics was applied by Glasser [227] to model the heat capacities, based on the microscopic ionic volumes V/N_A at an unspecified temperature. It was applied to 11 1-alkyl-3-methylimidazolium RTILs (C_2, C_4, and C_6) with a variety of anions and to $C_{14}mim^+NTF_2^-$ and C_4-2,3-dimethylimidazolium$^+$ PF_6^- with a linear correlation coefficient squared of 0.980 when forced through the origin. When data for 28 further imidazolium salts were added and the microscopic volumes from Tables 2.4 and 6.9 (rather than V/N_A) were employed, the slope was slightly larger than in [227] and the data followed the expression:

$$C_P/\text{J K}^{-1}\text{mol}^{-1} = (-9 \pm 5) + (1343 \pm 16)(v_+ + v_-)/\text{nm}^3 \quad (6.14)$$

Figure 6.3 a second degree power series in the temperature was modeled by Galinski et al. [156] for 19 imidazolium, pyridinium, and pyrrolidinium RTILs with a variety of anions by group contribution methods with three ion-specific parameters. When the C_P(298 K) values were plotted against the microscopic ionic volumes the slope was 1174 ± 54, in fair agreement with that of Eq. (6.14). Two further group contribution methods for obtaining the C_P of RTILs were published

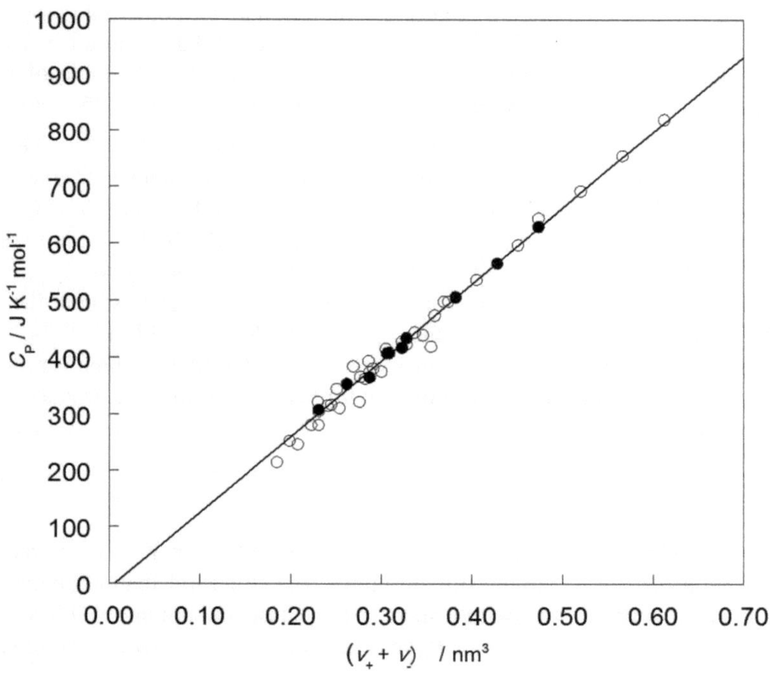

Fig. 6.3 The molar constant pressure heat capacity, C_P, of RTILs plotted against their microscopic molar volumes, $(v_+ + v_-)$, cf. Eq. (6.11): (●) according to Glasser and Jenkins [227], where $(v_+ + v_-) = V/N_A$, and (o) for many other imidazolium RTILs where v_+ and v_- are from Table 6.9 in the present book

recently by Farahani et al. [228] and by Sattari et al. [130], only the second one of these specifying the characteristics of the groups used.

6.2.6 Surface Tension

The surface tension, σ, at 25 °C of common RTILs is shown in Tables 6.2, 6.3, and 6.4. The surface tension has generally been measured over a temperature span of ~60 K and was invariably found to diminish linearly with increasing temperatures:

$$\sigma = a - bT \tag{6.15}$$

The values do decrease linearly with the temperature, within the temperature range of measurement (up to 120 °C according to Ghatee and Zolghadr [92]), notwithstanding the theoretical dependencies of the Eötvos and Guggenheim expressions, Eqs. (6.12) and (6.13) for fluids in general.

Values of σ of RTILs with less usual anions, such as $FeCl_4^-$, $GaCl_4^-$, $Au(CN)_3^-$, ICl_2^-, and Br_3^- were reported by Martino et al. [88] and with other inorganic anions by Larriba et al. [229]. The surface tensions of RTILs diminish as their molar volumes increase according to Jin et al. [204], and for 12 imidazolium and pyrrolidinium NTF_3^- RTILs the expression for 25 °C is: $\sigma/\mathrm{mN\,m^{-1}} = 22.77 + 3820/\left(V/\mathrm{cm^3mol^{-1}}\right) - 1090/\left(V/\mathrm{cm^3mol^{-1}}\right)^2$. In fact, linear dependencies: $\sigma = a - bV$ can be derived from the data in Table 6.2 for imidazolium RTILs with anions such as BF_4^-, PF_6^-, NO_3^-, and $CH_3CO_2^-$ with individual slopes b, but not, for instance for the anion NTF_2^-. Similarly, for 1-alkyl-3-methyl-imidazolium alkylsulfates σ diminishes monotonously (in most cases linearly) with the number of carbon atoms in the alkyl chains, both the cation and the anion being present at the surface, as found by Santos and Baldelli [96].

The old concept of the parachor introduced by Sugden [230], which is nowadays seldom invoked, has nevertheless found use for the prediction of the surface tension of RTILs. The parachor is defined as:

$$P_\sigma = M\sigma^{1/4}\rho^{-1} \tag{6.16}$$

and is only slightly temperature dependent, <0.03 % per K. The parachor is additive with respect to the constituent atoms and structural features (rings, etc.) [230] and should be additive with respect to the cation and anion of RTILs. For the division of experimental values of P_σ into the contributions of the individual ions the extra-thermodynamic assumption that $P_{\sigma-}/P_{\sigma+} = v_{\mathrm{cr}-}/v_{\mathrm{cr}+}$ was made by Ma et al. [231], where v_{cr} is the microscopic volume of the subscripted ion in crystals. For a family of RTILs with a common anion P_σ increases linearly with the number of carbon atoms in the alkyl chains. However, the increment in $P_\sigma/[(\mathrm{mN\,m^{-1}})^{1/4}$

mol cm^{-3}] per methylene group has been reported variously by different authors: 32.8 [232], 37.4 [231, 233], and 39.9 [234, 235]. Values of P_σ of various cations and anions have been reported in the two most recent references by Ma et al. and Soukova et al. [231, 232]. These supersede previous estimates by Gardas et al. [234], based on the proportionality of the parachor to the 5/6th power of the molar volume of the RTIL: $P_\sigma = 6.2V^{5/6}$ (with the commonly used units for these quantities).

In principle, therefore, Eq. (6.16) may be inverted to predict the surface tension of RTILs: $\sigma = (P_\sigma/V)^4$, but this procedure makes the predicted σ very sensitive to the values of the parachor (see the discrepancies in reported values mentioned above) and the molar volume, because of the fourth power dependence. Therefore this apparently simple method according to Lemraski and Zobeydi [236] is not recommended for the estimation of the unknown surface tension of RTILs.

Another predictive method, according to Shang et al. [237], uses the QSAR employing the topological index of the RTILs, the distance matrix between non-hydrogen atoms in the salt. Using a training set of 715 temperature dependent data points (for $\sigma = a - bT$) and 215 test data points the average deviation was ~4 % for the predicted values. However, the number of required structural parameters was very large (>30 numerical values are listed for various types of RTILs).

According to the hole theory of liquids (Sect. 3.3.1) the surface tension should be inversely proportional to the 2/3 power of the void volume. As applied by Larriba et al. to ionic liquids [229]:

$$\sigma = \xi^* k_B T v_v^{-2/3} \tag{6.17}$$

where $v_v = V/N_A - (v_+ + v_-)$ is the microscopic void volume, and $\xi^* = 0.674$. However, the theoretical coefficient had to be replaced by an empirical one for the 20 salts of imidazolium cations with inorganic anions considered [229]. This was also necessary for the purpose of this book for 50 RTILs with imidazolium and pyridinium cations and various anions for which surface tension data could be found, using the molar void volume $V_v = V - N_A(v_+ + v_-)$ (Sect. 6.3):

$$\sigma_{calc}/\text{mN m}^{-1} = 4.9\left(V_v/\text{cm}^3\text{mol}^{-1}\right)^{-2/3} = (1.005 \pm 0.036)\sigma_{exp}/\text{mN m}^{-1} \tag{6.18}$$

with a correlation coefficient $r_{corr} = 0.9590$, Fig. 6.4. The scatter arises from the definition of the void volume, V_v, which is a rather small difference between much larger values. Still, the inverse proportionality of the surface tension to the 2/3 power of the void volume that was found, that is, to the mean surface areas of the cavities, indicates that cavities of sizes commensurate with the ions of the RTILs do exist in them. Due to the appreciable scatter noted for Eq. (6.18), care should be used for the prediction of unknown surface tensions of RTILs from it.

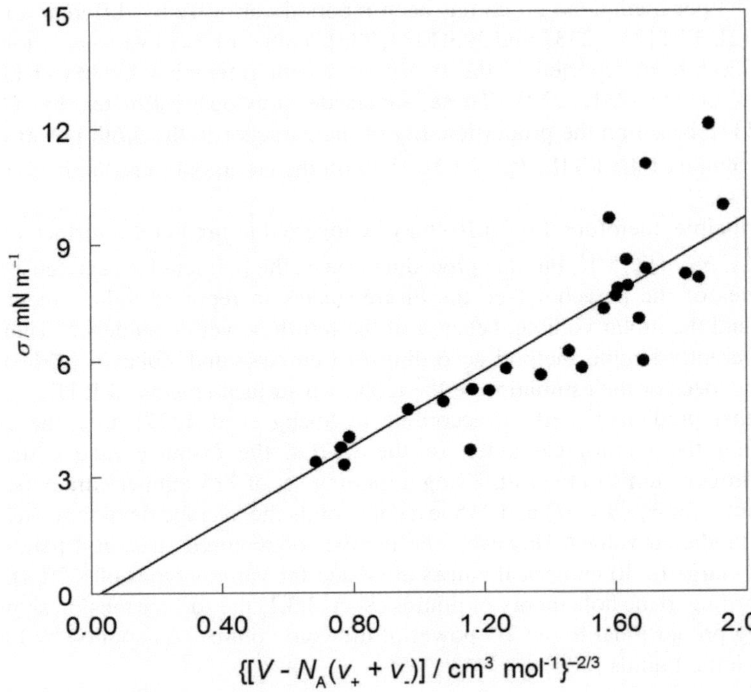

Fig. 6.4 The surface tension, σ, of RTILs plotted against the reciprocal of the surface area of the voids in them, $[V - N_A(v_+ + v_-)]^{-2/3}$, cf. Eq. (6.18)

6.3 Volumetric Data

6.3.1 Densities and Molar Volumes

The densities of RTILs and their temperature dependences are readily measured with good accuracy by a variety of methods with results that have recently been compiled by Marcus [238] and in the NIST database [56]. The densities are generally reported as ρ/g cm^{-3} to four decimals and the isobaric expansibilities as $10^3\alpha_P$/K^{-1} to three decimals. Values of ρ and $10^3\alpha_P$ at 25 °C for the common 1-alkyl-3-methylimidazolium RTILs with a large variety of anions are shown in Table 6.6, those for 1-alkylpyridinium RTILs in Table 6.7, and of quaternary ammonium and phosphonium ones in Table 6.8. Also shown in these three Tables are the molar volumes, V/cm^3 mol^{-1} = M/ρ at 25 °C, with the molar masses from Tables 6.2, 6.3, and 6.4.

The modeling of the (temperature dependent) densities is described briefly in Sect. 6.1.2, using equations of state (EoS) derived from various modifications of the statistical associated fluid theory (SAFT), the COSMO-RS model, the Sanchez-Lascomb lattice fluid model (SL), or the perturbed hard sphere model (PHS). Each

Table 6.6 The density, isobaric expansibility, molar volume, isothermal compressibility, and internal pressure at 25 °C of 1-alkyl-3-methylimidazolium ($C_n$mim) salts, from Farahani et al. [238] unless otherwise noted

Salt	ρ/g cm^{-3}	$10^3\alpha_P$/K^{-1}	V/cm^3 mol^{-1}	κ_T/GPa^{-1}	P_{int}/MPa[y]
C_1mim BF_4	1.306		140.9		
C_1mim PF_6	1.478		164.5		
C_1mim NTF2	1.5620	0.743[l]	241.5		
C_1mim Cl	1.1399[a]		116.3		
C_1mim SCN	1.1574	0.572	134.1		
C_1mim CH_3SO_4	1.3273	0.454	156.9		573
C_2mim BF_4	1.2803	0.639	154.7	0.337	496
C_2mim PF_6	1.413	0.784[n]	181.2		492
C_2mim NTF2	1.5183	0.662	257.7	0.380	366
C_2mim CH_3SO_4	1.2058	0.537	184.4	0.336[r]	
C_2mim $C_2H_5SO_4$	1.2390	0.541	190.8	0.262[s]	487
C_2mim $C_8H_{17}SO_4$	1.0948		292.7		
C_2mim CH_3SO_3	1.2409	0.532	166.3		
C_2mim $CH_3C_6H_4SO_3$	1.2253		230.5[q]		
C_2mim CF_3SO_3	1.3836	0.611	188.1	0.435[t]	397
C_2mim Cl	1.186[a]		123.6		
C_2mim Br	1.48[b]		129		
C_2mim I	1.69[b]		140		
C_2mim SCN	1.1162	0.545	151.6	0.299[u]	
C_2mim NO_3	1.2085[c]		143.3		
C_2mim HSO_4	1.3674	0.453	152.3		
C_2mim $AlCl_4$	1.2947	0.595	218.6		
C_2mim $N(CN)_2$	1.1016	0.608	160.9		
C_2mim $B(CN)_4$	1.0363[d]	0.764[d]	218.2		
C_2mim CH_3CO_2	1.1437		148.8		
C_3mim BF_4	1.237		171.4		
C_3mim PF_6	1.333[a]		202.5		
C_3mim NTF2	1.4757	0.685	274.6	0.482	355
C_3mim NO_3	1.184		158.2		
C_3mim CH_3SO_4	1.2031	0.533	196.3		
C_3mim CH_3CO_2	1.1190		164.5		
C_4mim BF_4	1.2015	0.631	188.1	0.360	448
C_4mim PF_6	1.3647	0.611	208.3	0.412	439
C_4mim NTF2	1.4371	0.660	291.8	0.498	351
C_4mim CH_3SO_4	1.2058	0.547	207.6	375[j]	427
C_4mim $C_2H_5SO_4$	1.2323[e]	0.629[e]	214.5		
C_4mim $C_8H_{17}SO_4$	1.0668	0.579	326.8	0.492	373
C_4mim CF_3SO_3	1.3006	0.730	221.7	0.432	401
C_4mim CH_3SO_3	1.2437[a]		188.4		
C_4mim Cl	1.0800	0.491[o]	161.7		470
C_4mim Br	1.2909[f]	0.511	169.7		
C_4mim I	1.4842[g]	0.234[g]	179.3		

(continued)

Table 6.6 (continued)

Salt	ρ/g cm^{-3}	$10^3\alpha_P$/K^{-1}	V/cm^3 mol^{-1}	κ_T/GPa^{-1}	P_{int}/MPa[y]
C_4mim SCN	1.0695	0.539	184.5		
C_4mim NO_3	1.1565[h]	0.553[h]	174.5		
C_4mim CH_3CO_2	1.0665	0.577	185.9		
C_4mim $FeCl_4$	1.3658[i]	0.605[i]	246.6		
C_4mim $N(CN)_2$	1.0588	0.433	193.9	0.415	
C_5mim BF_4	1.173		204.7		
C_5mim NTF_2	1.4045	0.676	308.5	0.471[v]	354
C_5mim NO_3	1.127		178.5		
C_5mim CH_3CO_2	1.0773		197.1		
C_6mim BF_4	1.1453	0.600[p]	221.9		402
C_6mim PF_6	1.2928	0.669	241.5	0.400[w]	420
C_6mim NTF2	1.3708	0.671	326.4	0.560[x]	351
C_6mim CH_3SO_4	1.2149[e]	0.611	229.4		
C_6mim $C_2H_5SO_4$	1.2014[e]	0.614[e]	243.4		
C_6mim $C_8H_{17}SO_4$	1.0676[j]	0.604[j]	352.7	481[j]	365[j]
C_6mim CH_3SO_3	1.1183	0.558	235.5		
C_6mim Cl	1.0393	0.546	194.9		
C_6mim Br	1.2292[g]	0.551[g]	201.1		
C_6mim I	1.3402[g]	0.552[g]	219.5		
C_6mim NO_3	1.1165[k]	0.614[k]	193.7		
C_6mim $B(CN)_4$	0.9906[d]	0.729[d]	284.8		
C_6mim $AlCl_4$	1.1953	0.786	281.2		
C_6mim $FeCl_4$	1.333		276.0		
C_6mim CH_3CO_2	1.0606		213.4		
C_7mim NTF2	1.3471	0.764	342.6	0.543	
C_8mim BF_4	1.1100	0.676	254.2	0.417	384
C_8mim PF_6	1.1191	0.666	304.1	0.458	387
C_8mim NTF2	1.3203	0.675	360.1	0.540	348
C_8mim $MeSO_4$	1.110[l]	0.588[l]	276.1		
C_8mim CH_3SO_3	1.0818	0.569	268.4		
C_8mim CF_3SO_3	1.142[l]		301.6		
C_8mim Cl	1.0096[j]	0.579[j]	228.4	389[j]	433[j]
C_8mim Br	1.169[l]	0.585[l]	235.4		
C_8mim I	1.2336[g]	0.666[g]	255.3		
C_8mim NO_3	1.065[l]	0.588[l]	241.3		
C_8mim $N(CN)_2$	1.013[m]		258.0	0.415	
C_8mim $FeCl_4$	1.280		309.4		
C_{10}mim BF_4	1.0697[k]	0.660[k]	289.3		
C_{10}mim NTF_2	1.274	0.691	395.2	0.334[r]	
C_{12}mim NTF_2	1.2447	0.699	339.4		
C_{14}mim NTF_2	1.1882	0.696	379.1		

[a][221], [b][239], [c][246], [d][256], [e][249], [f][251], [g][245], [h][242], [i][248], [j][243], [k][252], [l][99], [m][244], [n][250], [o][54], [p][247], [q][253], [r][255], [s][257] κ_S rather than κ_T, [t][234], [u][258] κ_S rather than κ_T, [v][240], [w][241], [x][254], [y][185]

Table 6.7 The density, isobaric expansibility, molar volume, isothermal compressibility, and internal pressure at 25 °C of 1-alkylpyridinium salts

Salt	ρ/g cm^{-3}	$10^3\alpha_P$/K^{-1}	V/cm^3 mol^{-1}	κ_T/GPa^{-1}	P_{int}/MPa[d]
C_1Py CH_3SO_4	1.3448[k]	0.504[k]	152.7		
C_2Py BF_4	1.3020[b]		149.8		
C_2Py $N(SO_2F)_2$	1.4587[n]	0.60[n]	219.7		
C_2Py NTF_2	1.5368[f]	0.632[f]	252.7		
C_2Py CH_3SO_4	1.2847[k]	0.545[k]	170.7		
C_2Py $C_2H_5SO_4$	1.2223[i]	0.530[i]	190.9		
C_2Py CF_3CO_2	1.273[s]		173.8		
C_2(3 M)Py NTF_2	1.4885[u]		250.8		
C_2(4 M)Py NTF_2	1.4920[i]	0.563[i]	250.2		
C_3Py BF_4	1.2529[t]	0.575[t]	166.9	0.333[t]	
C_3Py NTF_2	1.4935[h]	0.601[h]	269.5		
C_3(3 M)Py BF_4	1.3412[a]		166.4		
C_3(3 M)Py NTF_2	1.4475[o]	0.71[o]	287.7		
C_4Py NO_3	1.1861[p]	0.707[p]	167.2		
C_4Py BF_4	1.2144[e]	0.543[e]	183.7	0.250[e]	
C_4Py PF_6	1.2144[b]		231.5		
C_4Py $N(SO_2F)_2$	1.3694[n]	0.61[n]	254.5		
C_4Py NTF_2	1.4476[g]	0.663[g]	287.7		
C_4Py CF_3SO_3	1.3113[t]	0.587[t]	217.6	0.410[t]	
C_4(2 M)Py NTF2	1.476[r]	1.08[r]	291.7		
C_4(3 M)Py BF_4	1.1824[q]	0.584[q]	200.6	0.369	443[a]
C_4(3 M)Py CF_3SO_3	1.297[r]	1.16[r]	230.9		
C_4(3 M)Py NTF_2	1.4226[d]	0.639[d]	302.6		
C_4(3 M)Py $N(CN)_2$	1.0493[t]	0.581[t]	206.2	0.354[t]	
C_4(4 M)Py BF_4	1.189[c]		199.4		
C_4(4 M)Py NTF_2	1.4187[i]	0.633[i]	303.4		
C_4(4 M)Py CF_3CO_2	1.1700[b]		225.1		
C_5Py NTF_2	1.4214[l]	0.591[l]	302.9		
C_6Py BF_4	1.161[c]		216.3		
C_6Py $N(SO_2F)_2$	1.3019[n]	0.61[n]	289.2		
C_6Py NTF_2	1.3942[h]	0.645[h]	318.8		
C_6(3 M)Py NTF_2	1.3516[m]		339.3		
C_8Py BF_4	1.1127[f]	0.629[f]	250.9		
C_8Py NTF_2	1.3268[g]	0.663[g]	357.2		
C_8(3 M)Py BF_4	1.0952[t]	0.603[t]	267.4	0.430[t]	
C_{10}Py NTF_2	1.2835[g]	0.670[g]	390.0		
C_{12}Py NTF_2	1.2488[g]	0.681[g]	423.3		

[a][263], [b][220], [c][266], [d][242], [e][116], [f][269], [g][270], [h][118], [i][62], [j][185], [k][267], [l][135], [m][261], [n][117], [o][132], [p][268], [q][260], [r][259], [s][264], [t][265], [j][262]

Table 6.8 The density, isobaric expansibility, molar volume, isothermal compressibility, and internal pressure at 25 °C of quaternary ammonium and phosphonium salts

Salt	ρ/g cm^{-3}	$10^3\alpha_P$/K^{-1}	V/cm^3 mol^{-1}	κ_T/GPa^{-1}	P_{int}/MPa
$MeEt_3N$ CH_3SO_4	1.1725[o]	0.519[o]	194.1		
$MeBu_3N$ NTF_2	1.266[h]		379.9		
$MeOc_3N$ Cl	0.891[k]		454.1		
$MeOc_3N$ $N(CN)_2$	0.891[k]		488.4		
$MeOc_3N$ CF_3CO_2	0.961[e]		501.3		
$MeOc_3N$ NTF_2	1.0823[c]	0.462[c]	599.5		
$PrMe_3N$ NTF_2	1.44[h]		265		
$BuMe_3N$ NTF_2	1.3922[b]	0.641[b]	285.0	0.389[bb]	
$BuEt_3N$ BF_4	1.1397[y]		215.4		
$BuEt_3N$ PF_6	1.3662[y]		222.1		
$BuPrMe_2N$ NTF_2	1.3483[c]	0.460[c]	315.1		
$HxMe_3N$ NTF_2	1.3106[c]	0.458[c]	324.1		
$HxEt_3N$ BF_4	1.0935[y]		250.2		
$HxEt_3N$ PF_6	1.3513[y]		245.3		
$HxEt_3N$ NTF_2	1.2793[c]	0.578[c]	364.7		
$HxPrMe_2N$ NTF_2	1.2846[c]	0.553[c]	341.6		
$HxBu_3N$ NTF_2	1.15[h]		479		
$HpEt_3N$ NTF_2	1.2710[u]	0.702[u]	378.1		
$OcEt_3N$ BF_4	1.0653[y]		283.1		
$OcEt_3N$ PF_6	1.1902[y]		302.1		
$OcEt_3N$ NTF_2	1.2498[u]	0.660[u]	396.0		
$DcMe_3N$ NTF_2	1.2263[c]	0.628[c]	426.5		
$DcEt_3N$ NTF_2	1.2162[u]	0.651[u]	406.6		
$DcPrMe_2N$ NTF_2	1.2007[c]	0.541[c]	435.2		
$DoEt_3N$ NTF_2	1.1882[u]	0.655[u]	463.4		
$TdEt_3N$ NTF_2	1.1650[u]	0.661[u]	496.7		
EtMePyrr $C_2H_5SO_4$	1.1960[o]	0.525[o]	200.3		
PrMePyrr $N(FSO_2)_2$	1.3069[v]	0.60[v]	236.1		
PrMePyrr NTF_2	1.4274[z]	0.638[z]	286.2	0.516[cc]	395[g]
BuMePyrr CF_3SO_3	1.2527[w]		232.6		
BuMePyrr CH_3SO_4	1.1667[o]	0.534[o]	217.3		
BuMePyrr $N(FSO_3)_2$	1.3069[v]	0.601[v]	246.9		
BuMePyrr NTF_2	1.3946[i]	0.696[i]	302.7	0.447[bb]	388[g]
BuMePyrr $N(CN)_2$	1.020[e]		204.2		
BuMePyrr CH_3CO_2	1.0212[t]	0.588[t]	197.3		
BuMePyrr $PF_3(C_2F_5)_3$	1.5832[n]	1.001[n]	371.1		
BuEtPyrr $C_2H_5SO_4$	1.1187[o]	0.510[o]	239.2		
HxMePyrr NTF_2	1.32[aa]		342		
HxHxPyrr NTF_2	1.240[e]		419.8		
OcMePyrr NTF_2	1.296[e]		369.2		
DcMePyrr NTF_2	1.25[aa]		406		
PrMePip NTF_2	1.4121[f]	1.376[f]	299.3	0.485[f]	

(continued)

Table 6.8 (continued)

Salt	ρ/g cm^{-3}	$10^3\alpha_P$/K^{-1}	V/cm^3 mol^{-1}	κ_T/GPa^{-1}	P_{int}/MPa
$MeBu_3P$ NTF_2	1.28[d]		388.7		
$MeBu_3P$ $N(CN)_2$	0.96[l]		295		
$PrOc_3P$ Cl	0.8941[m]	0.638[m]	499.2		
$BuEt_3P$ $N(CN)_2$	1.00[l]		241		
$BuEt_3P$ $N(SO_2F)_2$	1.26[s]		282		
$BuOc_3P$ Cl	0.8969[m]	0.645[m]	516.5		
$PeEt_3P$ $N(SO_2F)_2$	1.24[s]		298		
$PeEt_3P$ NTF_2	1.32[h]		355		
$PeEt_3P$ $N(CN)_2$	0.99[l]		258		
$HxOc_3P$ Cl	0.8921[m]	0.661[m]	550.7		
$OcEt_3P$ NTF_2	1.26[h]		406		
$OcBu_3P$ BF_4	1.02[d]		395		
$OcBu_3P$ PF_6	1.12[d]		411		
$OcBu_3P$ NTF_2	1.18[d]		505		
$OcBu_3P$ SCN	0.95[d]		373		
$OcBu_3P$ $N(CN)_2$	0.97[l]		307		
$OcBu_3P$ CF_3SO_3	1.08[d]		430		
$OcBu_3P$ $MePhSO_3$	1.02[d]		477		
$DoEt_3P$ NTF_2	1.21[h]		469		
$DoBu_3P$ BF_4	0.97[d]		473		
$DoBu_3P$ NTF_2	1.13[d]		577		
$TdBu_3P$ $DoPhSO_3$	0.9384[c]	0.501[c]	773.1		
$TdHx_3P$ Cl	0.8916[a]	0.670[a]	582.4	0.586[a]	
$TdHx_3P$ Br	0.9552[q]	0.639[q]	590.5	0.561[r]	
$TdHx_3P$ BF_4	0.929[e]		614.3		
$TdHx_3P$ NTF_2	1.0666[a]	0.727[a]	716.3	0.612[a]	
$TdHx_3P$ $N(CN)_2$	0.8995[p]	0.669[p]	611.4	0.492[p]	
$TdHx_3P$ CH_3CO_2	0.8906[a]	0.662[a]	609.9	0.585[a]	
$TdHx_3P$ CH_3SO_3	0.9281[q]	0.657[q]	624.1	0.573[cc]	
$TdHx_3P$ CF_3SO_3	0.9824[j]	0.668[j]	644.6		
$TdHx_3P$ $PF_3(C_2F_5)_3$	1.1817[n]	0.997[n]	786.3		
$TdHx_3P$ $FeCl_4$	0.9888[x]	0.638[x]	689.6		

[a][271], [b][272], [c][89], [d][167], [e][259], [f][267], [g][185], [h][221], [i][273], [j][274], [k][244] at 293 K, [l][169], [m][275], [n][276], [o][277] for $MeEt_3NCH_3SO_4$ extrapolated to 298 K, for BuEtPyrr $C_2H_5SO_4$ at 328 K, [p][278], [q][279], [r][55], [s][170], [t][216], [u][280], [v][281], [w][282], [x][248], [y][283], [z][284], [aa][204], [bb][189], [cc][235]

of these models involves several substance-specific parameters obtained by fitting the (temperature-dependent) densities and can be used to extend the range of values to temperatures and pressures outside the range of measurement or to calculate other properties of the RTILs, such as viscosities or surface tensions.

Several authors, on the other hand, established schemes for the estimation of the densities, molar volumes, and expansibilities of RTILs from independent data concerning the ion constituents of the RTILs. Two kinds of approaches have been used for this purpose: group contributions and microscopic ionic volumes. Gardas and Coutinho [285] reviewed group contribution methods for the prediction of thermophysical properties of RTILs, including the isobaric thermal expansibility α_P and the isothermal compressibility κ_T, but not the density itself. Consideration of 109 RTILs with various cations and anions showed little variation of the $10^4\alpha_P/K^{-1}$ values among them, ranging from 4.48 to 7.44, but being mostly within 6.5 ± 0.5. Much fewer κ_T/GPa^{-1} values were considered ranging from 0.33 to 0.73. Valderrama et al. [221] used group contributions to calculate primarily the critical temperature T_c and critical molar volume V_c and the normal boiling points T_b of RTILs. However, there being no experimental values for these quantities, they checked their calculations by the round-about estimation of the densities of 200 RTILs from the following relationship:

$$\rho/\mathrm{g\,cm^{-3}}(T) = (0.01256 + 0.9533M/V_c)\left[(0.0039/M + 0.2987/V_c)V_c^{1.033}\right]^{\psi} \tag{6.19}$$

where $\psi = -[(1 - T/T_c)/(1 - T_b/T_c)]^{2/7}$ with an overall absolute deviation of 5.9 %. The use of the not readily available nor accurately known critical properties for the estimation of room temperature quantities such as the density appears to be unprofitable.

The use of the microscopic ionic volumes v_+ and v_- of the constituents of RTILs, as described in Sect. 2.2 with values listed in Table 6.9, produced much better justifiable results. (Table 2.4 shows ionic volumes pertinent to high-temperature melting salts, most of which are not relevant for the RTILs.) The molar volume of the RTILs at 298 K, $V(298\ \mathrm{K}) = M/\rho(298)$, might be expected to be represented by $N_A (v_+ + v_-) : V/\mathrm{cm^3 mol^{-1}} = 602.2(v_+ + v_-)/\mathrm{nm^3}$. Ye and Schreve [259] noted that too low values of $V(298\ \mathrm{K})$ are thereby obtained and proposed to add 0.025 nm^3 to the v_+ of the cations of imidazolium, pyridinium, pyrrolidinium, and quaternary ammonium and phosphonium RTILs and to v_- of one anion, NTF_2^- (but not to others) to correct for this. Slattery et al. [286] proposed instead for the densities of RTILs with BF_4^-, PF_6^-, $N(CN)_2^-$ and NTF_2^- anions the expression:

$$\rho/\mathrm{g\,cm^{-3}} = 1.4684\left[M(\mathrm{g\,mol^{-1}})/\left[(v_+ + v_-)/\mathrm{nm^3}\right]\right]^{1.0026} \tag{6.20}$$

valid at 295–297 K. Preiss et al. [49] improved on this approach by allowing for the temperature dependence of the density, albeit by a logarithmic expression $\ln\rho = -\alpha_P T + \beta$ (contrary to the empirical linear dependence of ρ on T), where α_P and β are substance-dependent parameters. This was subsequently adopted by Beichel et al. [287] with the same two universal parameters (1.4684, 1.0026) and three substance-dependent ones: α_P, and β, and the zero-point vibrational energy of the RTIL.

When the microscopic ionic volumes are not available in Table 6.9 they may be substituted by the corresponding van der Waals volumes divided by the factor 0.6785 according to Beichel et al. [287]: $(v_+ + v_-) = (v_{+\mathrm{vdW}} + v_{-\mathrm{vdW}})/(0.6785 \pm 0.0055)$.

Table 6.9 Ionic volumes of the ionic constituents of RTILs in the first column: from Marcus [288] (2nd column) and as noted and their polarizabilities from Bica et al. [289]

Ion	v/nm^3			α_{pol}/nm^3[i]
$C_n\text{mim}^+$	$0.106 + 0.0230n$	$0.111 + 0.024n$[a]	$0.103 + 0.027n$[i]	$0.00920 + 0.00178n$
$C_4\text{mmim}$	0.220	0.275[b]		
$C_n\text{Py}^+$	$0.096 + 0.0239n$	$0.102 + 0.024n$[a]	0.198 ($n = 4$)[a]	$0.00915 + 0.00182n$[n]
$C_4(4\text{ M})\text{Py}^+$	0.221	0.221[a]		
$C_n(C_m)_3N^+$	$0.034 + 0.024(n + 3m)$			$0.00080 + 0.00186(n + 3m)$[n]
$C_n\text{MePyrr}^+$	$0.119 + 0.0234n$			$0.00962 + 0.00181n$[n]
BuMePyrr^+		0.238[c]	0.221[a], 0.253[h]	0.01684[n]
EtMePip^+		0.192[d]		
$C_n(C_m)_3P^+$	$0.133 + 0.021(n + 3m)$			
Bu_4P^+		0.499[e]		
TdHx_3P^+		0.835[a]		0.06225[n]
Cl^-	0.047	0.047[f, g]	0.033[i]	0.00345
Br^-	0.056	0.056[f, g]		0.00442[n]
I^-	0.072	0.072[g]		0.00673[n]
NO_3^-	0.064	0.064[f, g]		
SCN^-	0.071	0.071[g]		0.00632[n]
$CH_3CO_2^-$	0.078	0.086[h]	0.071[i]	0.00465
$CF_3CO_2^-$	0.103	0.108[f]	0.103[i]	0.00523
BF_4^-	0.079	0.073[f]	0.078[i]	0.00280
PF_6^-	0.111	0.105[a,j]		0.00418
$AlCl_4^-$	0.156	0.156[j]		
$FeCl_4^-$	0.155	0.155[j]		
$N(FSO_2)_2^-$	0.139	0.139[k]		
NTF_2^-	0.230	0.232[a]	0.246[i]	0.01359

(continued)

Table 6.9 (continued)

Ion	v/nm^3			$\alpha_{pol}/\mathrm{nm}^{3i}$
$N(CN)_2^-$	0.089	0.089[a]	0.069[i]	0.00611
$B(CN)_4^-$	0.156	0.141[l]		0.01013[n]
$CH_3SO_3^-$	0.099	0.099[f]	0.096[i]	0.00630
$CF_3SO_3^-$	0.126	0.130[a,f]	0.128[i]	0.00688
$MePhSO_3^-$	0.196	0.208[i]		0.01555
$C_nSO_4^-$	$0.080 + 0.027n$	$0.080 + 0.027n$[i]		$0.00515 + 0.00179n$
$PF_3(C_2F_5)_3^-$	0.334	0.383[m]		

[a][286] ($2 \leq n \leq 8$, $m = 1$ or 2), [b][296], [c][292], [d][295], [e][290] (but $MeBu_3N^+$ has $v = 0.388\ \mathrm{nm}^3$), [f][259], [g][291], [h][216], [i][289] (for $C_n$mim $2 \leq n \leq 10$ and for C_nSO_4 $n = 1,2,8$), [j][39], [k][293], [l][294], [m][264], [n]From Eq. (6.31) and the listed R_D values in Table 6.10

The estimation of the molar volumes and densities of RTILs can be considerably simplified, requiring only two parameters valid for all the RTILs considered in this book [288]. A proportionality was found, shown for 162 RTILs with a large variety of anions in Fig. 6.5, between $N_A(v_+ + v_-)$ and the experimental V(298 K), with a squared correlation coefficient of $r_{corr}^2 = 0.9895$. Therefore, a good estimation of V (298 K) is:

$$V_{est}/\text{cm}^3\text{mol}^{-1} = 689.5\left[(v_+ + v_-)/\text{nm}^3\right] \qquad (6.21)$$

with a standard deviation of 3.8 % for the resulting densities $\rho(T) = M/V(T)$.

Equation 6.21 was used as a training set and densities of a test set of 20 RTILs with anions having v_- obtained from crystal unit cell data and not from density fitting, namely $AlCl_4^-$, $FeCl_4^-$, $B(CN)_4^-$, and $N(FSO_2)_2^-$, were estimated, resulting in an average deviation of only 1.2 % with regard to the experimental values [288].

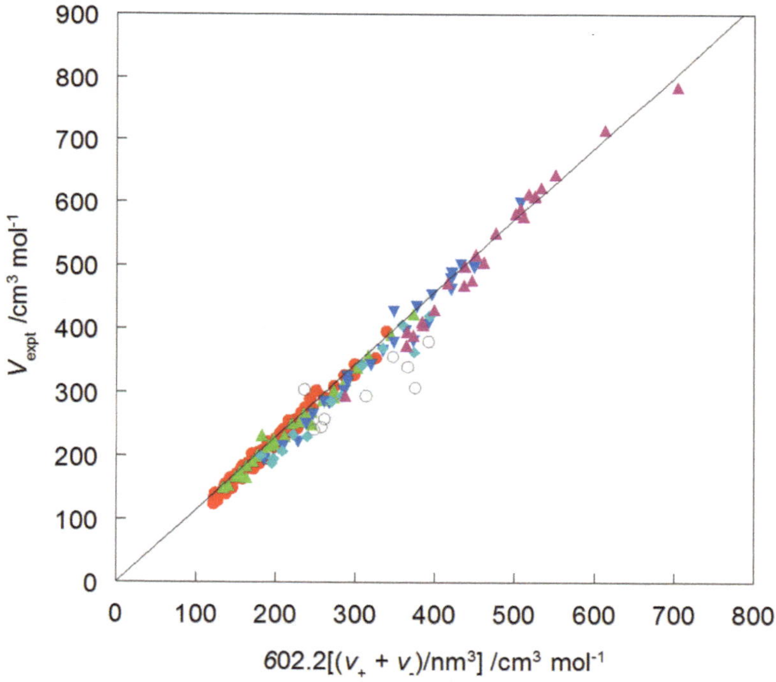

Fig. 6.5 A plot of the experimental molar volumes, V_{exp}, of 174 RTILs at 298.15 K against those calculated according to $V = N_A(v_+ + v_-)$ from the ionic volumes listed in Table 6.9: ● $C_n$mim$^+$ salts, ▲ C_nPy$^+$ salts, ▼ R_4N^+ salts, ◆ C_nMePyrr$^+$ salts, ■ R_4P^+ salts, and o are outliers (From Marcus [288] with permission from the publisher (Elsevier))

Following Gardas and Coutinho [297], the temperature dependence of the ionic volumes of RTILs at ambient pressures is estimated, as:

$$(v_+ + v_-)(T) = (v_+ + v_-)^0\left[1 + 0.666 \times 10^{-3}\{(T/\mathrm{K}) - 298.15\}\right] \quad (6.22)$$

where $(v_+ + v_-)^0$ represents the ionic values at the reference temperature $T = 298.15$ K listed in Table 6.9. Thus, with only two universal parameters, 689.5 cm^3 mol^{-1} nm^{-3} and 0.666×10^{-3}, the molar volumes $V(T)$ and densities $\rho(T) = M/V(T)$ can be well estimated up to at least 400 K at ambient pressures.

The discrepancy between $N_A(v_+ + v_-)$ and the experimental V(298 K) should be ascribed to the expansion of the RTIL on melting. The v_+ and v_- values in Table 6.9 were reported for unspecified temperatures, but should be construed as applying nominally for 298 K. There are only few data for the experimental densities of crystalline RTILs as functions of the temperature up to the melting point, hence the expansion on melting could be obtained for only 25 RTILs. The relative expansion data averaged as $9.9 \pm 3.3\,\%$, going a long way to explain [288] the discrepancy noted above. This expansion on melting results in the formation of cavities in the liquid salt, nominally represented by the void volume, V_v:

$$\begin{aligned} V_v(298\,\mathrm{K})/\mathrm{cm^3 mol^{-1}} &= (689.5 - 602.2)(v_+ + v_-)/\mathrm{nm^3} \\ &= 87.3(v_+ + v_-)/\mathrm{nm^3} \end{aligned} \quad (6.23)$$

The relationship between the surface tension σ of the RTIL (Tables 6.2, 6.3, and 6.4) and the average surface area of these cavities is dealt with in Sect. 6.2.

Several methods have been applied to calculate the void volume of RTILs (the fractional free volume, FFV), outstanding among these is the application of the COSMO-RS methodology by Palomar et al. and Shannon et al. [47, 298]. For 23 imidazolium RTILs (C_1mim to C_8mim) with a variety of anions a negative bias of $-1.6\,\%$ was found [47] between the COSMO-RS-calculated densities and the experimental ones, and a standard deviation of 2.6 % also resulted.

6.3.2 *Compressibilities and Internal Pressures*

The compressibilities of RTILs are available via two routes: high pressure density measurements, leading to the isothermal compressibilities κ_T and ultrasound speed measurements with ambient pressure densities, leading to the adiabatic compressibilities κ_S, which with some further data lead to the isothermal ones. The first route yields:

$$\kappa_T = \rho^{-1}(\partial\rho/\partial P)_T = V^{-1}(\partial V/\partial P)_T \quad (6.24)$$

In many cases the Tait expression is invoked for the pressure-dependence of the density:

$$\rho = \rho^* + C\ln[(B(T) + 0.1)/(B(T) + P)] \tag{6.25)}$$

where $\rho^* = \rho(298.15\ \text{K}, 0.1\ \text{MPa})$, C is a substance-dependent but temperature independent parameter, and $B(T) = B_0 + B_1T + B_2T^2 + \cdots$. Since $B(T)$ is of the order of several hundred MPa for RTILs, the derivative in Eq. (6.24) yields $\kappa_T(T) = C/B(T)$. The second route yields:

$$\kappa_S = \rho^{-1}u^{-2} \tag{6.26}$$

where u is the speed of ultrasound in the liquid RTIL. The conversion of κ_S values to κ_T ones:

$$\kappa_T = \kappa_S + V\alpha_p^2 T/C_P \tag{6.27}$$

requires the molar volume V, the isobaric expansibility α_P (from Tables 6.6, 6.7, and 6.8) and the isobaric molar heat capacity (from Tables 6.2, 6.3, and 6.4).

Perusal of Tables 6.6, 6.7, and 6.8 shows the available κ_T values to be of the order of 0.5 GPa^{-1}, with relatively little variation around this mean value. Grossly deviating values (by >0.2 GPa^{-1} on either side) are suspect as not being valid. In fact, considerable differences between values reported by various authors are common: for example, the values of κ_T/GPa^{-1} reported or deduced from the data for C_2mim^+ PF_6^- at 298.15 K are 0.378 [261], 0.394 [48], 0.418 [254], and 0.450 [299].

The product of the isothermal compressibility and the surface tension of RTILs has the dimension of length. When both quantities are available in Tables 6.2, 6.3, and 6.4 and Tables 6.6, 6.7, and 6.8 it is remarkably constant, namely $\kappa_T\sigma = 17.0 \pm 1.5$ nm for 29 items. Values that grossly depart from this product, say by >3 nm on either side, are suspected to have one of the factors wrong, probably the compressibility that is more difficult to measure. Thus, the value $\kappa_T = 0.733\ GPa^{-1}$ for C_2mim^+ $C_2H_5SO_4^-$ reported in [300] is excessively large. The constancy of the product $\kappa_T\sigma$ of RTILs contrasts with its clear dependence on the cohesive energy density, *ced*, of high-melting salts, for which $\kappa_T\sigma/\text{nm} = 14 + 0.276(ced/\text{GPa})$ as reported by Marcus [301].

The isothermal compressibility increases moderately with the chain-length of the alkyl groups for RTILs with a given anion, i. e., with the molar volumes. It increases in the series of anions $BF_4^- < PF_6^- < NTF_2^-$, but there are insufficient data to discern further trends.

A quantity closely related to the compressibility is the internal pressure as reviewed recently by Marcus [302]:

$$P_{\text{int}} = (\partial U/\partial V)_T = T(\partial P/\partial V)_T - P \approx T\alpha_P/\kappa_T \tag{6.28}$$

where U is the internal energy and the approximation is justified when the pressure (ambient or of the vapor) is small compared with the internal pressure. The values available for RTILs at ambient conditions (298.15 K and 0.1 MPa) are listed in Tables 6.6, 6.7, and 6.8. In view of the rather narrow ranges of the data for α_P and κ_T it is expected that also P_{int} values of RTILs have a narrow range, and indeed the available values for 48 RTILs are within 100 MPa of the mean value of 430 MPa. Values that are grossly outside this range, such as for $C_4Py^+BF_4^-$ (648 MPa), due to a small value of κ_T (250 GPa^{-1}), and for $PrMePip^+NTF_2^-$ (846 MPa), due to a large α_P (1.376×10^{-3} K^{-1}), suggests that these outlying values are incorrect.

The internal pressures of RTILs are commensurate with those of liquid organic substances on the one hand and are considerable smaller than those of high-temperature molten salts, the lowest value for such salts is 770 MPa at the melting point for KI [302].

6.4 Optical and Electric Properties

6.4.1 *Refractive Index and Molar Refractivity*

The refractive index of RTILs at the sodium D-line (589 nm), n_D, is available for a large number of the RTILs dealt with in this book. The data for 298.15 K are show in Table 6.10, rounded if necessary to four decimals, the variation between the reported data of diverse authors being often as large as ±0.0005. The values are within a narrow range: $1.37 \leq n_D \leq 1.57$, while RTILs with cyano-groups in the anion have $n_D > 1.5$ values. Special efforts are needed to obtain transparent and colorless RTILs with $n_D > 1.6$, using aromatic substituents, such as 1-benzyl-3-methylimidazolium bromide or triphenylhexylphosphonium dicyanamide, $n_D = 1.61$ according to Kayama et al. [303]. The refractive index of RTILs diminishes mildly and linearly with increasing temperatures:

$$n_D = a(n_D) - b(n_D)T \tag{6.29}$$

a typical slope being $b(n_D) \sim 2 \times 10^{-4} K^{-1}$.

Group contribution schemes have been proposed for the estimation of the refractive index of RTILs. Sattari et al. proposed for Eq. (6.29) that $a(n_D) = 1.5082 + \Sigma n_j a_j$ and $b(n_D) = -1.4207 \times 10^{-4} + \Sigma n_j b_j$ [330], where n_j is the number of functional groups of the j-th kind and a_j and b_j are their listed values.

The molar refractions of RTILs are given by the Lorentz-Lorenz expression:

$$R_D = V\left(n_D^2 - 1\right)/\left(n_D^2 + 2\right) \tag{6.30}$$

and have the dimension of a molar volume. If not reported by the authors that supplied the n_D data, then the R_D was calculated with the molar volume V data in

Table 6.10 The refractive index, the derived molar refractivity, and the static permittivity at 25 °C of RTILs

Salt	n_D	R_D/cm^3 mol^{-1}	ε_s
C_1mim NTF2	1.422[a]	61.4	
C_1mim CH_3SO_4	1.4827[b]	44.8	17.2[hh]
C_2mim BF_4	1.4142[c]	38.7	14.5[ff]
C_2mim NTF2	1.4225[d]	65.7	12.0[ee]
C_2mim $(SO_2F)_2N$	1.4475[pp]		
C_2mim CH_3SO_4	1.4813[tt]	52.5	
C_2mim $C_2H_5SO_4$	1.4794[e]	54.1	35.0[ee]
C_2mim $C_4H_9SO_4$			30.0[ii]
C_2mim $C_8H_{17}SO_4$	1.3869[f]	68.9	
C_2mim $CH_3C_6H_4SO_3$	1.5380[jj]	72.1	
C_2mim CH_3SO_3	1.4985[c]	48.8	
C_2mim CF_3SO_3	1.4332[a]	48.9	16.5[ee]
C_2mim SCN	1.5545[c]	48.6	
C_2mim HSO_4	1.4993[c]	44.7	18.4
C_2mim $N(CN)_2$	1.5160[c]	48.6	11.0
C_2mim $B(CN)_4$	1.4482[jj]	58.4	
C_2mim CH_3CO_2	1.5009[g]	43.8	
C_2mim CF_3CO_2	1.4405[a]	46.0	
C_2mim $PF_3(C_2F_5)_3$	1.3692[pp]		
C_3mim BF_4	1.4165[jj]	4301	
C_3mim NTF_2	1.3890[h]	47.9	13.3[ee]
C_3mim CH_3SO_4	1.4761[jj]	55.4	
C_3mim CH_3CO_3	1.4902[yy]	47.8	
C_4mim BF_4	1.4219[i]	47.8	13.9[ee]
C_4mim PF_6	1.4084[d]	51.5	14.0[ee]
C_4mim NTF2	1.4265[d]	74.9	14.0[ee]
C_4mim CH_3SO_4	1.4771[j]	58.5	14.8[hh]
C_4mim $C_8H_{17}SO_4$	1.3973[k]	78.8	
C_4mim CH_3SO_3	1.4792[jj]	53.4	
C_4mim CF_3SO_3	1.4376[c]	58.1	12.9[ee]
C_4mim Cl			15.0[hh]
C_4mim Br	1.5353[l]	52.9	
C_4mim I	1.5695[m]	58.8	
C_4mim SCN	1.5392[i]	57.8	13.7[ee]
C_4mim CH_3CO_2	1.4938[d]	54.1	
C_4mim $N(CN)_2$	1.5089[c]	57.9	11.3[ff]
C_5mim BF_4	1.4238[jj]	52.2	
C_5mim PF_6	1.4152[n]	55.4	
C_5mim NTF_2	1.4289[o]	79.5	15.0[ee]
C_5mim CH_3CO_2	1.4845[yy]	56.4	
C_6mim BF_4	1.4279[i]	57.1	12.0[ff]
C_6mim PF_6	1.4176[i]	60.8	10.1[gg]

(continued)

Table 6.10 (continued)

Salt	n_D	R_D/cm^3 mol^{-1}	ε_s
C_6mim NTF2	1.4302[c]	84.2	12.7[ff]
C_6mim $C_8H_{17}SO_4$	1.4700[xx]	98.4	
C_6mim CF_3SO_3	1.4393[c]	46.2	
C_6mim Cl	1.5095[p]	58.2	
C_6mim $N(CN)_2$	1.5042[c]	67.2	
C_6mim $B(CN)_4$	1.4531[qq]	77.0	
C_6mim CH_3CO_2	1.4829[yy]	60.9	
C_7mim PF_6	1.4203[rr]		
C_7mim NTF2	1.4317[o]	88.8	
C_8mim BF_4	1.4319[k]	65.9	7.5[hh]
C_8mim PF_6	1.4235[d]	70.3	8.8[gg]
C_8mim NTF2	1.4326[d]	93.6	6.5[hh]
C_8mim Cl	1.5062[jj]	67.9	7.0[hh]
C_9mim PF_6	1.4254[rr]		
C_{10}mim NTF_2	1.4356[d]	102.9	
C_{12}mim NTF_2	1.4376[d]	112.0	
C_1Py CH_3SO_4	1.5130[q]	45.9	
C_1(2 M)Py CH_3SO_4	1.5133[q]		
C_2Py $C_2H_5SO_4$	1.5053[jj]	56.7	
C_2Py NTF_2	1.4411[ss]	66.7	
C_2Py $N(SO_2F)_2$	1.4704[c]	61.3	
C_2Py $N(CN)_2$	1.5463[ww]		
C_2(2 M)Py NTF_2	1.4480[oo]		
C_2(3 M)Py NTF_2	1.4438[c]	66.6	
C_2(3 M)Py $C_2H_5SO_4$	1.5067[jj]		
C_2(3 M)Py NTF_2	1.4449[c]	66.7	
C_3Py BF_4	1.4445[s]	44.4	
C_3Py $N(CN)_2$	1.5407[ww]		
C_3(2 M)Py NTF_2	1.4492[oo]		
C_3(3 M)Py NTF_2	1.4449[c]	76.6	
C_4Py BF_4	1.4457[t]	49.0	
C_4Py $N(SO_2F)_2$	1.4712[c]	71.2	
C_4Py NTF_2	1.4438[u]	76.4	15.3[ee]
C_4Py CF_3SO_3	1.4583[v]	59.4	
C_4Py $N(CN)_2$	1.5345[ww]		
C_4(2 M)Py BF_4	1.4577	53.8[mm]	15.0[hh]
C_4(3 M)Py BF_4	1.4511[kk]	54.0	15.0[hh]
C_4(3 M)Py NTF_2	1.4460[c]	80.7	
C_4(3 M)Py CF_3SO_3	1.4616[jj]	63.4	
C_4(4 M)Py BF_4	1.4525[kk]	53.8	15.3[hh]
C_4(4 M)Py NTF_2	1.4465[jj]	81.0	
C_5Py $N(CN)_2$	1.5300[ww]		

(continued)

Table 6.10 (continued)

Salt	n_D	R_D/cm^3 mol^{-1}	ε_s
C_6Py $N(CN)_2$	1.5267[ww]		
C_6Py $N(SO_2F)_2$	1.4710[c]	80.8	
C_8(3 M)Py BF_4	1.4555[ll]	72.7[ll]	
C_8Py NTF_2	1.4472[u]	95.5	
C_{10}Py NTF_2	1.4486[u]	104.5	
C_{12}Py NTF_2	1.4497[u]	113.7	
$MeEt_3N$ CH_3SO_4	1.4644[nn]	53.6	
$MeBu_3N$ NTF_2	1.4264[w]	97.4	
$MeOc_3N$ NTF_2	1.4379[x]	157.3	
$PrMe_3N$ NTF_2	1.4049[uu]	64.9	
$BuMe_3N$ NTF_2	1.4082[w]	70.3	15.7[ee]
$PeMe_3N$ NTF_2	1.4108[uu]		
$PeEt_3N$ NTF_2			12.5[ii]
$HxMe_3N$ NTF_2	1.4131[uu]	80.8	
$HxEt_3N$ NTF_2	1.4260[y]	93.4	
$HxBu_3N$ NTF_2	1.4338[x]	124.5	
$HpEt_3N$ NTF_2	1.4271[y]	97.1	
$OcMe_3N$ NTF_2	1.4174[uu]		
$OcEt_3N$ NTF_2	1.4287[y]	102.0	
$DcEt_3N$ NTF_2	1.4317[y]	105.4	
$DoEt_3N$ NTF_2	1.4341[y]	120.7	
$TdEt_3N$ NTF_2	1.4359[y]	129.8	
EtMePyrr $N(CN)_2$			14.0[ff]
EtMePyrr $C_2H_5SO_4$	1.4728[nn]	56.2	
PrMePyrr $N(FSO_2)_2$	1.4452[c]	62.9	
PrMePyrr NTF_2	1.4233[c]	72.9	9.1[hh]
BuMePyrr CF_3SO_3	1.4329[c]	60.4	
BuMePyrr NTF_2	1.4457[v]	80.7	14.7[ee]
BuMePyrr $N(CN)_2$	1.4469[z]	54.5	18.0[ee]
BuMePyrr CH_3SO_4	1.4667[nn]	60.3	
PeMePyrr NTF_2			12.5[ii]
HxMePyrr NTF_2	1.425[dd]	87.4[dd]	
DcMePyrr NTF_2	1.431[dd]	105.1[dd]	
PrMePip NTF_2	1.4274[v]	76.9	
BuMePip NTF_2	1.4293[v]	81.6	
$MeBu_3P$ CH_3SO_4	1.4763[vv]		
$MeiBu_3P$ CH_3PhSO_3	1.5203[vv]		
$EtBu_3P$ $(C_2H_5)_2PO_4$	1.4662[vv]		
$OcBu_3P$ Cl	1.4939[jj]		
$TdHx_3P$ Cl	1.4831[aa]	166.4	
$TdHx_3P$ Br	1.4898[aa]	170.7	
$TdHx_3P$ NTF_2	1.4496[d]	192.6	

(continued)

Table 6.10 (continued)

Salt	n_D	R_D/cm^3 mol^{-1}	ε_s
$TdHx_3P\ N(CN)_2$	1.4834[bb]	174.7	
$TdHx_3P\ CH_3CO_2$	1.4818[d]	173.6	
$TdHx_3P\ CH_3SO_3$	1.4735[aa]	175.2	
$TdHx_3P\ CF_3SO_3$	1.4577[d]	175.7	

[a][59], [b][304], [c][284], [d][274], [e][305], [f][306], [g][76], [h][307], [i][308], [j][105], [k][309], [l][251], [m][310], [n][311], [o][312], [p][313], [q][263], [r][314], [s][315], [t][316], [u][117], [v][134], [w][218], [x][312], [y][280], [z][77], [aa][317], [bb][273], [dd][204], [ee][318], [ff][319], [gg][320], [hh][185], [ii][321], [jj][56], [kk][322], [ll][323], [mm][115], [nn][277], extrapolated to 298.15 K, [oo][113], [pp][266], [qq][256], [rr][324], [ss][325], [tt][186], [uu][326] suppl. Information, [vv][327], [ww] [231], [xx][328], [yy][329]

Tables 6.6, 6.7, and 6.8. The polarizability of an RTIL is proportional to its molar refraction:

$$\alpha = (3/4\pi N_A)R_\infty \approx 3.92 \times 10^{-31}\left(R_D/\text{cm}^3\text{mol}^{-1}\right) \quad (6.31)$$

The approximation is due to the use of R_D instead of the infinite frequency value, R_∞. The polarizabilities of imidazolium-based RTILs are additive in terms of the constituting ions, according to Bica et al. [289] and their values are shown in Table 6.9. The list provided by these authors is supplemented in the latter Table by the values obtained from Eq. (6.31) and the listed R_D values in Table 6.10. The cation polarizabilities increase with the lengths of the alkyl chains for imidazolium, pyridinium, 1-methylpyrrolidinium, and quaternary ammonium (RR'_3N^+) RTILs with a variety of anions rather uniformly by 0.0018 nm^3 per methylene group, but in the case of the imidazolium RTILs some anions (dicyanamide and alkylsulfates) were said by Bica et al. [289] to exhibit diminishing cation polarizabilities with increasing alkyl chain length, for no obvious reason.

6.4.2 *Static Permittivity*

The static permittivity (dielectric constant), ε_s, of RTILs cannot be measured directly, because these substances are highly conducting (Sect. 6.5). Therefore, the dielectric response is measured as a function of the frequency, and from these data the static permittivity is then derived. This path is not quite straight-forward, and therefore there are large discrepancies between the values reported by diverse authors. The values shown in Table 6.10 may, therefore be incorrect by as much as two units on either side. Even the trends with increasing lengths of the alkyl chains, whether ε_s increases or diminishes, are not clear, because authors differ on this point. For example, the ε_s of $C_n\text{mim}^+\text{NTF}_2^-$ RTILs show the following changes with the number of methylene groups in the 1-alkyl chain, Δn, from the value for the $C_2mim^+NTF_2^-$ salt: 12.25–0.27Δn [331], 11.5–.83Δn [185], 12.3 + 0.75Δn [332], 12.0 + 1.00Δn [318], 12.0 + 1.5Δn [321]. On the whole, values of the static permittivity in the range $9 \leq \varepsilon_s \leq 16$ are to be expected for RTILs at ambient conditions. Values grossly outside this range (e.g., the values for C_2mim^+ ethylsulfate, 35.0, and butylsulfate, 30.0 reported by Weingärtner [321] without comment on their large sizes) are unlikely to be valid.

6.5 Transport Properties

6.5.1 Viscosity

The viscosity of RTILs, which is in general considerably larger than that of ordinary organic solvents, is an important property that needs to be known for the many applications foreseen for these substances as 'green' solvents. The viscosity of RTILs diminishes rapidly with increasing temperatures. The temperature dependence of the dynamic viscosity, η, of RTILs generally does not follow the simple Arrhenius dependence but rather the Vogel-Fulcher-Tammann (VFT) one:

$$\eta = A_\eta \exp\left[B_\eta / \left(T - T_{0\eta}\right)\right] \tag{6.32}$$

The values of the coefficients A_η and B_η and of the temperature $T_{0\eta}$ of RTILs are shown in Tables 6.11, 6.12, and 6.13. Also shown in these Tables is the viscosity at 298.15 K, if the RTIL is liquid at this temperature. In those cases where the authors did not report the values of these parameters, they were calculated from the reported $\eta(T)$ data on the basis of arbitrarily setting $T_{0\eta} = 180$ K. This value is a fair average representative of the actually reported values, but the fitting is not very sensitive to its actual size. When the temperature of an RTIL is decreased below its melting temperature T_m to form a super-cooled liquid the viscosity increases up to near infinite values as T approaches $T_{0\eta}$, which may be construed as the glass transition temperature.

The viscosity is very sensitive to the presence of water in the RTILs, which diminishes it, and discrepancies between the reported values by diverse authors may be due to this factor. For example, the values of η(298 K)/(mPa s) for the dry and water-equilibrated PF_6^- salt of C_4mim^+ are 267 [252], 450 [70] or 397 [70], of C_6mim^+ they are 479 [252] or 585 [70] and 472 [70], and of C_8mim^+ they are 597 [252], 682 [70] or [357] or 506 [70]. Similar observations of the different viscosities of various dry and water-saturated salts of the $TdHx_3P^+$ cation were made by Neves et al. [279]. Many other cases of large discrepancies in the reported data are shown in Tables 6.11, 6.12, and 6.13.

Several expressions have been proposed for the temperature dependence of the viscosity of RTILs alternative to Eq. (6.32). The Litovitz expression, $\eta = A_\eta \exp\left(B_\eta / T^3\right)$ is empirical, and the third power of T does not have a physical significance.

Table 6.11 The viscosity at 25 °C and the VFT parameters of the viscosity of 1-alkyl-3-methylimidazolium ($C_n$mim) salts

Salt	η/mPa s	A_η/mPa s	B_η/K	$T_{0\eta}$/K
C_1mim NTF2	32.9[a], 34[b]	0.215	709.7	157.1
C_1mim CH_3SO_4	72.6[c]	0.2746	659.2	180
C_1mim $(CH_3)_2PO_4$	290.8[d]	0.1104	929.4	180
C_2mim BF_4	52.3[e], 61.7[f]	0.3185	602.6	180
C_2mim PF_6	15[g]			
C_2mim NTF2	34.0[h]	0.4397	513.7	180
C_2mim CH_3SO_4	78.8[i]	0.1917	801.2	165.1
C_2mim $C_2H_5SO_4$	101.3[h]	0.2547	707.2	180
C_2mim $C_8H_{17}SO_4$	601.6[i]	0.0933	1301.2	165.1
C_2mim CH_3SO_3	166.6[j]	0.0901	890.7	180
C_2mim CF_3SO_3	40.5[e], 30.7[f]	0.6122	595.4	180
C_2mim Cl	3530[k]	0.0124	1482	180
C_2mim SCN	22.2[d]	0.3490	490.4	180
C_2mim NO_3	60.7[e], 95.3[f]	0.3804	599.3	180
C_2mim HSO_4	1628.4[i]	0.0726	1200.5	165.1
C_2mim $N(CN)_2$	16.1[j], 21[b]	0.4712	416.4	180
C_2mim CH_3CO_2	143.6[j]	0.0624	916.4	180
C_2mim $FeCl_4$	18[l]			
C_3mim BF_4	74.0[m]	0.2223	686.3	180
C_3mim NTF2	43.7[n]	0.2701	601.3	180
C_4mim BF_4	114.9[e], 105.1[o]	0.2533	722.8	180
C_4mim PF_6	267.8[e]	0.2769	812.2	180
C_4mim NTF2	49.0[h]	0.3309	590.4	180
C_4mim CH_3SO_4	216.1[c]	0.1847	834.0	180
C_4mim $C_8H_{17}SO_4$	643.4[c]	0.2431	929.2	180
C_4mim CF_3SO_3	77.3[e]	0.4081	619.5	180
C_4mim Cl	18920[e]	0.0110	1696.4	180
C_4mim Br	1486.5[p]			
C_4mim I	647.4[q]	0.1115	946	189
C_4mim SCN	51.7[r]			
C_4mim NO_3	190.0[e], 166.4[f]	0.1430	849.7	180
C_4mim HSO_4	439.4[s]	0.1713	927.4	180
C_4mim CH_3CO_2	433.2[c]	0.0654	1040.6	180
C_4mim CF_3CO_2	68.6[c]	0.2065	684.8	180
C_4mim $FeCl_4$	41.0[t]	0.2315	682.1	166.3
C_4mim $N(CN)_2$	28.8[u]	0.0488[k]	935	180
C_5mim BF_4	158.1[f]			
C_5mim NTF_2	62.3[a]	0.156	790.6	166.2
C_5mim NO_3	257.6[f]			
C_6mim BF_4	227.2[e], 257.2[f]	0.2117	824.5	180
C_6mim PF_6	478.5[e]	0.2190	908.5	180
C_6mim NTF2	69.3[f, w]	0.2298	674.6	180

(continued)

Table 6.11 (continued)

Salt	η/mPa s	A_η/mPa s	B_η/K	$T_{0\eta}$/K
C_6mim CF_3SO_3	153.7[w]	0.1741	800.4	180
C_6mim Cl	10160[e]	0.0353	1485.2	180
C_6mim Br	3650[w]	0.0574	1306.7	180
C_6mim I	1432.4[q]	0.1117	1023	190
C_6mim NO_3	512.5[e], 363.8[f]	0.0745	1044.0	180
C_6mim $FeCl_4$	45[l]			
C_7mim NTF2	81.2[a]	0.119	883.5	162.8
C_8mim BF_4	313.7[e]	0.2141	861.7	180
C_8mim PF_6	596.8[e]	0.2188	934.7	180
C_8mim NTF2	90.2[c], 95.0[u]	0.2098	716.0	180
C_8mim CF_3SO_3	349.4[e]	0.2969	835.4	180
C_8mim Cl	18070[e]	0.0464	1520.9	180
C_8mim I	1423.4[q]	0.0320	1500	158
C_8mim NO_3	789.2[e]	0.0831	1082.1	180
C_8mim $N(CN)_2$	34[x]			
C_8mim $FeCl_4$	77[l]			
C_{10}mim BF_4	623.5[e]	0.1527	982.4	180
C_{10}mim PF_6	1303[e]	0.1000	1119.5	180
C_{10}mim NTF_2	119.2[a]	0.096	978.1	160.9
C_{10}mim CF_3SO_3	675.8[e]	0.2673	925.7	180
C_{10}mim SCN	683.7[u]			
C_{10}mim CH_3SO_4	71.7[s]	0.2999	646.5	180
C_{12}mim NTF_2	155.0[a]	0.087	1034.5	160.0
C_{14}mim NTF_2	206.3[a]	0.092	1040	163.4

[a][344], [b][153], [c][138], [d]]266], [e][252], [f][57] at 297 K, [g][338] at 353 K, the salt melts at 335 K, [h][272], [i][335], [j][76], [k][341] based on only three data at $T \geq 353$ K), [l][68], [m][345], [n][343], [o][336], [p][282], [q][334], [r][342], [s][346], [t][248], [u][333], [v][266], [w][340], [x][339], [y][337]

The expression $\Phi^{0.3} = a + bT$, where $\Phi = (1/\eta)$ is the fluidity of the RTIL, was suggested by Ghatee et al. [334, 358] with two substance-related parameters, and fits the data well. A mode coupling theory power law, $\eta = \eta_0[(T - T_x)/T_x]^{-\gamma}$ with three substance-related parameters has subsequently been proposed by Ghatee and Zare [138]. Here T_x is the crossover temperature at which the behavior of the fluid changes from fragile, according to Xu et al. [357], above it to strong, i.e., Arrhenius-type, below it. In fact, the later power law expression of the viscosity is readily transformed into the linear expression for the 0.3 power of the fluidity, $\Phi^{0.3}$, noted above [138]. The crossover temperatures T_x obtained from the viscosities of 43 RTILs of various kinds cover a narrow range: $223 \leq T_x/\mathrm{K} \leq 252$ (two 1-benzyl-3-imidazolium salts being outliers) and are between the glass-transition and melting temperatures of the RTILs.

Table 6.12 The viscosity at 25 °C and the VTF parameters of 1-alkylpyridinium salts

Salt	η/mPa s	A_η/mPa s	B_η/K	$T_{0\eta}$/K
$C_1Py\ CH_3SO_4$	116.2[a]	0.2072	771.0	175.8
$C_2Py\ NTF_2$	39.8[b]	0.010	1935	64.9
$C_2Py\ C_2H_5SO_4$	126.3[c]	0.1579	785.8	180
C_2(3 M)Py NTF_2	54.9[d]	0.000143	782.0	165.1
C_2(3 M)Py $C_2H_5SO_4$	150[e]	0.1538	807.1	180
C_2(4 M)Py NTF_2	32.8[f]	0.1768	721.8	160.2
$C_3Py\ BF_4$	119.5[g]	0.1409	804.2	178.8
C_3(3 M)Py NTF_2	55.8[h]	0.206	689	175
$C_4Py\ BF_4$	163.3[i], 160.3[j]	0.00253[i]	1086	166.2
$C_4Py\ NTF_2$	60.6[b]	0.137	808.2	165.2
$C_4Py\ CF_3SO_3$	126.0[j]	0.1358	867.8	171.2
C_4(2 M)Py BF_4	389.4[k]	0.0822	999.1	180
C_4(2 M)Py NTF2	95.7[b]	0.594	412.9	217.6
C_4(3 M)Py BF_4	169.5[b], 63.2[l]	0.3033[l]	628.7	180
C_4(3 M)Py CF_3SO_3	126.6[b]	0.001	3542	16.4
C_4(3 M)Py NTF_2	64.3[b]	0.446	528.7	191.7
C_4(3 M)Py $N(CN)_2$	35.6[z]	0.1881	619.6	180
C_4(4 M)Py BF_4	201.2[m], 237.0[n]	0.0971	901.1	180
C_4(4 M)Py NTF_2	55.1[f], 54.4[b]	0.0859	915.6	156.6
$C_6Py\ NTF_2$	96.2[o], 245.3[b]	0.264[r]	1098	136.1
C_6(3 M)Py NTF_2	853[z], 85[e]	0.3962	686.7	180
$C_8Py\ BF_4$	233.5[i]	0.00630[i]	949.7	174.2
$C_8Py\ NTF_2$	77.1[p], 114.3[q]	0.1478	747	180
C_8(3 M)Py BF_4	332[p], 505.5[r]	0.0892	970	180
C_8(4 M)Py BF_4	460[p]	0.0472	1092	180
C_8(4 M)Py NTF_2	105[p], 115.6[d]	0.1947	746	180
C_8(4 M)Py CF_3SO_3	365[p]	0.1289	939	180
$C_{10}Py\ NTF_2$	151.9[q]			
$C_{12}Py\ NTF_2$	200.5[q]			

[a][263], [b][132], [c][350], [d][348], [e][60], [f][116], [g][125], [h][134], [i][242], [j][123], [k][115], [l][347], [m][349], [n][351], [o][62], [p][120], [q][117], [r][323], [e][261]

A different approach relates the fluidity of RTILs to their molar volumes over a large temperature range, based on the Hildebrand-Lamoreaux expression [359], in analogy with its use for molten salts (Sect. 3.4) and many other fluids:

$$\Phi(T) = B[V(T) - V_0]/V_0 = -B + (B/V_0)V(T) \tag{6.33}$$

This expression is applicable to RTILs according to Marcus [360], provided the temperature is sufficient for the viscosity to be below 50 mPa s, because when the viscosity is larger (the fluidity is smaller than 0.02 s mPa^{-1}) the linearity of $\Phi(T)$

Table 6.13 The viscosity at 25 °C and the VTF parameters of the viscosity of quaternary ammonium and phosphonium salts

Salt	η/mPa s	A_η/mPa s	B_η/K	$T_{0\eta}$/K
$MeEt_3N\ CH_3SO_4$	409.3[a]	0.1309	949.6	180
$MeBu_3N\ NTF_2$	544.5[b]			
$MeBu_3N\ N(CN)_2$	410[c]			
$MeOc_3N\ CF_3CO_2$	1708[d]			
$PrMe_3N\ NTF_2$	77.7[hh]	0.2989	657.0	180
$BuMe_3N\ NTF_2$	99.0[e], 110.9	0.45	534	199
$PeEt_3N\ NTF_2$	161.6[f]	0.1308	839.9	180
$PeEt_3N\ N(CN)_2$	121[c]			
$HxMe_3N\ NTF_2$	192[g]	0.2777	771.2	180
$HxEt_3N\ NTF_2$	186.6[f]	0.1294	857.9	180
$OcEt_3N\ NTF_2$	221.4[f]	0.1375	871.5	180
$OcEt_3N\ N(CN)_2$	241[c]			
$DcEt_3N\ NTF_2$	281.9[f]	0.1031	934.1	180
$DoEt_3N\ NTF_2$	311.8[f]	0.1244	924.3	180
$EtMePyrr\ C_2H_5SO_4$	301.4[a]	0.2420	841.0	180
$PrMePyrr\ NTF_2$	54[h]	0.1991	785.7	158.1
$PrMePyrr\ N(CN)_2$	45			
$BuMePyrr\ CF_3SO_3$	30.0[i]	1.3810	407.9	165.6
$BuMePyrr\ CH_3SO_4$	467.5[a]	0.1083	988.3	180
$BuMePyrr\ N(FSO_2)_2$	53.2[j]	0.1823	905.2	138.7
$BuMePyrr\ NTF_2$	75.7[e], 110.9[k]	0.29	651	181
$BuMePyrr\ N(CN)_2$	34.7[l], 50[m]	0.232	737.6	150.9
$BuMePyrr\ PF_3(C_2F_5)_3$	33.3[i]	1.3765	388.6	176.2
$BuEtPyrr\ C_2H_5SO_4$	611.6[a]	0.1012	1027.4	180
$HxMePyrr\ NTF_2$	96[h]	0.1503	873.8	163.0
HxMePyrr N(CN)2	45[m]			
$DcMePyrr\ NTF_2$	150[h]	0.1038	1010.6	159.3
$MeBu_3P\ NTF_2$	207[n]			
$MeBu_3P\ N(CN)_2$	167[c]			
$MeBu_3P\ CH_3SO_4$	509[o]	0.00779	1862.6	130
$EtBu_3P\ (C_2H_5)_2PO_4$	306[o]	0.00388	1894.8	130
$PrOc_3P\ Cl$	2499[p]	0.00646	2095.6	135.1
$BuEt_3P\ N(CN)_2$	60[c]			
$BuEt_3P\ N(SO_2F)_2$	62[q]			
$BuOc_3P\ Cl$	2494[p]	0.00652	2095.6	135.1
$PeEt_3P\ N(SO_2F)_2$	70[q]			
$PeEt_3P\ NTF_2$	88[q]			
$PeEt_3P\ N(CN)_2$	72[c]			
$HxOc_3P\ Cl$	2164[p]	0.00781	2095.4	131.0
$OcEt_3P\ N(CN)_2$	104[c]			
$OcBu_3P\ BF_4$	1240[n]			
$OcBu_3P\ PF_6$	1720[n]			

(continued)

Table 6.13 (continued)

Salt	η/mPa s	A_η/mPa s	B_η/K	$T_{0\eta}$/K
$OcBu_3P\ NTF_2$	250[n]			
$OcBu_3P\ Cl$	4670[o]	0.00050	2697.9	130
$OcBu_3P\ SCN$	450[n]			
$OcBu_3P\ N(CN)_2$	245[c]			
$OcBu_3P\ CF_3CO_2$	453[n]			
$OcBu_3P\ CF_3SO_3$	778[n]			
$OcBu_3P\ MePhSO_3$	2435[n]			
$DoBu_3P\ BF_4$	1310[n]			
$DoBu_3P\ NTF_2$	303[n]			
$TdHx_3P\ Cl$	2729[r]	0.00121	1999.8	135.9
$TdHx_3P\ Br$	2988[r]	0.00125	1999.8	136.6
$TdHx_3P\ NTF_2$	336.7[r]	0.0304	1498.6	137.2
$TdHx_3P\ N(CN)_2$	438.6[r]	0.0249	1624.8	131.9
$TdHx_3P\ CH_3SO_3$	1379[r]	0.0133	1800.2	142.2
$TdHx_3P\ PF_3(C_2F_5)_3$	8.23[i], 336[s]	1.0869	616.9	143.1
$TdHx_3P\ FeCl_4$	1349[t]	0.0208	1786.1	136.5

[a][277] (for $MeEt_3N\ CH_3SO_4$ and BuEtPyrr $C_2H_5SO_4$ extrapolated to 298 K), [b][164], [c][169], [d][333], [e][352], [f][161], [g][354], [h][209], [i][276], [j][280], [k][272], [l][353], [m][153] at 293 K, [n][167], [o][356], [p][275], [q][170], [r][279], [s][355], [t][248]

with *V*(*T*) breaks down, Fig. 6.6. The volume V_0 is construed to mean that molar volume, at which the particles are so closely crowded together as to prevent viscous flow while maintaining rotational freedom [359]. The dangling alkyl chains of the RTIL cations prevent their free rotation at relatively low temperatures, hence the need for the sufficiently high temperature for Eq. (6.33) to be valid. The values of *B* and V_0 for some 120 RTILs of all kinds have been reported by Marcus [360], the V_0 being ~99 % of the molar volume at 298 K. The products BV_0 for series of imidazolium RTILs with given anions are approximately constant: 350 for BF_4^-, 250 for PF_6^-, and 670 for the NTF_2^- salts. The *B* coefficients tend to increase with increasing cohesive energies of the RTILs (as they do for molten salts), i. e., with the tightness of the electrostatic binding of the ions.

The pressure dependence of the viscosity of RTILs was studied by Harris et al. [361–363]. They applied the VFT expression (6.32) and reported the pressure dependences of the coefficients: $A_\eta = \exp(a' + b'P)$ and $B_\eta = c' + d'P + e'P^2$, leaving T_0 to remain pressure-independent. The viscosity increases quadratically with increasing pressures: for $C_8mim^+BF_4^-$, for example, $\eta/(\mathrm{mPa\,s}) = 344 + 3.77(P/\mathrm{MPa}) + 0.0403(P/\mathrm{MPa})^2$. Harris et al. reported values of the coefficients a', b', c', d', and e' and of T_0 for several imidazolium salts: $C_4mim^+PF_6^-$, $C_8mim^+BF_4^-$, and $C_8mim^+PF_6^-$ [361], $C_4mim^+BF_4^-$ [362], and $C_6mim^+PF_6^-$ and $C_4mim^+NTF_2^-$ [363].

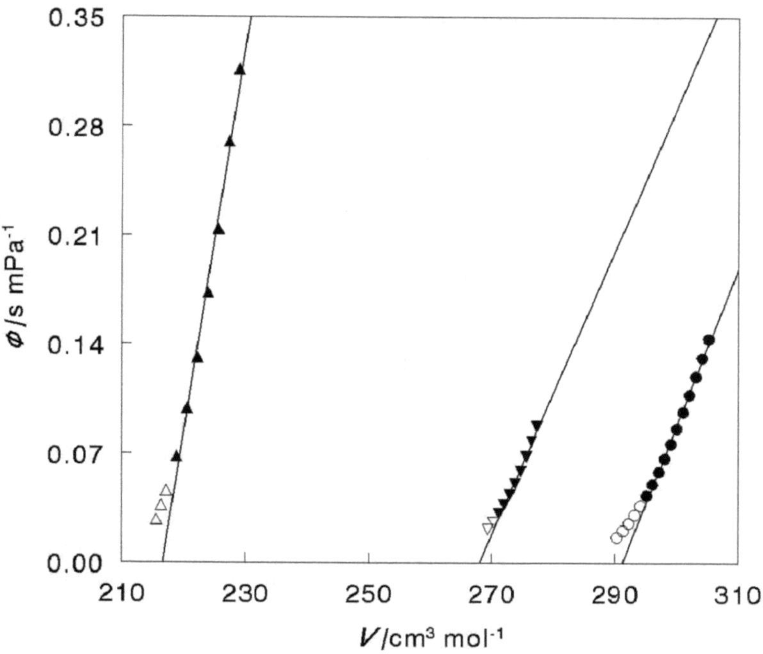

Fig. 6.6 Plots of the fluidity, Φ, against the molar volume, V, at a range of temperatures for Etmim $^+$ $B(CN)_4^-$ (▲), $Bumim^+$ NTF_2^- (▼), and Et(3Me)Py (●). Filled symbols conform to a straight line with a linear correlation coefficient ≥0.995, empty symbols do not, the viscosity being too large (From Marcus [360] by permission of the publisher (Elsevier))

Self-diffusion coefficients could be determined separately for cations and anions of some RTILs by the pulsed-gradient spin-echo NMR signals of ^1H and ^{19}F according to Noda et al. [364]. The authors applied this method to RTILs with the C_2mim^+ and C_4Py^+ cations and the BF_4^- and NTF_2^- anions, the self-diffusion coefficients obeying a VFT-type expression $D = A_D \exp[B_D/(T - T_{0\eta})]$. The Stokes-Einstein relationship:

$$D = k_B T / c\pi\eta r_{hd} \tag{6.34}$$

is obeyed by the individual ionic self-diffusion coefficients, but the constant c (between four for slip- and six for stick-conditions) and the hydrodynamic radii r_{hd} have not been specified. Individual self-diffusion coefficients for cations and anions can, of course, be obtained from computer simulations as carried out by Tsuzuki et al. [365]. The calculations for RTILs at 353 K did not agree with the experimental values at this temperature, but those calculated at 453 K did agree for $C_n mim^+$ NTF_2^- ($n = 1, 2. 4. 6, 8$) and C_4mim^+ with the anions $CF_3SO_3^-$, $CF_3CO_2^-$, BF_4^-, and PF_6^-.

6.5.2 Electrical Conductivity

The values of the molar electrical conductivity Λ of RTILs result from the direct determination of the specific conductivity κ and its multiplication by the molar volume V as $\Lambda = \kappa V$. On the whole, the specific conductivities of imidazolium RTILs are larger than those based on pyridinium, pyrrolidinium, and acyclic quaternary ammonium cations, but all are considerably lower than those of concentrated aqueous electrolytes used in batteries, as shown by Galinski et al. [156]. The values of Λ at 298 K are shown in Table 6.14 as are the parameters A_Λ, B_Λ, and $T_{0\Lambda}$ of the VFT expression:

$$\Lambda = A_\Lambda \exp[B_\Lambda/(T - T_{0\Lambda})] \tag{6.35}$$

It should be noted that contrary to the positive values of B_η for the viscosity (resulting from viscosities diminishing with increasing temperatures) the B_Λ values are negative, i.e., the conductivity increases with increasing temperatures. In those cases where the authors did not report the values of these parameters, they were calculated from the reported $\Lambda(T)$ data on the basis of arbitrarily setting $T_{0\Lambda} = 180$ K. This value is a fair average representative of the actually reported values, but the fitting is not very sensitive to its actual size. Insufficient $\Lambda(T)$ data for quaternary phosphonium RTILs were available for showing their VFT parameters.

The molar conductivity may be determined, in addition to its direct determination from the specific conductivity κ and the molar volume V (as $\Lambda = \kappa V$), also from the ionic self-diffusion coefficients, obtained from NMR measurements as mentioned above, according to the Nernst-Einstein equation $\Lambda = F^2(D_+ + D_-)/RT$, with results in fair agreement with the directly determined values according to Tokuda et al. [80, 352, 374]. Ionic self-diffusion coefficients can be estimated from molecular dynamics simulations and can then be transformed into the molar conductivity, as applied in [365, 387, 388], showing fair agreement with experimental values..

The pressure dependence of the molar conductivity can be expressed in a manner similar to that proposed for the viscosity, namely, allowing for the pressure dependence of the coefficients of the VFT expression: $A_\Lambda = \exp(a' + b'P)$ and $B_\Lambda = c' + d'P + e'P^2$, leaving $T_{0\Lambda}$ unchanged [371, 386, 389, 390]. The pressure and temperature dependences can also be expressed according to Lopez et al. [391] by a density-scaling equation:

$$\Lambda(T, V) = \Lambda_0 \exp\left[\left(A'_\Lambda/TV^\gamma\right)^\varphi\right] \tag{6.36}$$

with four substance-specific parameters: Λ_0, $A_\Lambda{}'$, γ, and φ, the latter two being similar to those for the corresponding equation for the viscosities, being generally between 2 and 3.

Table 6.14 The molar conductivity, Λ, at 25 °C and the VFT parameters of the molar conductivities of RTILs

Salt	Λ/S cm^2 mol^{-1}	A_Λ/S cm^2 mol^{-1}	B_Λ/K	$T_{0\Lambda}$/K
C_1mim NTF2	0.79[w], 0.212[gg]	12.50	−481.1	180
C_1mim I	0.532[j]			
C_2mim BF_4	2.00[n], 2.430[p]	95.4	−431.4	180
C_2mim PF_6	0.91[n]	97.2[p]	−569.3	180
C_2mim NTF2	2.15[n], 2.347[c]	190[jj]	−604	161
C_2mim $C_2H_5SO_4$	0.729[p]	194.1	−657.8	180
C_2mim CF_3SO_3	1.451[q], 0.167[ff]	9.78	−480.0	180
C_2mim Br	0.180[p]	380.4	−903.8	180
C_2mim I	0.678[j]			
C_2mim SCN	1.03[bb]			
C_2mim $N(CN)_2$	4.72[x]	213.4	−523.6	160.8
C_2mim $FeCl_4$	4.4[pp]			
C_3mim BF_4	1.06 [hh]			
C_3mim NTF2	1.459[dd]			
C_3mim I	0.471[j]			
C_4mim BF_4	0.824[p]	144.7	−615.0	180
C_4mim PF_6	0.31[u], 0.308[y]	380	−897	172
C_4mim NTF2	1.178[c], 1.096[l]	150	−605	175
C_4mim CF_3SO_3	0.82[u], 0.636[y]	270	−841	159
C_4mim I	0.472[j]			
C_4mim CF_3CO_2	0.37[w], 0.631[y]	230	−761	169
C_4mim Al_2Cl_7	3.012[k]	352.7	−807.3	128.4
C_4mim $FeCl_4$	2.2[pp]			
C_4mim $N(CN)_2$	0.221[v]	166.5[mm]	−517.4[mm]	180
C_4mim CF_3CO_2	0.692[mm]	153.9[mm]	−636.7[mm]	180
C_5mim NTF_2	0.917[aa]	6.83	−167.2	214.9
C_6mim BF_4	0.398[p], 0.267[ee]	270.5	−816.8	180
C_6mim PF_6	0.131[ee], 0.129[u]	548.9	−1086.9	168.0
C_6mim $PF_3(C_2H_5)_3$	0.518[q]	31.13[ll]	−666.8[ll]	180
C_6mim NTF2	0.40[w],0.072[gg]	13.93	−621.0	180
C_6mim Cl	0.0035[ee]	3675.4	−1635.1	180
C_6mim Al_2Cl_7	1.893[k]	180.0	−581.2	170.6
C_6mim $FeCl_4$	1.3[pp]			
C_7mim NTF2	0.558[dd]			
C_8mim BF_4	0.276[p], 0.155[ee]	246.9	−869.7	180
C_8mim PF_6	0.150[r]	444.0	−1072.7	163.8
C_8mim NTF2	0.29[w], 0.039[gg]	11.16	−668.7	180
C_8mim Cl	0.0028[ee]	2062.2	−1593.1	180
C_8mim $FeCl_4$	0.69[pp]			
C_2(4 M)Py NTF_2	2.06[h]	131.5	−490.2	180
C_3Py BF_4	0.676[i]	126.4	−505.1	201.6
C_3Py NTF_2	1.147[o]	123.6	−552.2	180

(continued)

Table 6.14 (continued)

Salt	Λ/S cm^2 mol^{-1}	A_Λ/S cm^2 mol^{-1}	B_Λ/K	$T_{0\Lambda}$/K
C_4Py BF_4	0.442[kk]	392.7	−799.7	180
C_4Py NTF_2	0.933[c], 0.943[l]	200	−675	172
C_4(3 M)Py BF_4	0.418[kk]	459.1	−824.6	180
C_4(3 M)Py NTF_2	0.823[g], 1.127[z]	227.5	−694.1	167.4
C_4(4 M)Py BF_4	0.365[kk]			
C_4(4 M)Py NTF_2	0.97[h]	152.6	−597.0	180
C_6Py NTF_2	0.532[o]	160.5	−672.9	180
C_6(3 M)Py NTF_2	0.498[z]	192.4	−686.2	178.4
C_6(4 M)Py NTF_2	0.628[z]	192.7	−721.4	176.9
$MeBu_3N$ NTF_2	0.1103[c]			
$MeBu_3N$ $N(CN)_2$	1.55[e]			
$BuMe_3N$ NTF_2	0.586[c], 0.605[l]	220	−731	174
$PeEt_3N$ $N(CN)_2$	0.67[e]			
$HxMe_3N$ NTF_2	0.14[ii]			
$HxEt_3N$ NTF_2	0.26[u], 0.321[nn]	138.0[nn]	−551.6[nn]	207.2[nn]
$HxBu_3N$ NTF_2	0.08[ii]			
$HpMe_3N$ NTF_2	0.14[ii]			
$HpEt_3N$ NTF_2	0.292[nn]	90.6[nn]	−470.8[nn]	216.1[nn]
$HpBu_3N$ NTF_2	0.08[ii]			
$OcMe_3N$ NTF_2	0.13[ii]			
$OcEt_3N$ NTF_2	0.245[nn]	230.1[nn]	−537.4[nn]	209.6[nn]
$OcEt_3N$ $N(CN)_2$	0.28[e]			
$OcBu_3N$ NTF_2	0.07[ii]			
$OcBu_3N$ CF_3SO_3	0.007[ii]			
$DcEt_3N$ NTF_2	0.198[nn]	104.1[nn]	−624.8[nn]	199.2[nn]
PrMePyrr NTF_2	0.392[m]	83.3[oo]	−443.2[oo]	170[oo]
PrMePyrr $N(FSO_2)_2$	0.800[oo]	36.6[oo]	−432.0[oo]	186[oo]
BuMePyrr $N(FSO_2)_2$	1.527[a]	202.7	−730.8	148.6
BuMePyrr NTF_2	0.660[m], 0.835[d]	213.5	−716.5	168.9
BuMePyrr $N(CN)_2$	2.43[cc]	50.1	−357.0	180
$MeBu_3P$ NTF_2	0.161[f]			
$MeBu_3P$ $N(CN)_2$	0.35[e]			
$BuEt_3P$ $N(CN)_2$	1.37[e]			
$BuEt_3P$ $N(SO_2F)_2$	1.27[b]			
$PeEt_3P$ $N(SO_2F)_2$	0.89[b]			
$PeEt_3P$ NTF_2	0.60[b]			
$PeEt_3P$ $N(CN)_2$	1.03[e]			
$OcEt_3P$ $N(CN)_2$	0.61[e]			
$OcBu_3P$ BF_4	0.0271[f]			
$OcBu_3P$ PF_6	0.0191[f]			
$OcBu_3P$ NTF_2	0.134[f]			
$OcBu_3P$ SCN	0.0694[f]			

(continued)

Table 6.14 (continued)

Salt	Λ/S cm^2 mol^{-1}	A_Λ/S cm^2 mol^{-1}	B_Λ/K	$T_{0\Lambda}$/K
$OcBu_3P\ N(CN)_2$	0.18[e]			
$OcBu_3P\ CF_3CO_2$	0.0548[f]			
$OcBu_3P\ CF_3SO_3$	0.0375[f]			
$OcBu_3P\ MePhSO_3$	0.0101[f]			
$DoBu_3P\ BF_4$	0.0221[f]			
$DoBu_3P\ NTF_2$	0.101[f]			
$TdHx_3P\ NTF_2$	0.081[s]			

[a][281], [b][170], [c] [164], [d][366], [e][169], [f][167], [g][347], [h][116], [i][125], [j][367], [k][260], [l][352], [m][368], [n][338], [o][62], [p][369], [q][370], [r][371], [s][137], [u][163], [v][372], [w][365] at 353 K and the experimental values reported are too large by a factor of 10, [x][373], [y][374], [z][283], [aa][375], [bb][376], [cc][377], [dd][312], [ee][378], [ff][379], [gg][380], [hh][381], [ii][382], [jj][80], [kk][383], [ll][384], [mm][385], [nn][280], [oo][386], [pp][68]

The product of the molar conductivity and the viscosity, the so-called Walden product, is generally expressed in terms of plots of ln Λ against ln$(1/\eta)$ for varying temperatures. For RTILs the *fractional* Walden product expression applies in the form:

$$\Lambda\eta^{\alpha} = const. \tag{6.37}$$

with α values somewhat smaller than unity. This results in the linear ln Λ against ln $(1/\eta)$ plots to be somewhat below, but parallel to, the so-called ideal Walden curve obtained for dilute aqueous KCl, where the electrolyte is completely dissociated to ions and $\alpha = 1$. The distance of the plotted line from the ideal line is construed as the fractional degree of ionic dissociation of the RTIL [164, 280, 373, 385, 392]. Values of $\alpha = 0.90 \pm 0.04$ are typical for RTILs, but it is difficult to discern clear trends with the natures of the cations and the anions.

6.5.3 Thermal Conductivity

A transport property of RTILs that is of consequence for several of their application is their thermal conductivity, i.e., the rate at which heat is dissipated from them. The thermal conductivity, λ_{th}, in units of W m^{-1} K^{-1}, decreases mildly and linearly with increasing temperatures:

$$\lambda_{th} = a_{th} + b_{th}T \tag{6.38}$$

The values of λ_t at 298 K and the coefficients a_{th} and (the negative) b_{th} are shown in Table 6.15. The thermal conductivity increases significantly when the RTIL contains water [393, 394] but only slightly when the pressure is increased, 0.1 % per

MPa [395–397]. Molecular dynamics simulations by Liu et al. [398] under-estimate considerably the thermal conductivity of RTILs relative to the experimental values, but when the ionic charges were reduced to 80 % in the force field used for the computation slight over-estimation resulted according to Theodorescu [251].

A simple correlation of the thermal conductivity of RTILs and their densities has been proposed by Fröba et al. [400], rewritten as:

$$\lambda_{\mathrm{th}}\rho/\left(\mathrm{W\,m^{-1}K^{-1}g\,cm^{-3}}\right) = 0.1130 + 0.02265/\left(M/\mathrm{kg\,mol^{-1}}\right) \tag{6.39}$$

However, the data in [400] yielded a different slope, 0.0321 rather than 0.02265, for this plot, Fig. 6.7.

A more complicated correlation, involving five universal coefficients, was proposed by Shojaee et al. [406]:

$$\lambda_{\mathrm{th}}T = a + b(M/T)^{\mathrm{c}} + d(T_{\mathrm{m}}/P)^{\mathrm{e}} \tag{6.40}$$

This expression considered both the temperature and the pressure dependences and was applied to 143 data points for the determination of the coefficients $a = 26.365$, $b = 7.233$, $c = -2.376$, $d = 11.040$, and $e = -0.019$ as a training set and was then applied to 66 data points as a test set.

The thermal conductivity can be estimated from the speed of sound, u, which can be measured accurately, and for which a database of 96 RTILs and its estimation from group contributions were established by Wu et al. [407], according to the following expression:

$$\lambda_{\mathrm{t}} = \rho^{2/3}M^{1/2}\left(k_0 M^{\alpha-1}/\rho^{\alpha} + k_1 T + k_2\right)u \tag{6.41}$$

requiring also the density ρ and the molar mass M and the universal coefficients (for RTILs) $k_0 = 1.1407 \times 10^{-4}$, $k_1 = 5.3458 \times 10^{-9}$, $k_2 = -9.7345 \times 10^{-6}$, and $\alpha = 0.7380$. A group contribution method for the estimation of the thermal conductivities of RTILs was proposed earlier by Wu et al. from the same laboratory [408].

6.6 Chemical Properties

6.6.1 Solvatochromic Parameters

The use of solvents for chemical reactions and separations depends on their abilities to solvate solutes and reaction intermediates to different extents, and these abilities concern RTILs too. Such abilities are commonly expressed in terms of linear solvation energy relationships (LSERs) in which a property of a solute in a solvent is expressed as the sum of terms that are products of suitable solute and solvent parameters and a coefficient. Properties such as solubility, toxicity, activity

Table 6.15 The thermal conductivity, λ_{th}, at 25 °C and the coefficients of its linear temperature dependence of RTILs

Salt	λ_{th} /W m^{-1} K^{-1}	$a_{\lambda th}$/W m^{-1} K^{-1}	$10^3 b_{\lambda th}$/W m^{-1} K^{-2}
C_2mim BF_4	0.200[a]	0.235	−0.121
C_2mim NTF2	0.121[b], 0.130[c]	0.138	−0.029
C_2mim $C_2H_5SO_4$	0.186[d], 0.181[c]	0.207, 0.206	−0.069, −0.082
C_2mim $C_8H_{17}SO_4$	0.167[d]	0.205	−0.125
C_2mim CH_3SO_3	0.178[e]		
C_2mim CF_3SO_3	0.16[f]		
C_2mim SCN	0.21[f]		
C_2mim $N(CN)_2$	0.194[e], 0.202[d]	0.257	−0.187
C_2mim CH_3CO_2	0.212[d]	0.290	−0.272
C_2mim CF_3CO_2	0.17[f]		
C_4mim BF_4	0.163[g], 0.186[h, a]	0.204	−0.061
C_4mim PF_6	0.137[b], 0.145[i]	0.158	−0.043
C_4mim NTF2	0.121[g], 0.127[c]	0.149	−0.071
C_4mim CF_3SO_3	0.141[i], 0.146[c]	0.161, 0.170	−0.068,−0.079
C_6mim BF_4	0.165[j], 0.157[i]	0.194	−0.012
C_6mim PF_6	0.129[b], 0.141[i]	0.163	−0.071
C_6mim $PF_3(C_2H_5)_3$	0.107[k]		
C_6mim NTF2	0.122[d, g], 0.127[c]	0.132, 0.140	−0.035, −0.043
C_8mim BF_4	0.163[j]		
C_8mim PF_6	0.123[b], 0.147[k]	0.152	−0.024
C_8mim NTF2	0.121[g], 0.128[c]	0.143	−0.050
C_{10}mim NTF_2	0.131[c]	0.152	−0.071
C_4Py BF_4	0.170[l]	0.185	−0.049
C_6Py BF_4	0.163[l]	0.178	−0.049
C_8Py BF_4	0.159[l]	0.174	−0.049
$MeOc_3N$ NTF_2	0.129[d]	0.156	−0.093
Et_4N BF_4	0.70[m]		
Et_4N PF_6	0.74[m]		
Et_4N SCN	0.75[m]		
Et_4N NO_3	0.97[m]		
$BuMe_3N$ NTF_2	0.115[n]	0.144	−0.100
BuMePyrr NTF_2	0.125[c], 0.118[k]	0.133	−0.029
BuMePyrr $PF_3(C_2F_5)_3$	0.106[c]	0.116	−0.032
$MeBu_3P$ CH_3SO_4	0.155[o]	0.172	−0.056
$TdHx_3P$ Cl	0.160[c]	0.190	−0.100
$TdHx_3P$ NTF_2	0.135[o], 0.143[c]	0.166, 0.167	−0.096, −0.079
$TdHx_3P$ $PF_3(C_2F_5)_3$	0.125[k]		

[a][393], [b][395], [c][394], [d][400], [e][404], [f][405] mean for $300 \leq T/K \leq 375$, [g][402], [h][399], [i][401], [j][396], [k][56], [l][397], [m][403], [n][398], [o][168]

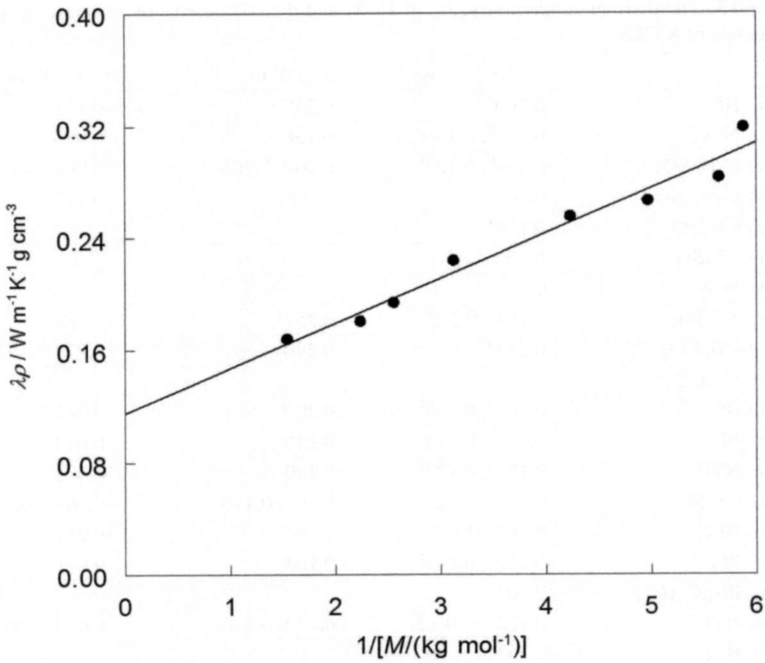

Fig. 6.7 The product of the thermal conductivity of RTILs, λ, and their densities, ρ, plotted against the reciprocal of their molar masses, Eq. (6.39), with data at 298 K from Fröba [400] supplemented with densities from Table 6.6

coefficient, etc. are thus expressed by means of LSERs. The relevant solvent properties in this respect are its polarity, Lewis acidity (or hydrogen bond donation (HBD) ability), Lewis basicity (or hydrogen bond acceptance (HBA) ability), and polarizability.

The general concept of solvent polarity is expressed by means of the normalized Reichardt polarity index E_T^N [409]:

$$E_T^N = [28591/(\lambda_{max}/\text{nm}) - 30.7]/32.4 \tag{6.42}$$

where λ_{max} is the wavelength of maximum light absorption of the betaine dye 2,6-diphenyl-4-(2,4,6-triphenylpyridinium-1-yl)phenolate in the solvent (RTIL), 30.7 is the value for tetramethylsilane with minimal polarity, and 32.4 is the normalizing factor to make $E_T^N = 1.000$ for water. Jessop et al. [410] compiled data from the literature for a large number of RTILs, many of which are shown in Table 6.16. It should be noted, however, that the E_T^N values do not pertain to just the polarity of the solvent, but has a considerable contribution from its HBD ability and also of its polarizability as shown by Marcus [411] for molecular solvents.

The Lewis acidity or HBD ability of solvents is described by the Kamlet-Taft α parameter, measured by means of ^{13}C NMR signals from pyridine-N-oxide [417] or N,N-dimethylbenzamide [418] as the probes suggested by Schneider et al. or according to the original Kamlet-Taft method, but then it includes contributions also from other properties. Values of the Kamlet-Taft α parameter, from the compilation by Jessop et al. [410] and from Kochly et al. [414] and Spange et al. [416], are shown in Table 6.16. An alternative measure of this property that has been applied by Schade et al. [413] to RTILs is Catalan's [412] *SA* parameter, using the solvatochromic probe bis(1,10-phenantrolino)dicyanoFe(II), and the resulting values are shown in Table 6.16. The *SA* values are linear with the α values: $SA = -0.032 + 0.612\alpha$. The Lewis acidity increases in a series of RTILs with a common anion: phosphonium < ammonium < pyrrolidinium < pyridinium < imidazolium, the latter having the largest α due to the hydrogen atom at the two-position of the imidazole ring according to Spange et al. [416].

The Lewis basicity or HBA ability of solvents is described by the Kamlet-Taft β parameter, measured by means of the couples of solvatochromic probes 4-nitrophenol compared to 4-nitroanisole or 4-nitroaniline compared to 4-nitro-N, N-diethyaniline, the second probe of the couple serving to eliminate the effects of the solvent polarity and polarizability on its first probe. The resulting values of β from the compilation of Jessop et al. [410], supplemented with data from Spange et al. [416] are shown in Table 6.16. An alternative measure of this property that has been applied by Schade et al. [413] to RTILs is Catalan's *SB* parameter, using the solvatochromic probe N.N-dimethyl-4-aminobenzodifuranone, and the resulting values are shown in Table 6.16. The *SB* values are linear with the β values: $SB = -0.126 + 1.056\beta$. The basicity decreases in a series of RTILs with a common anion: phosphonium > ammonium > pyrrolidinium > pyridinium > imidazolium, but more moderately than the acidity increases in the opposite direction shown above as shown by Spange et al. [416].

The polarizability of solvents, combined with their polarity, are measured by the Kamlet-Taft π^* values, obtained most profitably with the solvatochromic probe 4-nitroanisole and shown, as compiled by Jessop et al. [410] and supplemented with values from Kochly et al. [414] and Spange et al. [416] in Table 6.16. The π^* values are in a narrow range, 0.95 ± 0.15, i.e., the RTILs are all very polar as expected for ionic liquids. The disadvantage of having both these properties in one parameter is removed by the separate Catalan parameters *SP* for polarizability and *SdP* for dipolarity. These use results from two solvatochromic probes: a thiazoline one and 2-(4-(N,N-dimethylamino)benzylidene)malononitrile according to Schade et al. [413], and the values of *SP* and *SdP* are also shown in Table 6.16. The polarizability of RTILs is not affected much by the acidity or basicity of its constituent ions, but their polarity does depend on these properties.

Table 6.16 Solvatochromic parameters[a] and Catalan parameters[b] of RTILs

Salt	E_T^{Na}	α^a	SA^d	β^a	SB^d	π^{*a}	SP^d	SdP^d
C_1mim NTF2	0.670	0.66, 0.60[c]		0.36[c]		0.99, 0.80[c]		
C_2mim BF_4	0.71							
C_2mim PF_6	0.676	0.66		0.20		0.99		
C_2mim $PF_3(C_2F_5)_3$		0.83[c]	0.476	0.20[c]	0.191	0.71[c]	0.755	0.756
C_2mim NTF2	0.685	0.66		0.28		0.90		
C_2mim $C_2H_5SO_4$				0.71				
C_2mim $C_8H_{17}SO_4$	0.63	0.65		0.77		0.93		
C_2mim $MePhSO_3$		0.43[c]		0.78[c]		1.13[c]		
C_2mim NO_3	0.642	0.48		0.66		1.13		
C_2mim $N(CN)_2$	0.648	0.53, 0.51[c]		0.35, 0.61[c]		1.08, 0.98[c]		
C_3mim NTF2	0.654							
C_4mim BF_4	0.664	0.63, 0.52[c]	0.288	0.38, 0.55[c]	0.466	1.05, 0.96[c]	0.800	1.003
C_4mim PF_6	0.669	0.63, 0.63[d]	0.296	0.19, 0.21[d]	0.364	1.04, 1.03[d]	0.808	0.934
C_4mim $PF_3(C_2F_5)_3$		0.59[c]	0.329	0.25[c]	0.233	0.78[c]	0.747	0.829
C_4mim NTF2	0.645	0.62, 0.61[d]	0.304	0.25, 0.24[d]	0.372	0.90, 0.98[d]	0.793	0.867
C_4mim CH_3SO_4		0.53	0.191	0.66	0.799	1.06	0.837	1.062
C_4mim $C_8H_{17}SO_4$	0.64	0.69, 0.41[c]	0.223	0.79, 0.77[c]	0.691	0.89, 0.96[c]	0.769	1.032
C_4mim CF_3SO_3	0.667	0.63, 0.50[c]	0.271	0.46, 0.57[c]	0.482	1.00, 0.90[c]	0.808	0.934
C_4mim CH_3SO_3		0.44, 0.36[c]		0.77, 0.85[c]		1.02, 1.04[c]		
C_4mim Cl		0.47, 0.32[c]	0.167	0.87, 0.95[c]	0.869	1.10, 1.13[c]	0.833	1.172
C_4mim Br		0.36[c]	0.191	0.87[c]	0.774	1.14[c]	0.873	1.120
C_4mim I		0.41[c]		0.75[c]		1.13[c]		
C_4mim SCN		0.43[c]		0.71[c]		1.06[c]		
C_4mim NO_3		0.38[c]	0.215	0.81[c]	0.664	1.05[c]	0.868	1.034
C_4mim CH_3CO_2	0.611	0.43, 0.36[c]		1.05, 0.85[c]		0.96, 1.06		

(continued)

Table 6.16 (continued)

Salt	E_T^{Na}	α^a	SA^d	β^a	SB^d	π^{*a}	SP^d	SdP^d
C_4mim CF_3CO_2	0.623	0.56, 0.43[c]		0.74[c]		0.56, 0.90[c]		
C_4mim $N(CN)_2$	0.639	0.54, 0.44[c]		0.71, 0.64[c]		1.05, 0.98[c]		
C_5mim NTF_2		0.63, 0.51[c]		0.26, 0.44[c]		0.97, 0.86[c]		
C_6mim BF_4		0.44[c]	0.239	0.60[c]	0.495	0.96[c]	0.800	1.003
C_6mim PF_6	0.66	0.57, 0.51[c]	0.279	0.58, 0.50[c]	0.463	1.02, 0.93[c]	0.820	0.954
C_6mim $PF_3(C_2F_5)_3$		0.55[c]	0.304	0.27[c]	0.206	0.80[c]	0.717	0.906
C_6mim NTF2	0.651	0.65		0.25		0.98		
C_6mim CF_3SO_3	0.673	0.67, 0.47[c]	0.255	0.52, 0.61[c]	0.515	0.98, 0.92[c]	0.798	0.959
C_6mim Cl	0.59	0.48, 0.31[c]	0.159	0.94, 0.97[c]	0.996	1.02, 1.06[c]	0.787	1.139
C_6mim Br	0.61	0.45, 0.36[c]	0.191	0.74, 0.88[c]	0.786	1.09, 1.09[c]	0.861	1.101
C_6mim NO_3		0.40[c]		0.76[c]		1.01[c]		
C_6mim $N(CN)_2$	0.630	0.51, 0.44[c]	0.239	0.69[c]	0.596	1.05, 1.00[c]	0.866	0.990
C_8mim BF_4	0.65	0.62, 0.45[c]	0.247	0.41, 0.63[c]	0.525	0.98, 0.93[c]	0.788	0.983
C_8mim PF_6	0.633	0.58, 0.52[c]	0.288	0.46, 0.53[c]	0.392	0.88, 0.92[c]	0.798	0.959
C_8mim $PF_3(C_2F_5)_3$		1.03[c]		0.31[c]				
C_8mim NTF2	0.627	0.60, 0.48[c]	0.263	0.28, 0.47[c]	0.373	0.97, 0.86[c]	0.774	0.918
C_8mim CF_3SO_3		0.48[c]	0.263	0.64[c]	0.590	0.90[c]	0.744	0.992
C_8mim Cl		0.31[c]	0.159	0.98[c]	1.001	1.03[c]	0.784	1.096
C_8mim Br		0.35[c]	0.183	0.89[c]	0.810	1.04[c]	0.806	1.091
C_8mim I		0.39[c]		0.81[c]		1.07[c]		
C_8mim NO_3		0.36[c]		0.80[c]		0.98[c]		
C_8mim $N(CN)_2$		0.43[c]	0.231	0.71[c]	0.592	0.97[c]	0.822	0.999
C_{10}mim Cl		0.31[c]	0.159	0.98[c]	0.932	0.97[c]	0.696	1.115
C_{10}mim Br		0.34[c]	0.175	0.91[c]	0.883	1.00[c]	0.740	1.105
C_{10}mim NO_3		0.36[c]		0.81[c]		0.96[c]		

(continued)

Table 6.16 (continued)

Salt	E_T^{Na}	α^a	SA^d	β^a	SB^d	π^{*a}	SP^d	SdP^d
C_{10}mim BF_4		0.47^c	0.255	0.65^c	0.560	0.90^c	0.744	0.992
C_{10}mim PF_6		0.48^c	0.263	0.55^c	0.408	0.89^c	0.754	0.968
C_{10}mim NTF_2	0.627	0.48^c	0.263	0.49^c	0.396	0.86^c	0.774	0.918
C_{10}mim CF_3SO_3		0.48^c	0.263	0.65^c	0.565	0.87^c	0.764	0.943
C_{10}mim $N(CN)_2$		0.44^c	0.239	0.71^c	0.605	0.96^c	0.800	1.003
C_2(3 M)Py BF_4	0.670							
C_3Py BF_4	0.661							
C_4Py BF_4	0.639	0.53		0.21		1.08		
C_4Py PF_6			0.231		0.310		0.818	0.909
C_4Py NTF_2	0.593	0.48, 0.51^d	0.279	0.12, 0.28^d	0.339	1.01, 0.98^d	0.815	0.863
C_4Py CF_3SO_3		0.41^c	0.215	0.57^c	0.477	0.92^c	0.830	0.930
C_4Py $N(CN)_2$		0.43^c	0.231	0.65^c	0.534	1.01^c	0.918	0.957
C_4Py NO_3		0.44^c	0.239	0.73^c	0.613	1.05^c	0.920	1.001
C_4(3 M)Py NTF_2	0.602	0.54		0.28		0.97		
C_4(4 M)Py BF_4	0.630							
C_4(4 M)Py NTF_2	0.588	0.51		0.29		0.98		
C_6Py CF_3SO_3		0.41^c		0.57^c		0.90^c		
C_6Py NTF_2	0.593	0.50, 0.50^c	0.271	0.07, 0.44^c	0.379	0.98, 0.86^c	0.805	0.888
C_6Py $N(CN)_2$		0.41^c	0.215	0.69^c	0.571	1.03^c	0.878	1.010
C_6Py NO_3		0.39^c	0.207	0.78^c	0.654	1.04^c	0.868	1.034
C_6(3 M)Py NTF_2	0.608	0.54		0.31		0.98		
C_8Py BF_4	0.606	0.54		0.34		0.97		
C_8Py NTF_2	0.588	0.51		0.28		0.99		
C_8(2 M)Py NTF_2	0.554	0.48		0.35		0.95		
C_8(3 M)Py BF_4	0.60	0.51		0.44		1.02		
C_8(3 M)Py NTF_2	0.565	0.46		0.28		0.97		
$MeBu_3N$ NTF_2	0.774^b	0.85^b, 0.41^c	0.215	0.33^b, 0.46^c	0.350	0.92^b, 0.89^c	0.786	0.938

(continued)

Table 6.16 (continued)

Salt	E_T^N[a]	α[a]	SA[d]	β[a]	SB[d]	π^*[a]	SP[d]	SdP[d]
$MeBu_3N\ N(CN)_2$		0.25[c]	0.121	0.73[c]	0.644	1.03	0.815	1.067
$MeHx_3N\ NTF_2$		0.70[c]		0.51[c]		0.90[c]		
$MeHx_3N\ N(CN)_2$		0.24[c]		0.8[c]		1.00[c]		
$MeOc_3N\ N(CN)_2$		0.25[c]	0.121	0.80[c]	0.687	0.94[c]	0.863	0.946
$MeOc_3N\ NTF_2$	0.469	0.33, 0.39[c]	0.207	0.23, 0.55[c]	0.396	0.87, 0.87[c]	0.827	0.884
$BuMe_3N\ NTF_2$	0.574[a], 0.968[b]	1.27[b]		0.26[b]		0.97[a], 0.90[b]		
$EtMePyrr\ N(CN)_2$	0.556	0.38				1.03		
$PrMePyrr\ NTF_2$	0.669	0.74		0.34		0.91		
$BuMePyrr\ NTF_2$	0.540	0.41		0.23		0.95		
$HxMePyrr\ NTF_2$	0.657	0.76		0.06		0.81		
$OcMePyrr\ NTF_2$	0.651	0.80		0.08		0.73		
$PrMePip\ NTF_2$	0.564	0.52		0.35		0.93		
$BuMePip\ NTF_2$	0.553	0.50, 0.43[c]		0.36, 0.42[c]		0.92, 0.82[c]		
$BuMePip\ N(CN)_2$	0.552	0.31		0.49		1.13		
$OcMePip\ NTF_2$	0.539	0.48		0.37		0.91		
$MeBu_3P\ NTF_2$		0.51[c]	0.279	0.48[c]	0.362	0.87[c]	0.827	0.884
$MeBu_3P\ N(CN)_2$		0.25[c]	0.121	0.75[c]	0.659	1.03[c]	0.878	1.010
$MeBu_3P\ Cl$		0.19[c]	0.083	1.11[c]	1.009	1.10[c]	0.664	1.297
$TdHx_3P\ Cl$		0.09[c]	0.023	1.14[c]	1.096	0.93[c]	0.664	1.145
$TdHx_3P\ NO_3$				0.89[c]		0.93[c]		
$TdHx_3P\ PF_6$			0.060		0.400		0.798	0.959
$TdHx_3P\ PF_3(C_2F_5)_3$		0.64[c]		0.45[c]		0.87[c]		
$TdHx_3P\ NTF_2$	0.475	0.37, 0.36[c]	0.191	0.27, 0.58[c]	0.447	0.83, 0.87[c]	0.827	0.884
$TdHx_3P\ N(CN)_2$		0.19[c]	0.083	0.85[c]	0.731	0.93[c]	0.800	1.003
$TdHx_3P\ PF_3(C_2F_5)_3$		0.64[e]	0.363	0.45[c]	0.264	0.87[c]	0.827	0.884

[a][410], [b][415], [c][416], [d][414], [e][413]

6.6.2 Mutual Solubility with Water

The hydrophilic character due to the iconicity of RTILs with not too long alkyl chains is expected to promote their solubility in water and of water in them. Still, even relatively hydrophobic RTILs that have small water solubilities are hygroscopic and absorb water from the atmosphere. For example, according to Seddon et al. [419] C_4mim^+ PF_6^- reached a mole fraction of 0.16 of water when exposed to the laboratory atmosphere.

These mutual solubilities at ambient conditions are reported in Table 6.17, in terms of mass fractions, w(W in RTIL) and w(RTIL in W), and mole fractions x (W in RTIL) and x(RTIL in W). RTILs with short alkyl chains and small anions are miscible with water, for example C_1mim^+ $CF_3CO_2^-$, C_4mim^+ BF_4^-, $C_2(2E)Py^+$ $C_2H_5SO_4^-$, $BuMePyrr^+$ $N(CN)_2^-$, and even C_8mim^+ Cl^-, so it is difficult to predict how short the alkyl chain of the RTIL should be for complete miscibility with water. Certain anions, such as BF_4^-, $CF_3SO_3^-$ and $N(CN)_2^-$ seem to confer hydrophilicity (water miscibility) on RTILs whereas PF_6^- and NTF_2^- confer hydrophobicity according to Chapeaux et al. [420]. However, conflicting results have been reported: C_4mim^+ $N(CN)_2^-$ and C_6mim^+ $N(CN)_2^-$ are water-miscible according to Gomez et al. [343], but C_2mim^+ $N(CN)_2^-$ with a shorter alkyl chain has only a limited solubility in water according to Cho et al. [421].

The mutual solubility of water and RTILs increases with the temperature [423, 430, 434] and a consolute temperature might eventually be reached, and indeed has been reached already at ambient temperatures for those RTILs that are miscible with water. For the solution of RTILs in water the molar heats of solution range from 6.5 to 12.6 kJ mol^{-1} according to Freire et al. [430].

The solubilities of RTILs in water (in terms of g L^{-1}) were predicted by Cho et al. [421] by means of linear solvation energy relationships with three to eight parameters in several alternative equations (the parameters were, but the equations were not provided in the paper) with a standard deviation of ± 0.14.

When the alkyl chains are long, then these mutual solubilities of RTILs and water diminish and a two-liquid-phases system results at equilibrium that can be utilized for solvent extraction processes. Only a few examples can be shown here, one being the extraction from water of the long-lived, hence hazardous, strontium and cesium fission products, the RTIL employed being methyl-tributylammonium$^+$ NTF_2^-. Suitable ionophores, a crown ether and a calixarene, were used to carry the metal ions from the aqueous to the RTIL phase according to Chen [429]. A series of imidazolium RTILs: C_nmim^+ PF_6^- $(4 \leq n \leq 9)$ was used by Chun et al. [443] with a crown ether ionophore to extract alkali metal ions from aqueous solutions. The applicability of RTILs as the solvents or diluents of active extractants for the extraction of actinide and lanthanide elements from aqueous solutions was investigated by Kolarik [445]. The extraction of polar organic solutes that are typical water contaminants (hexanoic acid, toluene, 1-nonanol, for example) into C_nmim^+ NTF_2^- ($n = 4$, 6, and 8), C_4mim^+ PF_6^-, C_4MPyrr^+ NTF_2^-, $TdHx_3P^+$ $CH_3SO_3^-$, $TdBu_3P^+$ $DoPhSO_3^-$, and $TdHx_3P^+$ $DoPhSO_3^-$ was studied by McFarlane

Table 6.17 Mass fraction (w) and mole fraction (x) solubilities of water in RTILs and of RTILs in water and their n-octanol/water partition constants $\log P^{o}_{w}$ at ambient temperatures (specified in the references)

Salt	$w_{W\ in\ RTIL}$	$w_{RTIL in\ W}$	$x_{W\ in\ RTIL}$	$10^3 x_{RTIL in\ W}$	$\log P^{o}_{w}$
C_1mim CF_3SO_3	Miscible[a]				
C_1mim NTF_2	0.025[a]	0.0194[f]			−1.35[t]
C_1mim CF_3CO_2	Miscible[a]				
C_2mim BF_4					−2.66[t], −2.57[x]
C_2mim PF_6	0.051[e]	0.055[e]			−2.36[t], −1.82[y]
C_2mim CF_3SO_3	Miscible[a]			−2.75[x]	
C_2mim NTF_2	0.020[i]	0.017[i]			−1.05[v], −1.18[x]
C_2mim CH_3CO_2					−2.53[w]
C_2mim CF_3CO_2	Miscible[a]			−2.3[w]	
C_2mim $N(CN)_2$		0.035[t]			
C_2mim $B(CN)_4$					−0.77[i]
C_3mim BF_4					−1.74[y]
C_3mim NTF_2	0.0078[j]	0.0051[t]			−0.88[t]
C_4mim BF_4	Miscible[d]			−2.52[v], −1.44[y]	
C_4mim PF_6	0.023[b]	0.020[b]	0.26[b,k]	1.29[b], 1.32[g]	−1.72[t], −1.66[x],−2.39[s]
C_4mim NTF2	0.014[a]	0.0207[f]		0.288[h], 1.21[k]	−0.50[t], +0.11[x]
C_4mim Cl	Miscible[d]			−2.40[w],−2.77[t]	
C_4mim Br				−2.48[v], −2.65[z]	
C_4mim I	Miscible[d]				
C_4mim NO_3					−2.42[w],−2.90[s]
C_4mim CH_3CO_2					−2.77[s]
C_4mim CF_3CO_2	Miscible[a]				
C_4mim CF_3SO_3	Miscible[a,q]			−1.61[t]	
C_4mim $N(CN)_2$	Miscible[q]			−2.32[w]	

(continued)

Table 6.17 (continued)

Salt	$w_{W\ in\ RTIL}$	$w_{RTIL in\ W}$	$x_{W\ in\ RTIL}$	$10^3 x_{RTIL in\ W}$	$\log P^o{}_w$
C_5mim NTF_2		0.00019[t]			−0.11[t]
C_5mim BF_4					−1.09[y]
C_5mim PF_6	0.0125[c]	0.0123[bb]			
C_5mim NTF_2				0.34[cc]	−0.11[w]
C_6mim BF_4					−1.58[t], −0.71[y]
C_6mim PF_6	0.0075[c]	0.0075[bb]	0.229[k]	0.444[g]	−1.20[t]
C_6mim CF_3CO_2					−2.3[w]
C_6mim $N(CN)_2$	Miscible[q]				
C_6mim NTF2	0.0034[f]	0.0109[f]		0.137[h],0.434[k]	+0.19[v],+0.64[x]
C_6mim Cl	Miscible[d]			−1.71[t]	
C_7mim NTF_2					+0.57[t]
C_7mim PF_6	0.0037[c]	0.0037[bb]			
C_8mim BF_4	0.108[b]	0.018[b]	0.63[b]	1.17[b]	−0.68[t], −1.34[x]
C_8mim PF_6	0.013[b]	0.007[b]	0.20[b,k]	0.350[b],0.135[g]	−0.35[t]
C_8mim NTF_2	0.0021[f]	0.0082[f]		0.085[h],0.127[k]	+0.92[v],+1.05[w],+0.79[t]
C_8mim Cl	Miscible[d]			−0.60[t], −0.27[aa]	
C_8mim Br				−1.95[z]	
C_8mim SCN	Miscible[l]				
C_8mim $N(CN)_2$	Miscible[l]				
C_9mim PF_6		0.0015[bb]			
C_9mim Cl					−0.13[w]
C_{10}mim NTF_2			0.0047[cc]		
C_{10}mim $N(CN)_2$	Miscible[l]				
C_{10}mim CF_3SO_3	Miscible[l]				
C_{10}mim CF_3CO_2	Miscible[l]				

(continued)

Table 6.17 (continued)

Salt	$w_{\mathrm{W\ in\ RTIL}}$	$w_{\mathrm{RTIL in\ W}}$	$x_{\mathrm{W\ in\ RTIL}}$	$10^3 x_{\mathrm{RTIL in\ W}}$	$\log P^{o}_{\mathrm{w}}$
C_{10}mim Cl				+0.31[t], −0.29[aa]	
C_{10}mim SCN	Miscible[l]				
C_{12}mim Cl				−0.14[dd]	
C_{12}mim Br				−1.26[z]	
C_1Py CH_3SO_4	Miscible[q]				
C_2Py Cl				−3.55[t]	
C_2Py NTF_2		0.311[r]		−0.9[w]	
C_2Py CF_3CO_2				−2.57[x]	
C_2Py Cl				−3.55[w]	
C_2(2 M)Py NTF_2		0.255[r]			
C_2(3 M)Py NTF_2		0.279[r]			
C_2(3 M)Py NTF_2		0.308[r]			
C_2(2E)Py $C_2H_5SO_4$	Miscible[q]				
C_3Py NTF_2			0.236[k]	0.57[p]	−1.40[p]
C_3(4 M)Py NTF_2			0.236[k]	0.375[p]	
C_4Py Cl					−2.82[t]
C_4Py Br					−2.43[z]
C_4Py NTF_2			0.255[r]		−0.26[w]
C_4(2 M)Py Cl					−2.78[t]
C_4(3 M)Py Cl					−2.62[t]
C_4(3 M)Py NTF_2		0.0020[t]	0.222[r]		+0.21[i]
C_4(4 M)Py Cl					−2.57[t]
C_4(4 M)Py CF_3SO_3	Miscible[q]				
C_4(4 M)Py NTF_2	0.011[i]	0.0047[i]	0.249[r]		+0.21[i]
C_6Py NTF_2	0.010[i]	0.0028[i]			+0.77[i]

(continued)

Table 6.17 (continued)

Salt	$w_{\mathrm{W\ in\ RTIL}}$	$w_{\mathrm{RTIL in\ W}}$	$x_{\mathrm{W\ in\ RTIL}}$	$10^3 x_{\mathrm{RTIL in\ W}}$	$\log P^{o}{}_{\mathrm{w}}$
C_5(3 M)Py NTF_2					+1.07[i]
C_5(3 M)Py Cl					+1.58[i]
C_5(4 M)Py Cl					−1.65[w]
C_6(4 M)Py NTF_2	0.0088[i]	0.0017[i]		0.064[p]	1.07[i], 0.14[p]
C_6(4 M)Py CF_3SO_3					−0.64[w]
C_8Py Cl					−0.72[t],−0.82[w]
C_8Py Br					−1.74[z]
C_8Py NTF_2	0.0086[m]				
C_8(3 M)Py NTF_2					+1.38[i]
C_8(4 M)Py BF_4	0.113[m]	0.0104[t]			
C_8(4 M)Py CF_3SO_3	0.172[m]	0.0081[t]			
C_8(4 M)Py $N(CN)_2$	0.549[m]				
C_8(4 M)Py NTF_2	0.0074[i]	0.0015[i]			+1.38[i]
$MeBu_3N$ NTF_2	0.0027[j]				
$BuMe_3N$ NTF_2				0.45[p]	
Me_2Pyrr NTF_2				0.84[p]	
PrMePyrr NTF_2	0.0132[n]	0.0099[n]	0.233[k]		
BuMePyrr NTF_2	0.0068[j]	0.0061[n]	0.211[k]	0.27[p]	
BuMePyrr CF_3SO_3	Miscible[q]				
BuMePyrr $N(CN)_2$	Miscible[q]				
BuMePyrr $PF_3(C_2F_5)_3$		0.00004[t]			
BuMePyrr Br					−2.92[z]
OcMePyrr NTF_2	0.0084[n]	0.00054[n]			
OcMePyrr Br					−2.26[z]
PrMePip NTF_2	0.0117[n]	0.0080[n]	0.212[k]	0.375[k]	

(continued)

Table 6.17 (continued)

Salt	$w_{W\ in\ RTIL}$	$w_{RTIL in\ W}$	$x_{W\ in\ RTIL}$	$10^3 x_{RTIL in\ W}$	$\log P^o_w$
$TdHx_3P$ Cl	0.1334[o]				
$TdHx_3P$ Br	0.0665[o]				
$TdHx_3P$ NTF_2	0.0023[o]				
$TdHx_3P$ CH_3SO_3	0.1181[o]				
$TdHx_3P$ $N(CN)_2$	0.0309[o]				

[a][422] at 20 °C, [b][423] at 22 °C, [c][424], [d][70], [e][425] at 45 °C, [f][426], [g][427], [h][428], [i][420] at 23.5 °C, [j][429], [k][430] at 25 and data also at 15, 20, 30, 35, 40 and 45 °C, [l][339], [m][120], [n][431], [o][317], [p][432] at 22 °C, [q][433], [r][434] at 24 °C, also data at 20–70 °C, [s][435], [t][421], [u][436], [v][437], [w][438], [x][439], [y][440], [z][441], [aa][442], [bb][443], [cc][444] in 0.1 M HNO3

et al. [428]. The partitioning of compounds of biological significance (such as caffeine, Thymine, and amine-, hydroxide- and carboxylate-substituted aromatic compounds) from aqueous solutions into C_4mim^+ PF_6^-, C_6mim^+ PF_6^-, and C_8mim^+ BF_4^- was studied by Padro et al. [446], having been measured and predicted by means of linear solvation energy relationships.

6.6.3 Hydrophilic/Hydrophobic Balance

The hydrophilic/hydrophobic balance of RTILs is of importance in the evaluation of their impact on the environment and their toxicity. The ionic character of RTILs makes them considerably hydrophilic, unless they carry long alkyl chains that turn them to be hydrophobic. The common measure of this hydrophilic/hydrophobic balance is the logarithm of the 1-octanol/water partition coefficient, $\log P^{O}{}_{W}$, for dilute solutions of the substance in question in the mutually saturated 1-octanol/ water system. The values for RTILs are shown in Table 6.18. The values of $\log P^{O}{}_{W}$ are concentration dependent, and this explains to some extent the discrepancies noted in Table 6.18 between the values reported by different authors: ideally the values should pertain to infinite dilution in both phases.

The effect of the length of the alkyl chain of the RTIL is clear: the longer it is the less hydrophilic is the salt and the less negative is $\log P^{O}{}_{W}$. The effect of the anion is more subtle, and in the study by Cho et al. [421] a list of the additive anion contribution to $\log P^{O}{}_{W}$ is reported: NTF_2^- 1.50, $CF_3SO_3^-$ 0.53, PF_6^- 0.19, BF_4^- –0.20, Br^- –0.34, Cl^- –0.44, NO_3^- –0.52, and $CH_3CO_2^-$ –0.63. A six-parameter expression is reported in [421] for the prediction of the $\log P^{O}{}_{W}$ values of RTILs from independent descriptors of these salts with moderate success (a standard deviation of 0.18).

6.6.4 Carbon Dioxide Solubility

RTILs have been suggested as effective media for the absorption of carbon dioxide from fuel combustion or other processes and its separation from mixtures of gases. The mole fraction solubility of CO_2 in RTILs at 1 MPa pressure and at two operation temperatures, 303 and 333 K, was measured in a variety of imizazolium, pyridinium, and pyrrolidinium RTILs by Galan-Sanches [447] as quoted by Torralba-Calleja et al. [448]. Similar data for 2 MPa pressure and 298 K were supplied by Yokozeki et al. [449]. The CO_2 contents at saturation under these conditions are shown in Table 6.19 and are appreciable, x_{CO2} ranging from ~0.1 to ~0.2 at 1 MPa pressure, increasing with the pressure as a second power expression, and diminishing with increasing temperatures.

A different manner for expressing the solubility of CO_2 in RTILs is by means of the Henry law constant K_H. One definition of it is the ratio of pressure of the solute

Table 6.18 The logarithm of the 1-octanol/water partition coefficients of RTILs at ambient conditions from Lee and Lin [438] unless otherwise noted

Salt	$\log P^{O}{}_{W}$
C_1mim NTF_2	−1.35
C_2mim BF_4	−2.66, −2.57[a]
C_2mim PF_6	−2.36, −1.82[b]
C_2mim NTF_2	−1.0, −1.18[a]
C_2mim $B(CN)_4$	−0.77[c]
C_2mim CF_3SO_3	−2.75[a]
C_2mim CH_3CO_2	−2.53
C_3mim BF_4	−1.74[b]
C_3mim NTF_2	−0.88
C_4mim BF_4	−2.52, −1.44[b]
C_4mim PF_6	−1.72, −2.39[d], −1.66[a]
C_4mim NTF_2	−0.5, +0.11[a]
C_4mim Cl	−2.40, −2.77[e]
C_4mim Br	−2.48, −2.65[f]
C_4mim NO_3	−2.42, −2.90[d]
C_4mim $N(CN)_2$	−2.32
C_4mim CF_3SO_3	−1.61
C_4mim CH_3CO_2	−2.77
C_5mim BF_4	−1.09[b]
C_5mim NTF_2	−0.11
C_6mim BF_4	−1.58, −0.71[b]
C_6mim PF_6	−1.2
C_6mim NTF2	+0.16, +0.64[a]
C_6mim Cl	−1.73
C_6mim CF_3CO_2	−2.30[a]
C_7mim NTF_2	+0.57
C_8mim BF_4	−0.68, −1.34[a]
C_8mim PF_6	−0.35
C_8mim NTF2	+1.05, 0.79[e]
C_8mim Cl	−0.6, −0.27[g]
C_8mim Br	−1.95[f]
C_9mim Cl	−0.13
C_{10}mim Cl	−0.29[g]
C_{10}mim Cl	+0.31
C_{12}mim Cl	−0.14[g]
C_{12}mim Br	−1.26[f]
C_2Py Cl	−3.55
C_2Py NTF_2	−0.90
C_2Py CF_3CO_2	−2.57[a]
C_4Py Cl	−2.82
C_4Py Br	−2.43[f]
C_4Py NTF_2	−0.26[a]

(continued)

Table 6.18 (continued)

Salt	$\log P^{O}{}_{W}$
C_4(3 M)Py NTF_2	+0.21[c]
C_4(2 M)Py Cl	−2.78[e]
C_4(3 M)Py Cl	−2.62[e]
C_4(4 M)Py Cl	−2.57
C_6Py NTF_2	+0.77[c]
C_6(3 M)Py NTF_2	+1.07[c]
C_6(3 M)Py Cl	−1.58[e]
C_6(4 M)Py Cl	−1.65
C_8Py Cl	−0.82, −0.72[e]
C_8Py Br	−1.74[f]
C_8(3 M) NTF_2	+1.38[c]
C_4MPyrr Br	−2.92[f]
C_8MPyrr Br	−2.26[f]

[a][439], [b][440], [c][420], [d][435], [e][421], [f][441], [g][442], [b][437]

in the gas phase to its mole fraction in the solution $K_H = p_{CO2(g)}/x_{CO2(RTIL)}$. Such K_H values at 298 K, obtained by the application of the COSMO-RS method and from experiments by Zhang et al. [450] are shown in Table 6.19, supplemented by values obtained by other authors. The mole fractions of CO_2 in the RTIL at saturation can be readily calculated from the ratio of its (partial) pressure to the Henry constant: $x_{CO2(RTIL)} = p_{CO2}/K_H$. Thus, the larger K_H is, the smaller the saturation mole fraction of CO_2 at a given pressure.

Imidazolium-based RTILs with a given anion show increasing absorption (smaller K_H values) as the 1-alkyl chain length increases. For a given cation the K_H values diminish in the series $PF_6^- > CH_3SO_4^- > CF_3SO_3^- > BF_4^- > NTF_2$.

The question ‘why is CO_2 soluble in imidazolium-based RTILs’ was raised by Cadena et al. [457] and further discussed by Huang and Rüther [462]. Strategies for improving the sorption of CO_2 by RTILs (consisting mainly of $TdHx_3P^+$ cations with a variety of anions) were dealt with by Wang et al. [463]. The use of aprotic heterocyclic anions (such as 1,2,4-triazolide) in the imidazolium-based RTILs for maximizing CO_2 absorption was recently suggested by Seo et al. [464]. A partial answer to these questions and suggestions is the interaction of the CO_2 with the anion of the RTIL, the cation playing a minor role. Still, a comparison by Cadena et al. [457] of 1-alkyl-3-methylimidaolium with 1-alkyl-2,3-dimethylimidazolium cations, in which the hydrogen in the two-position of the former is replaced by the methyl group, shows that since the former cation is capable of hydrogen bonding to the CO_2, it is a somewhat better CO_2 absorber.

The solubility of CO_2 in RTILs was estimated by Sistla et al. [197] by means of their Hildebrand solubility parameters, δ_H, that ought to match as closely as possible that of CO_2 ($\delta_H = 17.85$ MPa$^{1/2}$) for maximizing the solubility and separating it from CH_4, for example ($\delta_H = 14.0$ MPa$^{-1/2}$). Such a close matching is achieved by $TdHx_3P^+$ $PF_3(C_2F_5)_3^-$ from among the RTILs dealt with here ($\delta_H = 19.8$ MPa$^{-1/2}$).

Table 6.19 Mole fraction solubilities of CO_2 in RTILs at 1.0 MPa and at 303 K and 333 K[a] and at 2.0 MPa at 298 K,[b] estimated by means of the COSMO –RS program in the second and third columns, experimental in the fourth, and the Henry's law constants K_H at 298 K in the fifth and sixth columns

Salt	x_{CO2} 303 K 1 MPa[a]	x_{CO2} 333 K 1 MPa[a]	x_{CO2} 298 K 2 MPa[e]	K_H/MPa 298 K[g]	K_H/MPa 298 K
C_1mim BF_4				6.4	
C_1mim PF_6				8.6	
C_1mim CF_3SO_3				7.0	
C_1mim CH_3SO_4				7.2	
C_1mim NTF_2				4.9	
C_1mim $PF_3(C_2F_5)_3$				2.3	
C_2mim BF_4			0.202[f]	5.9	8.0[h]
C_2mim PF_6				6.0	
C_2mim CH_3SO_3			0.198[c]		
C_2mim CF_3SO_3	0.187[b]			5.8	7.3[h]
C_2mim CH_3SO_4		0.060[c]	0.242[c]		
C_2mim $C_2H_5SO_4$		0.074[d]		5.9	1.50[i], 8.82[j]
C_2mim $PF_3(C_2F_5)_3$		0.079[c]	0.340[c]	2.2	
C_2mim NTF_2	0.2257	0.1446	0.3900	4.2	3.56[h, k]
C_2mim SCN			0.182[c]		
C_2mim HSO_4		0.133[c]			
C_2mim CH_3CO_2			0.4280		
C_2mim CF_3CO_2			0.2820		
C_2mim $N(CN)_2$					7.8[h], 0.99[i]
C_3mim NTF_2					3.7[l]
C_3mim PF_6					5.2[l]
C_4mim BF_4	0.1461	0.0895		5.4	5.90[h], 5.65[k]
C_4mim PF_6	0.1662	0.1012, 0.140[d]	0.264[f]	4.4	5.34[h, k]
C_4mim NTF2				3.7	3.30[h], 3.7[l]
C_4mim CH_3SO_4	0.1190	0.0733		5.0	
C_4mim SCN	0.0978	0.0664			
C_4mim NO_3		0.097[d]			
C_4mim CH_3CO_2			0.4550		
C_4mim CF_3CO_2			0.3010		
C_4mim CF_3SO_3	0.228[b]		0.424[c]	5.0	1.21[i]
C_4mim $C_2H_5CO_2$			0.3900		
C_4mim $(CH_3)_3C$ CO_2			0.4310		
C_4mim $N(CN)_2$	0.1434	0.0997			
C_6mim BF_4			0.304[f]	4.8	
C_6mim PF_6			0.222[f]	3.7	
C_6mim CF_3SO_3	0.230[b]			4.5	
C_6mim $PF_3(C_2F_5)_3$			0.4930, 0.4559[g]	2.0	2.52[h], 2.37[g]
C_6mim NTF2			0.4330	3.4	3.16[h], 3.5[l]

(continued)

Table 6.19 (continued)

Salt	x_{CO2} 303 K 1 MPa[a]	x_{CO2} 333 K 1 MPa[a]	x_{CO2} 298 K 2 MPa[e]	K_H/MPa 298 K[g]	K_H/MPa 298 K
C_8mim BF_4	0.1873	0.1213, 0.110[d]			
C_8mim PF_6		0.139[k]			
C_8mim NTF_2					3.0[h, l]
C_4Py BF_4		0.089[d]			
C_4(3 M)Py NTF_2					3.3[h]
C_4(3 M)Py $N(CN)_2$	0.1436	0.0683			
C_4(4 M)Py BF_4	0.1443	0.0961		4.2	
C_4(4 M)Py PF_6				3.6	
C_4(4 M)Py CF_3SO_3				4.1	
C_4(4 M)Py NTF_2				3.3	
C_6Py CF_3SO_3	0.0962	0.0652			
C_6(3 M)Py NTF_2					3.28[d]
BuMePyrr SCN	0.0971	0.0608			
BuMePyrr BF_4				3.9	
BuMePyrr PF_6				3.6	
BuMePyrr NTF_2				3.1	
BuMePyrr CF_3SO_3				3.7	
BuMePyrr $N(CN)_2$	0.1204	0.0613			
BuMePyrr CF_3CO_2	0.1674	0.1030			
BuMePyrr $PF_3(C_2F_5)_3$			0.436[g]	1.9	2.85[l]
Bu_4P HCO_2			0.3480		
$TdHx_3$P Cl					1.38[m]
$TdHx_3$P NTF_2					2.14[m]
$TdHx_3$P CH_3SO_3					1.61[n]
$TdHx_3$P $DoPhSO_3$					1.86[n]

[a][447, 448], [b][461] interpolated, [c][459] interpolated, [d][458], [e][449], [f][460] extrapolated, [g][450], [h][451], [i][455], [j][456], [k][457], [l][452], [m][453], [n][454]

References

1. Greaves TL, Drummond CJ (2008) Protic ionic liquids: properties and applications. Chem Rev 108:206–237
2. Walden P (1914) Molecular weights and electrical conductivity of several fused salts. Bull Acad Imp Sci (St Peersburg): 405–422
3. Greaves TL, Weerawardena A, Fong C, Krodkiewska I, Drummond CJ Jr (2006) Protic ionic liquids: solvents with tunable phase behavior and physicochemical properties. J Phys Chem B 110:22479–22487, Correction: 26506
4. Song X, Kanzaki R, Ishiguro S-I, Umebayashi Y (2012) Physicochemical and acid-base properties of a series of 2-hydroxyethylammonium-based protic ionic liquids. Anal Sci 28:469–480
5. Hayes R, Imberti S, Warr GG, Atkin R (2013) the nature of hydrogen bonding in protic ionic liquids. Angew Chem Int Ed 52:4623–4627

6. Arancibia EL, Castells RC, Nardillo AG (1987) Thermodynamic study of the behavior of two molten organic salts as stationary phases in gas chromatography. J Chromatogr 398:21–29
7. Evans FD, Chen S-H, Schriver GW, Arnett EM (1981) Thermodynamics of solution of nonpolar gases in a fused salt. Hydrophobic bonding behavior in a nonaqueous system. J Am Chem Soc 103:481–482
8. Henderson WA, Fylstra P, De Long HC, Trulova PC, Parsons S (2012) Crystal structure of the ionic liquid EtNH3NO3 – insights into the thermal phase behavior of protic ionic liquids. Phys Chem Chem Phys 14:16041–16046
9. Biquard M, Letellier P, Fromon M (1985) Vapor pressure in the water-ammonium nitrate mixture at 298.15 K. Thermodynamic properties of the water-fused salt system. Can J Chem 63:3587–3595
10. Emel'yanenko VN, Boeck G, Verevkin SP, Ludwig R (2014) volatile times for the very first ionic liquid: understanding the vapor pressures and enthalpies of vaporization of ethylammonium nitrate. Chem Eur J 20:11640–11645
11. Weingärtner H, Knocks A, Schrader W, Kaatze U (2001) dielectric spectroscopy of the room temperature molten salt ethylammonium nitrate. J Phys Chem A 105:8646–8650
12. Allen M, Evans DF, Lumry R (1985) Thermodynamic properties of the ethylammonium nitrate + water system: partial molar volumes, heat capacities, and expansivities. J Solution Chem 14:549–558
13. Perron G, Hardy A, Justice J-C, Desnoyers JE (1993) Model system for concentrated electrolyte solutions: thermodynamic and transport properties of ethylammonium nitrate in acetonitrile and in water. J Solution Chem 22:1159–1171
14. Evans DF, Yamaguchi A, Roman R, Casassa EZ (1982) Micelle formation in ethylammonium nitrate, a low-melting fused salt. J Colloid Interf Sci 88:89–93
15. Hadded M, Bahri H, Letellier P (1986) Surface tensions of water-ethylammonium nitrate binary mixtures at 298 K. J Chim Phys 83:419–426
16. Gramstad T, Haszeldine RN (1957) Perfluoroalkyl derivatives of sulfur. VII. Alkyl trifluoromethanesulfonates as alkylating agents, trifluoromethanesulfonic anhydride as a promoter for esterification, and some reactions of trifluoromethanesulfonic acid. J Chem Soc 4069–4079
17. Corkum R, Milne J (1978) The density, electrical conductivity, freezing point, and viscosity of mixtures of trifluoromethanesulfonic acid and water. Can J Chem 56:1832–1835
18. Sarada T, Granata RD, Foley RT (1978) Properties of trifluoromethanesulfonic acid monohydrate pertinent to its use as a fuel cell electrolyte. J Electrochem Soc 125:1899–1906
19. Barthel J, Buchner R, Hölzl CG, Conway BE (1998) Dynamics of molten CF3SO3H · H2O probed by temperature dependent dielectric spectroscopy. J Chem Soc Faraday Trans 94:1953–1958
20. Barthel J, Maier R, Conway BE (1999) Density, viscosity, and specific conductivity of trifluoromethanesulfonic acid monohydrate from 309.15 K to 408.15 K. J Chem Eng Data 44:155–156
21. Hardacre C, Holbrey JD, McMath SEJ, Nieuwenhuyzen M (2002) Small-angle scattering from long-chain alkylimidazolium-based ionic liquids. ACS Symp Ser 818:400–412
22. Adya AK (2005) Nanoscopic structure of ionic liquids by neutron and X-ray diffraction. J Indian Chem Soc 82:1197–1225
23. Crozier ED, Alberding N, Sundheim BR (1983) EXAFS study of bromomanganate ions in molten salts. J Phys Chem 79:939–943
24. Carmichael AJ, Hardacre C, Holbrey JD, Nieuwenhuyzen M, Seddon KR (1999) A method for studying the structure of low-temperature ionic liquids by XAFS. Anal Chem 71:4572–4574
25. Takahashi S, Suzuya K, Kohara S, Koura N, Curtiss LA, Saboungi M-L (1999) Structure of 1-ethyl-3-methylimidazolium chloroaluminates. Neutron diffraction measurements and ab initio calculations. Z Phys Chem (Munich) 209:209–221

26. Hardacre C, Holbrey JD, McMath SEJ, Bowron DT, Soper AK (2003) Structure of molten 1,3-dimethylimidazolium chloride using neutron diffraction. J Chem Phys 118:273–279
27. Hardacre C, McMath SEJ, Nieuwenhuyzen M, Bowron DT, Soper AK (2003) Liquid structure of 1, 3-dimethylimidazolium salts. J Phys Condens Matter 15:S159–S166
28. Hagiwara R, Matsumoto K, Tsuda T, Ito Y, Kohara S, Suzuya K, Matsumoto H, Miyazaki Y (2002) The structures of alkylimidazolium fluorohydrogenate molten salts studied by high-energy X-ray diffraction. J Non-chryst Solids 312–314:414–418
29. Bradley AE, Hardacre C, Holbrey JD, Johnston S, McMath SEJ, Nieuwenhuyzen M (2002) small-angle x-ray scattering studies of liquid crystalline 1-alkyl-3-methylimidazolium salts. Chem Mater 14:629–635
30. Mizuhata M, Maekawa M, Deki S (2007) Ordered structure in room temperature molten salts containing aliphatic quaternary ammonium ions. ECS Trans 3:89–95
31. Triolo A, Russina O, Fazio B, Appetecchi GB, Carewska M, Passerini S (2009) Nanoscale organization in piperidinium-based room temperature ionic liquids. J Chem Phys 130:164521/1–6
32. de Andrade J, Böes ES, Stassen H (2002) A force field for liquid state simulations on room temperature molten salts. 1-Ethyl-3-methylimidazolium tetrachloro-aluminate. J Phys Chem B 106:3546–3548
33. Salanne M, Siqueira LJA, Seitsonen AP, Madden PA, Kirchner B (2012) From molten salts to room temperature ionic liquids: Simulation studies on chloroaluminate. systems. Faraday Disc 154:171–188
34. Canongia Lopes JNAC, Padua AAH (2006) Nanostructural organization in ionic liquids. J Phys Chem B 110:3330–3335
35. Dupont J (2011) From molten salts to ionic liquids: a "nano" journey. Acc Chem Res 44:1223–1231
36. Lind JE Jr, Abdel-Rehim HAA, Rudich SW (1966) Structure of organic melts. J Phys Chem 70:3610–3619
37. Ueno K, Tokuda H, Watanabe M (2010) Ionicity in ionic liquids: correlation with ionic structure and physicochemical properties. Phys Chem Chem Phys 12:1649–1658
38. Krossing I, Slattery JM, Daguenet C, Dyson PJ, Oleinikova A, Weingärtner H (2006) Why are ionic liquids liquid? A simple explanation based on lattice and solvation energies. J Am Chem Soc 128:13427–13434
39. Jenkins HDB, Roobottom HK, Passmore J, Glasser L (1999) Relationships among ionic lattice energies, molecular (formula unit) volumes, and thermochemical radii. Inorg Chem 38:3609–3620
40. Jenkins HDB, Glasser L (2003) Standard absolute entropy, S 298° values from volume or density. 1. Inorganic materials. Inorg Chem 42:8702–8708
41. Klamt A, Schürmann G (1993) COSMO: a new approach to dielectric screening in solvents with explicit expressions for the screening energy and its gradient. J Chem Soc Perkin Trans 2:799–805
42. Guzman O, Lara JER, del Rio F (2015) Liquid-vapor equilibria of ionic liquids from a saft equation of state with explicit electrostatic free energy contributions. J Phys Chem B 119:5864–5872
43. Johnson JK, Muller EA, Gubbins KE (1994) Equation of state for lennard-jones chains. J Phys Chem 98:6413–6419
44. Oliveira MB, Llovell F, Coutinho JAP, Vaga LF (2012) Modeling the [NTf2] pyridinium ionic liquids family and their mixtures with the soft statistical associating fluid theory equation of state. J Phys Chem B 116:9089–9100
45. Mac Dowell N, Llovell F, Sun N, Hallett JP, George A, Hunt PA, Welton T, Simmons BA, Vega LF (2014) new experimental density data and soft-SAFT models of alkylimidazolium ([CnC1im]+) chloride (Cl-), methylsulfate ([MeSO4]-), and dimethylphosphate ([Me2PO4]-) based ionic liquids. J Phys Chem B 118:6206–6221

46. Ji X, Held C, Sadowski G (2012) Modeling imidazolium-based ionic liquids with ePC-SAFT. Fluid Phase Equilib 335:64–75
47. Palomar J, Ferro VR, Torrecilla JS, Rodriguez F (2007) density and molar volume predictions using cosmo-rs for ionic liquids. An approach to solvent design. Ind Eng Chem Res 46:6041–6048
48. Machida H, Sato Y, Smith RL Jr (2008) Pressure-volume-temperature (PVT) measurements of ionic liquids ([bmim+][PF6−], [bmim+][BF4−], [bmim+][OcSO4−]) and analysis with the Sanchez-Lacombe equation of state. Fluid Phase Equilib 264:147–155
49. Preiss UPRM, Slattery JM, Krossing I (2009) In silico prediction of molecular volumes, heat capacities, and temperature-dependent densities of ionic liquids. Ind Eng Chem Res 48:2290–2296
50. Hosseini SM, Moghadasi J, Papari MM, Nobandegani FF (2011) Modeling the volumetric properties of mixtures involving ionic liquids using perturbed hard-sphere equation of state. J Mol Liq 160:67–71
51. Hosseini SM, Papari MM, Moghadasi J, Nobandegani FF (2012) Performance assessment of new perturbed hard-sphere equation of state for molten metals and ionic liquids: application to pure and binary mixtures. J Non-Cryst Solids 358:1753–1758
52. Ma J, Li J, Fan D, Peng C, Liu H, Hu Y (2011) Modeling pVT properties and vapor-liquid equilibrium of ionic liquids using cubic-plus-association equation of state. Chin J Chem Eng 19:1009–1016
53. Hosseini SM, Alavianmehr MM, Moghadasi J (2013) Density and isothermal compressibility of ionic liquids from perturbed hard-dimer-chain equation of state. Fluid Phase Equilib 356:185–192
54. Machida H, Taguchi R, Sato Y, Smith RL Jr (2011) Measurement and correlation of high pressure densities of ionic liquids, 1-ethyl-3-methylimidazolium l-lactate ([emim][lactate]), 2-hydroxyethyl-trimethylammonium l-lactate ([(C2H4OH)(CH3)3 N][Lactate]), and 1-butyl-3-methylimidazolium chloride ([bmim][Cl]). J Chem Eng Data 56:923–928
55. Tome LIN, Gardas RL, Carvalho PJ, Pastoriza-Gallego MJ, Pineiro MM, Coutinho JAP (2011) Measurements and correlation of high-pressure densities of phosphonium based ionic liquids. J Chem Eng Data 56:2205–2217
56. IL Thermo Database (2013) Natl Inst Stand Technol No. 147
57. Shirota H, Mandai T, Fukazawa H, Kazo T (2011) comparison between dicationic and monocationic ionic liquids: liquid density, thermal properties, surface tension, and shear viscosity. J Chem Eng Data 56:2453–2459
58. Fredlake CP, Crosthwaite JM, Hert DG, Aki SNVK, Brennecke JF (2004) Thermophysical properties of imidazolium-based ionic liquids. J Chem Eng Data 49:954–964
59. Berthod A, Ruiz-Angel MJ, Carda-Broch S (2008) Ionic liquids in separation techniques. J Chromatogr A 1184:6–18
60. Crosthwaite JM, Muldoon MJ, Dixon JK, Anderson JL, Brennecke JF (2005) Phase transition and decomposition temperatures, heat capacities and viscosities of pyridinium ionic liquids. J Chem Thermodyn 37:559–568
61. Blesic M, Swadzba-Kwasny M, Belhocine T, Nimal Guanarantew HQ, Canongia Lopes JN, Costa Gomes MF, Padua AAH, Seddon KR, Rebelo LPN (2009) 1-Alkyl-3-methylimidazolium alkanesulfonate ionic liquids, [C(n)H(2)(n)(+1)mim][C(k)H(2)k)(+1)SO(3)]: synthesis and physicochemical properties. Phys Chem Chem Phys 11:8939–8948
62. Liu Q-S, Yang MF, Li P-P, Sun S-S, Weiz-Biermann U, Tan Z-C, Zhang Q-G (2011) Physicochemical properties of ionic liquids [C3py][NTf2] and [C6py][NTf2]. J Chem Eng Data 56:4094–4101
63. Liu Q-S, Yang MF, Yan P-F, Liu X-M, Tan Z-C, Weiz-Biermann U (2010) Density and surface tension of ionic liquids [Cnpy][NTf2] (n=2, 4, 5). J Chem Eng Data 55:4928–4930
64. Huo Y, Xia S, Zhang Y, Ma P (2009) Group contribution method for predicting melting points of imidazolium and benzimidazolium ionic liquids. Ind Eng Chem Res 48:2212–2217

65. Torrecilla JS, Rodriguez F, Bravo JL, Rothenberg G, Seddon KR, Lopez-Martin I (2008) Optimising an artificial neural network for predicting the melting point of ionic liquids. Phys Chem Chem Phys 10:5826–5831
66. Luo H, Huang J-F, Dai S (2008) studies on thermal properties of selected aprotic and protic ionic liquids. Sep Sci Technol 43:2473–2488
67. Lopez-Martin I, Burello E, Davey PN, Seddon KR, Rothenberg G (2007) Anion and cation effects on imidazolium salt melting points: a descriptor modelling study. ChemPhysChem 8:690–695
68. Yoshida Y, Saito G (2006) Influence of structural variations in 1-alkyl-3-methylimida-zolium cation and tetrahalogenoferrate(III) anion on the physical properties of the paramagnetic ionic liquids. J Mater Chem 16:1254–1262
69. Zhou Z-B, Takeda M, Ue M (2004) New hydrophobic ionic liquids based on perfluoroalkyltrifluoroborate anions. J Fluor Chem 125:471–476
70. Hudleston JG, Visser AE, Reichert WM, Willauer HD, Broker GA, Rogers RD (2001) Characterization and comparison of hydrophilic and hydrophobic room temperature ionic liquids incorporating the imidazolium cation. Green Chem 3:156–164
71. Wasserscheid P, van Hal R, Bosmann A (2002) 1-n-Butyl-3-methylimidazolium ([bmim]) octylsulfate – an even 'greener' ionic liquid. Green Chem 4:400–404
72. Lashkarbolooki M, Zeinolabedini A, Ayatollahi S (2012) Artificial neural network as an applicable tool to predict the binary heat capacity of mixtures containing ionic liquids. Fluid Phase Equilib 324:102–107
73. Troncoso J, Cerdeirina CA, Sanmaned YA, Romani L, Rebelo LPN (2006) Thermodynamic properties of imidazolium-based ionic liquids: densities, heat capacities, and enthalpies of fusion of [bmim][PF6] and [bmim][NTf2. J Chem Eng Data 51:1856–1859
74. Gardas RL, Coutinho JAP (2008) A group contribution method for heat capacity estimation of ionic liquids. Ind Eng Chem Res 47:5751–5757
75. Kabo GJ, Paulechka YU, Kabo AG, Blokhin AV (2010) Experimental determination of enthalpy of 1-butyl-3-methylimidazolium iodide synthesis and prediction of enthalpies of formation for imidazolium ionic liquids. J Chem Thermodyn 42:1292–1297
76. Freire MG, Teles ARR, Rocha MAA, Schröder B, Neves CMSS, Carvalho PJ, Evtuguin DV, Santos LMNBF, Coutinho JAP (2011) Thermophysical characterization of ionic liquids able to dissolve biomass. J Chem Eng Data 56:4813–4822
77. Gomez E, Calvar N, Dominguez A, Macedo EA (2013) Thermal analysis and heat capacities of 1-alkyl-3-methylimidazolium ionic liquids with NTf2-, TFO-, and DCA- anions. Ind Eng Chem Res 52:2103–2110
78. Verevkin SP, Zaitsau DH, Emel'yanenko VN, Ralys RV, Yermalayeu AV, Schick C (2013) Does alkyl chain length really matter? Structure-property relationships in thermochemistry of ionic liquids. Thermochem Acta 562:84–95
79. Xie Y, Zhang Y, Lu X, Ji X (2014) Energy consumption analysis for CO2 separation using imidazolium-based ionic liquids. Appl Energy 136:325–335
80. Tokuda H, Hayamizu K, Ishii K, Susan MABH, Watanabe M (2005) Physicochemical properties and structures of room temperature ionic liquids. 2. Variation of alkyl chain length in imidazolium cation. J Phys Chem B 109:6103–6110
81. Zhu J, Bau L, Chen B, Fei W (2009) Thermodynamical properties of phase change materials based on ionic liquids. Chem Eng J 147:58–62
82. Preiss U, Verevkin AP, Koslowski T, Krossing I (2011) Going full circle: phase-transition thermodynamics of ionic liquids. Chem Eur J 17:6508–6517
83. Paulechka YU (2010) Heat capacity of room-temperature ionic liquids: a critical review. J Phys Chem Ref Data 39:033108/1–24
84. Zhu Q, Gao Y, Xiao J, Xie GJ (2012) Preconcentration and determination of aromatic amines with temperature-controlled ionic liquid dispersive liquid phase microextraction in combination with high performance liquid chromatography. AOAC Int 95:1534–1538

85. Zhang Y, Maginn EJ (2014) Molecular dynamics study of the effect of alkyl chain length on melting points of [CnMIM][PF6] ionic liquids. Phys Chem Chem Phys 16:13489–13499
86. Verevkin AP, Zaitsau DH, Emel'yanenko VN, Yermalayeu AV, Schick C, Liu H, Maginn EJ, Bulut S, Krossing I, Kalb R (2013) Making sense of enthalpy of vaporization trends for ionic liquids: new experimental and simulation data show a simple linear relationship and help reconcile previous data. J Phys Chem B 117:6473–8486
87. Law G, Watson PR (2001) Surface tension measurements of n-alkylimidazolium ionic liquids. Langmuir 17:6138–6141
88. Martino W, de la Mora JF, Yoshida Y, Saito G, Wilkes J (2006) Surface tension measurements of highly conducting ionic liquids. Green Chem 8:390–397
89. Kilaru P, Baker GA, Scovazzo P (2007) Density surface tension measurements of imidazolium-, quaternary phosphonium-, and ammonium-based room-temperature ionic liquids: data and correlations. J Chem Eng Data 52:2306–2314
90. Pereiro AB, Verdia P, Tojo E, Rodriguez A (2007) Physical properties of 1-butyl-3-methylimidazolium methyl sulfate as a function of temperature. J Chem Eng Data 52:377–380
91. Fröba AP, Kremer H, Leipertz A (2008) Density, refractive index, interfacial tension, and viscosity of ionic liquids [EMIM][EtSO4], [EMIM][NTf2], [EMIM][N(CN)2], and [OMA][NTf2] in dependence on temperature at atmospheric pressure. J Phys Chem B 112:12420–12430
92. Ghatee MH, Zolghadr AR (2008) Surface tension measurements of imidazolium-based ionic liquids at liquid-vapor equilibrium. Fluid Phase Equilib 263:168–175
93. Tong J, Liu Q-S, Xu W-G, Fang F-W, Yang J-Z (2008) Estimation of physicochemical properties of ionic liquids 1-alkyl-3-methylimidazolium chloroaluminate. J Phys Chem B 112:4381–4386
94. Klomfar J, Součkova M, Patek J (2009) Surface tension measurements for four 1-alkyl-3-methylimidazolium-based ionic liquids with hexafluorophosphate anion. J Chem Eng Data 54:1389–1394
95. Součkova M, Klomfar J, Patek J (2011) Surface tension of 1-alkyl-3-methylimidazolium based ionic liquids with trifluoromethanesulfonate and tetrafluoroborate anion. Fluid Phase Equilib 303:184–190
96. Santos CS, Baddelli S (2009) Alkyl chain interaction at the surface of room temperature ionic liquids: systematic variation of alkyl chain length (R = C1-C4, C8) in both cation and anion of [RMIM][R-OSO3] by sum frequency generation and surface tension. J Phys Chem B 113:923–933
97. Domanska U, Krolikowska M (2010) Effect of temperature and composition on the surface tension and thermodynamic properties of binary mixtures of 1-butyl-3-methylimidazolium thiocyanate with alcohols. J Colloid Interf Sci 348:661–667
98. Klomfar J, Součkova M, Patek J (2010) Surface tension measurements with validated accuracy for four 1-alkyl-3-methylimidazolium based ionic liquids. J Chem Thermodyn 42:323–329
99. Kolbeck C, Lehmann J, Lovelock KRJ et al (2010) density and surface tension of ionic liquids. J Phys Chem B 114:17025–17036
100. Anantharaj R, Benerjee T (2011) Phase behavior of 1-ethyl-3-methylimidazolium thiocyanate ionic liquid with catalytic deactivated compounds and water at several temperatures: experiments and theoretical predictions. Int J Chem Eng 209435/1–13
101. Guan W, Ma X-X, Li L, Tong J, Fang D-W, Yang J-Z (2011) Ionic parachor and its application in acetic acid ionic liquid homologue 1-alkyl-3-methylimidazolium acetate {[$C_n$mim][OAc](n = 2,3,4,5,6)}. J Phys Chem B 115:12915–12920
102. Klomfar J, Součkova M, Patek J (2011) Temperature dependence of the surface tension and density at 0.1 MPa for 1-ethyl- and 1-butyl-3-methylimidazolium dicyanamide. J Chem Eng Data 56:3454–3462

103. Ruso JW, Hoffmann M (2011) Measurements of surface tension and chemical shift on several binary mixtures of water and ionic liquids and their comparison for assessing aggregation. J Chem Eng Data 56:3703–3710
104. Anantharaj R, Banerjee T (2013) Thermodynamic properties of 1-ethyl-3-methylimidazolium methanesulphonate with aromatic sulphur, nitrogen compounds at T = 298.15-323.15 K and P = 1 bar. Can J Chem Eng 97:245–256
105. Beigi AAN, Abdouss M, Yousefi M, Pourmortazavi AM, Vahid A (2013) Investigation on physical and electrochemical properties of three imidazolium based ionic liquids (1-hexyl-3-methylimidazolium tetrafluoroborate, 1-ethyl-3-methylimidazolium bis(trifluoromethylsulfonyl) imide and 1-butyl-3-methylimidazolium methylsulfate). J Mol Liq 177:361–368
106. Cao Y, Mu T (2014) Comprehensive investigation on the thermal stability of 66 ionic liquids by thermogravimetric analysis. Ind Eng Chem Res 53:8651–8664
107. Gruzdev MS, Ramenskaya IM, Chervonova UV, Kumeev RS (2009) Preparation of 1-butyl-3-methylimidazolium salts and study of their phase behavior and intramolecular interactions. Russ J Gen Chem 79:1720–1727
108. Yan C, Han M, Wan H, Guan G (2010) QSAR correlation of the melting points for imidazolium bromides and imidazolium chlorides ionic liquids. Fluid Phase Equilib 292:104–109
109. Domanska U, Morawski P (2007) Influence of high pressure on solubility of ionic liquids: experimental data and correlation. Green Chem 9:361–368
110. Domanska U, Krolikowski M (2010) Phase equilibria study of the binary systems (1-butyl-3-methylimidazolium tosylate ionic liquid + water, or organic solvent). J Chem Thermodyn 42:355–362
111. Zhang ZH, Sun LX, Tan ZC, Xu F, Lu XC, Zeng JL, Sawada Y (2007) Thermodynamic investigation of room temperature ionic liquid. Heat capacity and thermodynamic functions of BPBF4. J Therm Anal Calorim 89:289–294
112. Pacholec F, Poole CF (1983) Stationary phase properties of the organic molten salt ethylpyridinium bromide in gas chromatography. Chromatographia 17:370–376
113. Garcia-Mardones M, Bandres I, Lopez MC, Gascon I, Lafuente C (2012) Experimental theoretical study of two pyridinium-based ionic liquids. J Solution Chem 41:1836–1852
114. Calvar N, Gomez E, Macedo EA, Dominguez A (2013) Thermal analysis and heat capacities of pyridinium and imidazolium ionic liquids. Thermochem Acta 565:178–182
115. Bandres I, Pera G, Martin S, Castro M, Lafuente C (2009) Thermophysical study of 1-butyl-2-methylpyridinium tetrafluoroborate ionic liquid. J Phys Chem B 113:11936–11942
116. Liu Q-S, Li P-P, Weiz-Biermann U, Liu V, Chen J (2012) Density, electrical conductivity, and dynamic viscosity of n-alkyl-4-methylpyridinium bis(trifluoromethylsulfonyl)imide. J Chem Eng Data 57:2999–3004
117. Yunus NM, Abdul Mutalib MI, Man Z, Bustam MA, Murugesan T (2010) Thermophysical properties of 1-alkylpyridinium bis(trifluoromethylsulfonyl)imide ionic liquids. J Chem Thermodyn 42:491–495
118. Guerrero H, Martin S, Perez-Gregorio V, Lafuente C, Bandres I (2012) Volumetric characterization of pyridinium-based ionic liquids. Fluid Phase Equilib 317:102–109
119. Domanska U, Krolikowski M, Pobudkowska A, Letcher TM (2009) Phase equilibria study of the binary systems (n-butyl-4-methylpyridinium tosylate ionic liquid + organic solvent, or water). J Chem Eng Data 54:1435–1441
120. Papaiconomou N, Salminen J, Lee J-M, Prausnitz JM (2007) Physicochemical properties of hydrophobic ionic liquids containing 1-octylpyridinium, 1-octyl-2-methylpyridinium, or 1-octyl-4-methylpyridinium cations. J Chem Eng Data 52:833–840
121. Pereiro AB, Rodriguez A, Blesic M, Shimizu K, Lopes JNC, Rebelo LPN (2011) Mixtures of pyridine and nicotine with pyridinium-based ionic liquids. J Chem Eng Data 56:4356–4363
122. Tong B, Liu Q-S, Tan Z-C, Welz-Biermann U (2010) Thermochemistry of alkyl pyridinium bromide ionic liquids: calorimetric measurements and calculations. J Phys Chem A 114:3782–3787

123. Bandres I, Royo FM, Gascon I, Castro M, Lafuente C (2010) Anion influence on thermophysical properties of ionic liquids: 1-butylpyridinium tetrafluoroborate and 1-butylpyridinium triflate. J Phys Chem B 114:3601–3607
124. Iken H, Guillen F, Chaumat H, Mazieres M-R, Plaquevent J-C, Tzedakis T (2012) Scalable synthesis of ionic liquids: comparison of performances of microstructured and stirred batch reactors. Tetrahedron Lett 53:3474–3477
125. Bandres I, Lopez MC, Castro M, Barbera J, Lafuente C (2012) Thermophysical properties of 1-propylpyridinium tetrafluoroborate. J Chem Thermodyn 44:148–153
126. Garcia-Miaja G, Troncoso J, Romani L (2007) Density and heat capacity as a function of temperature for binary mixtures of 1-butyl-3-methylpyridinium tetrafluoroborate plus water, plus ethanol, and plus nitromethane. J Chem Eng Data 52:2261–2265
127. Pinto AM, Rodriguez H, Arce A, Soto A (2013) Carbon dioxide absorption in the ionic liquid 1-ethylpyridinium ethylsulfate and in its mixtures with another ionic liquid. Intl J Greenh Gas Control 18:296–304
128. Farhani N, Gharagheizi F, Mirkhani SA, Tumba K (2013) A simple correlation for prediction of heat capacities of ionic liquids. Fluid Phase Equilib 337:73–82
129. Sashina ES, Kashirskii DA, Janowska G, Zaborski M (2013) Thermal properties of 1-alkyl-3-methylpyridinium halide-based ionic liquids. Thermochem Acta 568:185–188
130. Sattari M, Gharagheizi F, Ilani-Kashkouli P, Mohammadi AH, Ramjugernath DJ (2014) Development of a group contribution method for the estimation of heat capacities of ionic liquids. J Therm Anal Calorim 115:1863–1882
131. Domanska U, Krolikowski M, Pobudkowska A, Bochenska P (2012) Solubility of ionic liquids in water and octan-1-ol and octan-1-ol/water, or 2-phenylethanol/water partition coefficients. J Chem Thermodyn 55:225–233
132. Bittner B, Wrobel RJ, Milchert E (2012) Physical properties of pyridinium ionic liquids. J Chem Thermodyn 55:159–165
133. Garcia-Mardones M, Cea P, Gascon I, Lafuente C (2014) Thermodynamic study of the surface of liquid mixtures containing pyridinium-based ionic liquids and alkanols. J Chem Thermodyn 78:234–240
134. Bhattacharjee A, Carvalho PJ, Coutinho JAP (2014) Fluid Phase Equilib 375:80–88
135. Wang J-Y, Zhang X-j, Hu Y-q, Qi G-d, Liang L-y (2012) Properties of n-butylpyridinium nitrate ionic liquid and its binary mixtures with water. J Chem Thermodyn 45:43–47
136. Li H, Zhao G, Liu, Zhang S (2013) Physicochemical characterization of MFm – based ammonium ionic liquids. J Chem Eng Data 58:1505–1515
137. Ballantyne AD, Brisdon AK, Dryfe RA (2008) Immiscible electrolyte systems based on asymmetric hydrophobic room temperature ionic liquids. Chem Commun 4980–4982
138. Ghatee MH, Zare M (2011) Power-law behavior in the viscosity of ionic liquids: existing a similarity in the power law and a new proposed viscosity equation. Fluid Phase Equilib 311:76–82
139. Davey TW, Ducker WA, Hayman AR, Simpson J (1998) Krafft temperature depression in quaternary ammonium bromide surfactants. Langmuir 14:3210–3213
140. Scurto MA, Newton E, Weikrl RR, Draucker L, Hallett J, Liotta CL, Leitner W, Eckert CA (2008) Melting point depression of ionic liquids with CO2: phase equilibria. Ind Eng Chem Res 47:493–501
141. Köhler S, Liebert T, Heinze T (2009) Ammonium-based cellulose solvents suitable for homogeneous etherification. Macromol Biosci 9:836–841
142. Pas SJ, Pringle JM, Forsyth M, MacFarlane DR (2004) Thermal physical properties of an archetypal organic ionic plastic crystal electrolyte. Phys Chem Chem Phys 6:3721–3725
143. Krieger BM, Lee HY, Emge TJ, Wishart JF, Castner EW Jr (2010) Ionic liquids and solids with paramagnetic anions. Phys Chem Chem Phys 12:8919–8925
144. Henderson WA, Young VG Jr, Passerini S, Trulove PC, De Long HC (2006) Plastic phase transitions in N-Ethyl-N-methylpyrrolidinium Bis(trifluoromethanesulfonyl)imide. Chem Mater 18:934–938

145. Kim K, Cho Y-H, Shin H-C (2013) 1-Ethyl-1-methylpiperidinium bis(trifluoromethanesulfonyl)imide as a co-solvent in Li-ion batteries. J Power Sources 225:113–118
146. Xu M, Ivey DG, Xie Z, Qu W, Dy E (2013) The state of water in 1-butyl-1-methylpyrrolidinium bis(trifluoromethanesulfonyl)imide and its effect on Zn/Zn(II) redox behavior. Electrochim Acta 97:289–295
147. Fu S, Gong S, Liu C, Zheng L, Feng, Nie J, Zhou Z (2013) Ionic liquids based on bis(2,2,2-trifluoroethoxysulfonyl)imide with various oniums. Electrochim Acta 94:229–237
148. Le M-L-P, Alloin F, Strobel P, Lepretre J-C, Cointeaux L, del Valle CP (2012) Electrolyte based on fluorinated cyclic quaternary ammonium ionic liquids. Ionics 18:817–827
149. Sun I-W, Wang HP, Teng H, Su S-G, Lin Y-C, Kuo C-W, Chen P-R, Wu T-Y (2012) Cyclic ammonium-based ionic liquids as potential electrolytes for dye-sensitized solar cells. Int J Elechtrochem Sci 7:9748–9754
150. Furlani M, Albinsson I, Mellander B-E, Appetecchi GB, Passerini S (2011) Annealing protocols for pyrrolidinium bis(trifluoromethylsulfonyl)imide type ionic liquids. Electrochim Acta 57:220–227
151. Domanska U (2010) Physico-chemical properties and phase behaviour of pyrrolidinium-based ionic liquids. Int J Mol Sci 11:1825–1841
152. Forsyth SA, Fraser KJ, Howlett PC, MacFarlane DR, Forsyth M (2006) N-Methyl-N-alkylpyrrolidinium nonafluoro-1-butanesulfonate salts: ionic liquid properties and plastic crystal behavior. Green Chem 8:256–261
153. MacFarlane DR, Forsyth SA, Golding J, Deacon GB (2002) Ionic liquids based on imidazolium, ammonium and pyrrolidinium salts of the dicyanamide anion. Green Chem 4:444–448
154. Carvalho PJ, Ventura SPM, Batista MLS, Schröder B, Gonçalves F, Esperança J, Mutelet F, Coutinho JAP (2014) Understanding the impact of the central atom on the ionic liquid behavior: phosphonium vs. ammonium cations. J Chem Phys 140:064505/1-9
155. Matsumoto H, Sakaebe H, Tatsumi K (2005) Preparation of room temperature ionic liquids based on aliphatic onium cations and asymmetric amide anions and their electrochemical properties as a lithium battery electrolyte. J Power Sources 146:45–50
156. Galinski M, Lewandowski A, Stepniak I (2006) Ionic liquids as electrolytes. Electrochim Acta 51:5567–5580
157. Gonzalez EJ, Gonzalez B, Macedo EA (2013) Thermophysical properties of the pure ionic liquid 1-Butyl-1-methylpyrrolidinium dicyanamide and its binary mixtures with alcohols. J Chem Eng Data 58:1440–1448
158. Shimizu Y, Ohte Y, Yamamura Y, Tsuzuki S, Saito K (2012) Comparative study of imidazolium- and pyrrolidinium-based ionic liquids: thermodynamic properties. J Phys Chem B 116:5406–5413
159. Domanska U, Lrolikowska M (2011) Fluid Phase Equilib 308:55–63
160. Sanchez LG, Espel JR, Onink F, Meindersma GW, de Haan AB (2009) Density, viscosity, and surface tension of synthesis grade imidazolium, pyridinium, and pyrrolidinium based room temperature ionic liquids. J Chem Eng Data 54:2803–2812
161. Ghatee MH, Bahrami M, Khanjari N (2013) Measurement and study of density, surface tension, and viscosity of quaternary ammonium-based ionic liquids ([N222(n)]Tf2N). J Chem Thermodyn 65:42–52
162. Taggougui M, Diaw M, Carre B, Willmann P, Lemordant D (2008) Solvents in salt electrolyte: benefits and possible use as electrolyte for lithium-ion battery. Electrochim Acta 53:5496–5502
163. O'Mahony AM, Silvester DS, Aldous L, Hardacre C, Compton RG (2008) Effect of water on the electrochemical window and potential limits of room-temperature ionic liquids. J Chem Eng Data 53:2884–2891
164. Pan Y, Boyd LE, Kruplak JF, Cleland WE Jr, Wilkes JS, Hussey C (2011) Physical and transport properties of Bis(trifluoromethylsulfonyl)imide-based room-temperature inic

liquids: application to the diffusion of Tris(2,2′-bipyridyl)ruthenium(II). J Electrochem Soc 158:F1–F9
165. Klomfar J, Souckova M, Patek J (2014) Low temperature densities from (218 to 364) K and up to 50 MPa in pressure and surface tension for Trihexyl(tetradecyl)phosphonium Bis (trifluoromethylsulfonyl)imide and dicyanamide and 1-Hexyl-3-methylimidazolium hexafluorophosphate. J Chem Eng Data 59:2263–2274
166. Olivera MB, Dominguez-Perez M, Cabeza O, Lopes-da-Silva JA, Freire MG, Coutinho JAP (2013) Surface tensions of binary mixtures of ionic liquids with bis(trifluoromethylsulfonyl) imide as the common anion. J Chem Thermodyn 64:22–27
167. Tsunashima K, Sugiya M (2007) Physical and electrochemical properties of room temperature ionic liquids based on quaternary phosphonium cations. Electrochemical 75:734–736
168. Ferreira AGM, Simoes PN, Ferreira AF, Fonseca MA, Oliviera MSA, Trino ASM (2013) Transport thermal properties of quaternary phosphonium ionic liquids and IoNanofluids. J Chem Thermodyn 64:80–92
169. Tsunashima K, Kodama S, Sugiya M, Kunugi Y (2010) Physical electrochemical properties of room-temperature dicyanamide ionic liquids based on quaternary phosphonium cations. Electrochim Acta 56:762–766
170. Tsunashima K, Kawabara A, Matsumiya M, Kodama S, Enomoto R, Sugiya M, Kunugi Y (2011) Low viscous and highly conductive phosphonium ionic liquids based on bis (fluorosulfonyl)amide anion as potential electrolytes. Electrochem Commun 13:178–181
171. Ferreira AGM, Simoes PN, Ferreira AF (2012) Quaternary phosphonium-based ionic liquids: thermal stability and heat capacity of the liquid phase. J Chem Thermodyn 45:16–27
172. Sun N, He X, Dong K, Zhang X, Lu X, He H, Zhang S (2006) Prediction of the melting points for two kinds of room temperature ionic liquids. Fluid Phase Equilib 246:137–142
173. Bini R, Ciappe C, Duce C, Micheli A, Solaro R, Starita A, Tine MR (2008) Ionic liquids: prediction of their melting points by a recursive neural network model. Green Chem 10:306–309
174. Ludwig R, Kragl U (2007) Do we understand the volatility of ionic ligands? Angew Chem Intl Ed 46:6582–6584
175. Zaitsau DH, Kabo GJ, Strechan AA, Paulechka YU, Tscherisch A, Verevkin SP, Heintz A (2006) Experimental vapor pressures of 1-alkyl-3-methylimidazolium bis(trifluoromethylsulfonyl)imides and a correlation scheme for estimation of vaporization enthalpies of ionic liquids. J Phys Chem A 110:7303–7306
176. Esperança JMSS, Lopes JNC, Tariq M, Santos LMNBF, Magee JW, Rebelo LPN (2010) Volatility of aprotic ionic liquids – a review. J Chem Eng Data 55:3–12
177. Santos LMNBF, Lopes JN, Coutinho JAP, Esperanca JMSS, Gomes LR, Manucho IM, Rebelo LPNJ (2007) Ionic liquids: first direct determination of their cohesive energy. J Am Chem Soc 129:284–285
178. Rocha MAA, Lima CFRAC, Gomes LR, Schroder B, Coutinho JAP, Marrucho IM, Esperança JMSS, Rebelo LPN, Shimizu K, Lopes JNC, Santos LMNBF (2011) High-accuracy vapor pressure data of the extended [Cnmim]NTF2] ionic liquid series: trend changes and structural shifts. J Phys Chem B 115:10919–10926
179. Rane KS, Errington JR (2014) Saturation properties of 1-alkyl-3-methylimidazolium based ionic liquids. J Phys Chem B 118:8734–8743
180. Paduszynski K, Domanska U (2012) Thermodynamic modeling of ionic liquid systems: development and detailed overview of novel methodology based on the PC-SAFT. J Phys Chem B 116:5002–5018
181. Greaves TL, Drummond CJ (2013) Solvent nanostructure, the solvophobic effect and amphiphile self-assembly in ionic liquids. Chem Soc Rev 42:1096–1120
182. Preiss U, Zaitsau DH, Beichel W, Himmel D, Higelin A, Merz K, Caesa N, Verevkin SP (2015) Estimation of lattice enthalpies of ionic liquids supported by hirshfeld analysis. ChemPhysChem 16:2890–2898

183. Rocha MAA, Ribeiro FMS, Schröder B, Coutinho JAP, Santos LMNBF (2014) Volatility study of [C1C1im][NTf2] and [C2C3im][NTf2] ionic liquids. J Chem Thermodyn 68:317–321
184. Weerachanchai P, Chen Z, Leong SSJ, Chang MW, Lee J-M (2012) Hildebrand solubility parameters of ionic liquids: effects of ionic liquid type, temperature and DMA fraction in ionic liquid. Chem Eng J 213:356–362
185. Singh T, Kumar A (2008) Static dielectric constant of room temperature ionic liquids: internal pressure and cohesive energy density approach. J Phys Chem B 112:12968–12972
186. Ren N-n, Gong Y-h, Lu Y-z, Meng H, Li C-x (2014) Surface tension measurements for seven imidazolium-based dialkylphosphate ionic liquids and their binary mixtures with water (methanol or ethanol) at 298.15 K and 1 atm. J Chem Eng Data 59:189–196
187. Liu Z, Wu X, Wang WA (2006) Novel united-atom force field for imidazolium-based ionic liquids. Phys Chem Chem Phys 8:1096–1104
188. Schröder B, Coutinho JAP (2014) Predicting enthalpies of vaporization of aprotic ionic liquids with COSMO-RS. Fluid Phase Equilib 370:24–33
189. Jahangiri S, Toghikhani M, Behnejad H, Ahmadi S (2008) Theoretical investigation of imidazolium based ionic liquid/alcohol mixture: a molecular dynamic simulation. Mol Phys 106:1015–1023
190. Lovelock KRJ, Armstrong JP, Licence P, Jones RG (2007) Vapourisation of ionic liquids. Phys Chem Chem Phys 9:982–990
191. Kilaru PK, Scovazzo P (2008) Correlations of low-pressure carbon dioxide and hydrocarbon solubilities in imidazolium-, phosphonium-, and ammonium-based room-temperature ionic liquids. Part 2. Using activation energy of viscosity. Ind Eng Chem Res 47:910–919
192. Lovelock KRJ, Armstrong JP, Licence P, Jones RG (2014) Vaporisation and thermal decomposition of dialkylimidazolium halide ion ionic liquids. Phys Chem Chem Phys 16:1339–1353
193. Jaquemin J, Nancarrow P, Rooney DW, Gomes MFC, Husson P, Majer V, Padua AAH, Hardacre C (2008) Prediction of ionic liquid properties. II. Volumetric properties as a function of temperature and pressure. J Chem Eng Data 53:2133–2143
194. Marciniak A (2010) The solubility parameters of ionic liquids. Int J Mol Sci 11:1973–1990
195. Marciniak A (2011) The Hildebrand solubility parameters of ionic liquids – part 2. Int J Mol Sci 12:3553–3575
196. Tong J, Yang HX, Liu RJ, Li C, Xia LX, Yang JZ (2014) Determination of the enthalpy of vaporization and prediction of surface tension for ionic liquid 1-alkyl-3-methylimidazolium propionate [Cnmim][Pro](n = 4, 5, 6). J Phys Chem B 118:12972–12978
197. Sistla YS, Jain L, Khanna A (2012) Validation prediction of solubility parameters of ionic liquids for CO2 capture. Sep Purif Technol 97:51–64
198. Lee SH, Lee SB (2005) The Hildebrand solubility parameters, cohesive energy densities and internal energies of 1-alkyl-3-methylimidazolium-based room temperature ionic liquids. Chem Commun 3469–3471
199. Swiderski K, McLean A, Gordon CM, Vaughan DH (2004) Estimates of internal energies of vaporisation of some room temperature ionic liquids. Chem Commun 2178–2179
200. Rocha MAA, Santos LMNBF (2013) First volatility study of the 1-alkylpyridinium based ionic liquids by Knudsen effusion. Chem Phys Lett 585:59–62
201. Chandran A, Prerkash K, Senepati S (2010) Structure and dynamics of acetate anion-based ionic liquids from molecular dynamics study. Chem Phys 374:46–54
202. Xu A, Wang J, Zhang Y, Chen Q (2012) Effect of alkyl chain length in anions on thermodynamic and surface properties of 1-butyl-3-methylimidazolium carboxylate ionic liquids. Ind Eng Chem Res 51:3458–3465
203. Vilas M, Rocha MAA, Fernandes AM, Tojo E, Santos LMNBF (2015) Novel 2-alkyl-1-ethylpyridinium ionic liquids: synthesis, dissociation energies and volatility. Phys Chem Chem Phys 17:2560–2572
204. Jin H, O'Hare B, Dong J, Arzhantzev S, Baker GA, Wishart JF, Benesi AJ, Maroncelli M (2008) Physical properties of ionic liquids consisting of the 1-butyl-3-methylimidazolium

cation with various anions and the bis(trifluoromethyl-sulfonyl)imide anion with various cations. J Phys Chem B 112:81–92
205. Zaitsau DH, Yermalayeu AV, Emel'yanko VN, Heintz A, Verevkin SP, Schick C, Berdzinski S, Strehmel V (2014) Structure-property relationships in ILs: vaporization enthalpies of pyrrolidinium based ionic liquids. J Mol Liq 192:171–176
206. Paduszynski K, Domanska U (2013) Experimental and theoretical study on infinite dilution activity coefficients of various solutes in piperidinium ionic liquids. J Chem Thermodyn 60:169–178
207. Requejo PF, Gonzalez EJ, Mecedo EA, Dominguez A (2014) Effect of the temperature on the physical properties of the pure ionic liquid 1-ethyl-3-methylimidazolium methylsulfate and characterization of its binary mixtures with alcohols. J Chem Thermodyn 74:193–200
208. Derecskei B, Derecskei-Kovacs A (2008) Molecular modelling simulations to predict density and solubility parameters of ionic liquids. Mol Simul 34:1167–1175
209. Greaves TL, Drummond CJ (2008) Ionic liquids as amphiphile self-assembly media. Chem Soc Rev 37:1709–1726
210. Greaves TL, Weerawardena A, Krodkiewska I, Drummond CJ (2008) Protic ionic liquids: physicochemical properties and behavior as amphiphile self-assembly solvents. J Phys Chem B 112:896–905
211. Rebelo LPN, Lopes JNC, Esperança JMSS, Filipe E (2005) On the critical temperature, normal boiling point, and vapor pressure of ionic liquids. J Phys Chem B 109:6040–6045
212. Weiss VC (2010) Guggenheim's rule and the enthalpy of vaporization of simple and polar fluids, molten salts, and room temperature ionic liquids. J Phys Chem B 114:9183–9194
213. Rai N, Maginn EJ (2012) Critical behaviour and vapour-liquid coexistence of 1-alkyl-3-methylimidazolium bis(trifluoromethylsulfonyl)amide ionic liquids via Monte Carlo simulations. Faraday Disc 154:53–69
214. Wu T-Y, Chen B-K, Kuo C-W, Hao L, Peng Y-C, Sun I-W (2012) Standard entropy, surface excess entropy, surface enthalpy, molar enthalpy of vaporization, and critical temperature of bis(trifluoromethanesulfonyl)imide-based ionic liquids. J Taiwan Inst Chem Eng 43:860–867
215. Weiss VC, Heggen B, Muller-Plathe F (2010) Critical parameters and surface tension of the room temperature ionic liquid [bmim][PF6]: a corresponding-states analysis of experimental and new simulation data. J Phys Chem C 114:3599–3608
216. Almeida HFD, Teles ARR, Lopes-da-Silva JA, Freire MG, Coutinho JAP (2012) Influence of the anion on the surface tension of 1-ethyl-3-methylimidazolium-based ionic liquids. J Chem Thermodyn 54:49–54
217. Almeida HFD, Passos H, Lopes-da-Silva JA, Fernandes AM, Freire MG, Coutinho JAP (2012) Thermophysical properties of five acetate-based ionic liquids. J Chem Eng Data 57:3005–3013
218. Bhattacharjee A, Luis A, Santos JH, Lopes-da-Silva JA, Freire MG, Carvalho PJ, Coutinho JAP (2014) Thermophysical properties of sulfonium- and ammonium-based ionic liquids. Fluid Phase Equilib 381:36–45
219. Bhattacharjee A, Luis A, Lopes-da-Silva JA, Freire MG, Coutinho JAP, Carvalho PJ (2014) Thermophysical properties of phosphonium-based ionic liquids. Fluid Phase Equilib 400:103–113
220. Valderrama JO, Robles PA (2007) Critical properties, normal boiling temperatures, and acentric factors of fifty ionic liquids. Ind Eng Chem Res 46:1338–1344
221. Valderrama JO, Sanga WW, Lazzus JA (2008) Critical properties, normal boiling temperature, and acentric factor of another 200 ionic liquids. Ind Eng Chem Res 47:1318–1330
222. Valderrama JO, Rojas RE (2009) Critical properties of ionic liquids. Ind Eng Chem Res 48:6890–6900 (revisited)
223. Valderrama JO, Forero LA, Rojas RE (2012) Critical properties and normal boiling temperature of ionic liquids. Update and a new consistency test. Ind Eng Chem Res 51:7838–7844

224. Valderrama JO, Forero LA, Rojas RE (2015) Extension of a group contribution method to estimate the critical properties of ionic liquids of high molecular mass. Ind Eng Chem Res 54:3490–3497
225. Ge R, Hardacre C, Jacquemin J, Nancarrow P, Rooney DW (2008) Heat capacities of ionic liquids as a function of temperature at 0.1 MPa. Measurement and prediction. J Chem Eng Data 53:2148–2153
226. Sanmamed YA, Navia P, Gonzalez-Salgado D, Troncoso J, Romani LJ (2010) Pressure temperature dependence of isobaric heat capacity for [Emim]BF4] [Bmim]BF4] [Hmim] BF4] and [Omim]BF4]. Chem Eng Data 55:600–604
227. Glasser L, Jenkins HBD (2011) Ambient isobaric heat capacities, cp,m, for ionic solids and liquids: an application of volume-based thermodynamics (VBT). Inorg Chem 50:8565–8560
228. Farahani N, Gharagherzi F, Mirkhani SA, Tumba K (2013) A simple correlation for prediction of heat capacities of ionic liquids. Fluid Phase Equilib 337:73–82
229. Larriba C, Yoshida Y, de la Mora JF (2008) Correlation between surface tension and void fraction in ionic liquids. J Phys Chem B 112:12401–12407
230. Sugden S (1924) A relation between surface tension, density and chemical composition. J Chem Soc Trans 125:1177–1189
231. Ma X-X, Wei J, Guan W, Pan Y, Zheng L, Wu Y, Yang J-Z (2015) Ionic parachor and its application to pyridinium-based ionic liquids of {[CnPy]DCA] (n = 2, 3, 4, 5, 6). J Chem Thermodyn 89:51–59
232. Souckova M, Klomfar J, Patek J (2015) Surface tension and 0.1 MPa density data for 1-Cn-3-methylimidazolium iodides with n = 3, 4, and 6, validated using a parachor and group contribution model. J Chem Thermodyn 83:52–60
233. Xu W-G, Li L, Ma X-X, Wei J, Duan W-B, Guan W, Yang J-Z (2012) Density surface tension, and refractive index of ionic liquids homologue of 1-alkyl-3-methylimidazolium tetrafluoroborate [Cnmim]BF4] (n = 2,3,4,5,6). J Chem Eng Data 57:2177–2184
234. Gardas RL, Coutinho JAP (2008) Applying a QSPR correlation to the prediction of surface tensions of ionic liquids. Fluid Phase Equilib 265:57–65
235. Gardas RL, Rooney DW, Hardacre C (2009) Development of a QSPR correlation for the parachor of 1,3-dialkyl imidazolium based ionic liquids. Fluid Phase Equilib 283:31–37
236. Lemraski EG, Zobeydi R (2014) Applying parachor method to the prediction of ionic liquids surface tension based on modified group contribution. J Mol Liq 193:204–209
237. Shang Q, Yan F, Xia S, Wang Q, Ma P (2013) Predicting the surface tensions of ionic liquids by the quantitative structure property relationship method using a topological index. Chem Eng Sci 101:266–270
238. Marcus Y (2015) Volumetric behavior of room temperature ionic liquids: Chapter 19. In: Wilhelm E, Letcher T (eds) Volumetric properties. Royal Society of Chemistry, Cambridge, pp 512–525
239. Every HA, Bishop AG, MacFarlane DR, Orädd G, Forsyth M (2004) Transport properties in a family of dialkylimidazolium ionic liquids. Phys Chem Chem Phys 6:1758–1765
240. Esperança JMSS, Visak ZP, Plechkova NV, Seddon KR, Guedes HJR, Rebelo LPNJ (2006) Density, speed of sound, and derived thermodynamic properties of ionic liquids over an extended pressure range. 4. [C3mim]NTf2] and [C5mim]NTf2]. J Chem Eng Data 51:2009–2015
241. Gardas RL, Freire MG, Carvalho PJ, Marrucho IM, Fonseca IM, Ferreira AGM, Coutinho JAP (2007) High-pressure densities and derived thermodynamic properties of imidazolium-based ionic liquids. J Chem Eng Data 52:80–88
242. Mokhtarani B, Sharifi A, Mortaheb HR, Mirzaei M, Mafi M, Sadeghian F (2009) Density viscosity of 1-butyl-3-methylimidazolium nitrate with ethanol, 1-propanol, or 1-butanol at several temperatures. J Chem Thermodyn 41:1432–1438
243. Singh T, Kumar A (2009) Temperature dependence of physical properties of imidazolium based ionic liquids: internal pressure and molar refraction. J Solution Chem 38:1043–1053

244. Carrera GVSM, Afonso CAM, Branco LC (2010) Interfacial properties, densities, and contact angles of task specific ionic liquids. J Chem Eng Data 55:609–615
245. Sastry NV, Vaghela NM, Macwan PM (2013) Densities, excess molar and partial molar volumes for water + 1-butyl- or, 1-hexyl- or, 1-octyl-3-methylimid-azolium halide room temperature ionic liquids at T = (298.15 and 308.15) K. J Mol Liq 180:12--18
246. Zhao FY, Liang LY, Wang JY, Hu YQ (2012) Density surface tension of binary mixtures of 1-ethyl-3-methylimidazolium nitrate with alcohols. Chin Chem Lett 23:1295–1298
247. Akbar MM, Murugesan T (2013) Thermophysical properties of 1-hexyl-3-methylimidazolium tetrafluoroborate [hmim]BF4]+N-methyldiethanolamine (MDEA) at temperatures (303.15 to 323.15) K. J Mol Liq 177:54–59
248. Cruz MM, Borges RP, Godinho M, Marques CS et al (2013) Thermophysical and magnetic studies of two paramagnetic liquid salts: [C4mim]FeCl4] and [P66614]FeCl4]. Fluid Phase Equilib 350:43–50
249. Matkowsaka D, Hofman T (2013) Volumetric properties of the ionic liquids: [C6mim] MeSO4] [C6mim]EtSO4] [C4mim]EtSO4] and their mixtures with methanol or ethanol. J Mol Liq 177:301–305
250. Neves CMSS, Kurnia KA, Shimizu K, Marrucho IM et al (2014) The impact of ionic liquid fluorinated moieties on their thermophysical properties and aqueous phase behaviour. Phys Chem Chem Phys 16:21340–21348
251. Teodorescu M (2014) Isothermal vapor + liquid equilibrium and thermophysical properties for 1-butyl-3-methylimidazolium bromide + 1-butanol binary system. Ind Eng Chem Res 53:13522–13528
252. Seddon KR, Stark A, Torres M-J (2002) Viscosity density of 1-alkyl-3-methylimidazolium ionic liquids. ACS Symp Ser 819:34–49
253. Domanska U, Krolikowska M, Krolikowski M (2010) Phase behaviour and physico-chemical properties of the binary systems {1-ethyl-3-methylimidazolium thiocyanate, or 1-ethyl-3-methylimidazolium tosylate + water, or + an alcohol}. Fluid Phase Equilib 294:72–83
254. de Azevedo R, Esperança JMSS, Szyslowski J, Visak ZP, Pires PF, Guedes HJR, Rebelo LPN (2005) Thermophysical thermodynamic properties of ionic liquids over an extended pressure range: [bmim]NTf] and [hmim]NTf2]. J Chem Thermodyn 37:888–899
255. Tome LIN, Carvalho PJ, Freire MG, Marrucho IM, Fonseca IMA, Ferreira AGM, Coutinho JAP, Gardas RL (2008) Measurements correlation of high-pressure densities of imidazolium-based ionic liquids. J Chem Eng Data 53:1914–1921
256. Koller T, Rausch MH, Ramos J, Schulz PS, Wasserscheid P, Ecomonou IG, Fröba AP (2013) Thermophysical properties of the ionic liquids [EMIM]B(CN)4] and [HMIM]B(CN)4]. J Phys Chem B 117:8512–8523
257. Kozlov DN, Kiefer J, Seeger T, Fröba AP, Leipertz A (2011) Determination of physico-chemical parameters of ionic liquids and their mixtures with solvents using laser-induced gratings. J Phys Chem B 115:8528–8533
258. Singh S, Bahadur I, Redhi GG, Ramjugemath D, Ebenso EE (2014) Density and speed of sound measurements of imidazolium-based ionic liquids with acetonitrile at various temperatures. J Mol Liq 200:160–167
259. Ye C, Shreeve JM (2007) Rapid accurate estimation of densities of room-temperature ionic liquids and salts. J Phys Chem A 111:1456–1461
260. Zheng Y, Dong K, Wang Q, Zhang J, Lu X (2013) Density viscosity, and conductivity of Lewis acidic 1-butyl- and 1-hydrogen-3-methylimidazolium chloroaluminate ionic liquids. J Chem Eng Data 58:32–42
261. Gu Z, Brennecke JF (2002) Volume expansivities and isothermal compressibilities of imidazolium and pyridinium-based ionic liquids. J Chem Eng Data 47:339–345
262. Gonzales B, Calvar N, Gomez E, Macedo EA, Dominguez A (2008) Synthesis and physical properties of 1-ethyl 3-methylpyridinium ethylsulfate and its binary mixtures with ethanol and water at several temperatures. J Chem Eng Data 53:1824–1828

263. Gomez E, Calvar N, Dominguez A, Macedo EA (2010) Synthesis and temperature dependence of physical properties of four pyridinium-based ionic liquids: influence of the size of the cation. J Chem Thermodyn 42:1324–1329
264. Liu QS, Tong J, Tan ZC, Welz-Biermann U, Yang JZ (2010) Density surface tension of ionic liquid [C2mim]PF3(CF2CF3)3] and prediction of properties [Cnmim]PF3(CF2CF3)3] (n = 1, 3, 4, 5, 6). J Chem Eng Data 55:2586–2589
265. Deng Y, Husson P, Delort V, Bess-Hoggan P, Sancelme M, Costa Gomes MF (2011) Influence of an oxygen functionalization on the physicochemical properties of ionic liquids: density, viscosity, and carbon dioxide solubility as a function of temperature. J Chem Eng Data 56:4194–4202
266. Seki S, Tsuzuki S, Hayamizu K, Umebayashi Y, Serizawa N, Takei K, Miyashiro H (2012) Comprehensive refractive index property for room-temperature ionic liquids. J Chem Eng Data 57:2211–2216
267. Gardas RL, Costa HF, Freire MG, Carvalho PJ, Marrucho IM, Fonseca IMA, Ferreira AGM, Coutinho JAP (2008) Densities derived thermodynamic properties of imidazolium-pyridinium- pyrrolidinium- and piperidinium-based ionic liquids. J Chem Eng Data 53:805–811
268. Safarov J, Kul I, El-Awady WA, Shahverdiyev A, Hassel E (2011) Thermodynamic properties of 1-butyl-3-methylpyridinium tetrafluoroborate. J Chem Thermodyn 43:1315–1322
269. Zhao H, Malhorta SV, Luo RG (2003) Phys Chem Liq 41:487–492
270. Gonzalez B, Corderi S, Santamaria AG (2013) Application of 1-alkyl-3-methylpyridinium bis(trifluoromethylsulfonyl)imide ionic liquids for the ethanol removal from its mixtures with alkanes. J Chem Thermodyn 60:9–14
271. Esperança JMSS, Guedes HJR, Blesic M, Rebelo LPN (2006) Densities derived thermodynamic properties of ionic liquids. 3. Phosphonium-based ionic liquids over an extended pressure range. J Chem Eng Data 51:237–242
272. Jaquemin J, Husson P, Padua AAH, Majer V (2006) Density and viscosity of several pure and water-saturated ionic liquids. Green Chem 8:172–180
273. Pereiro AB, Veiga HIM, Esperança JMSS, Rodriguez A (2009) Effect of temperature on the physical properties of two ionic liquids. J Chem Thermodyn 41:1419–1423
274. Tariq M, Forte PAS, Gomes MFC, Lopes JNC, Rebelo LPN (2009) Densities refractive indices of imidazolium- and phosphonium-based ionic liquids: effect of temperature, alkyl chain length, and anion. J Chem Thermodyn 41:790–798
275. Adamova G, Gardas RL, Rebelo LPN, Robertson AJ, Seddon KR (2011) Alkyltrioctylphosphonium chloride ionic liquids: synthesis and physicochemical properties. Dalton Trans 40:12750–12764
276. Gacino FM, Reguiera T, Lugo L, Comunas MJP, Fernandez J (2011) Influence of molecular structure on densities and viscosities of several ionic liquids. J Chem Eng Data 56:4984–4999
277. Gonzalez B, Gomez E, Dominguez A, Vilas M, Tojo E (2011) Physicochemical characterization of new sulfate ionic liquids. J Chem Eng Data 56:14–20
278. Gonçalvez FAMM, Costa CSMF, Ferreira CE, Bernardo JCS, Johnson I, Fonseca IMA (2011) Pressure-volume-temperature measurements of phosphonium-based ionic liquids and analysis with simple equations of state. J Chem Thermodyn 43:914–923
279. Neves CMSS, Carvalho PJ, Freire MG, Coutinho JAP (2011) Thermophysical properties of pure and water-saturated tetradecyltrihexylphosphonium-based ionic liquids. J Chem Thermodyn 43:948–957
280. Machanova K, Boisset A, Sedlakova Z, Anouti M, Bendova M, Jaquemin J (2012) Thermophysical properties of ammonium-based bis{(trifluoromethyl)sulfonyl}imide ionic liquids: volumetric and transport properties. J Chem Eng Data 57:2227–2235
281. Makino T, Kanakubo M, Umecky T, Suzuki A, Nishida T, Takano J (2012) Electrical conductivities, viscosities, and densities of n-methoxymethyl- and n-butyl-n-methylpyrrolidinium ionic liquids with the bis(fluorosulfonyl)amide anion. J Chem Eng Data 57:751–755

282. Kim K-S, Shin B-K, Lee H (2012) Physical electrochemical properties of 1-butyl-3-methylimidazolium bromide, 1-butyl-3-methylimidazolium iodide, and 1-butyl-3-methylimidazolium tetrafluoroborate. Korean J Chem Eng 21:1010–1014
283. Liu Q-S, Li P-P, Welz-Biermann U, Liu X-X, Chen L (2013) Density, dynamic viscosity, and electrical conductivity of pyridinium-based hydrophobic ionic liquids. J Chem Thermodyn 66:88–94
284. Seoane RG, Corderi S, Gomez E, Calvar N, Gonzalez EJ, Macedo EA, Dominguez A (2012) Temperature dependence and structural influence on the thermophysical properties of eleven commercial ionic liquids. Ind Eng Chem Res 51:2492–2504
285. Gardas RL, Coutinho JAP (2009) Group contribution methods for the prediction of thermophysical and transport properties of ionic liquids. AIChE J 55:1274–1290
286. Slattery JM, Daguenet C, Dyson PJ, Schubert TJS, Krossing I (2007) How to predict the physical properties of ionic liquids: a volume-based approach. Angew Chem Int Ed 46:5384–5388
287. Beichel W, Preiss UP, Verevkin SP, Koslowski T, Krossing I (2014) Empirical description and prediction of ionic liquids' properties with augmented volume-based thermodynamics. J Mol Liq 192:3–8
288. Marcus Y (2015) Ionic and molar volumes of room temperature ionic liquids. J Mol Liq 209:289–293
289. Bica K, Deetlefs M, Schröder C, Seddon KR (2013) Polarisabilities of alkylimidazolium ionic liquids. Phys Chem Chem Phys 15:2703–2711
290. Gardas RL, Ge R, Goodrich P, Hardacre C, Hussain A, Rooney DW (2010) Thermophysical properties of amino acid-based ionic liquids. J Chem Eng Data 55:1505–1515
291. Marcus Y, Jenkins HBD, Glasser L (2002) Ion volumes: a comparison. J Chem Soc Dalton Trans 3795–3798
292. Xie T, Brockner W, Gjikaj M (2010) New ionic liquid compounds based on tantalum pentachloride $TaCl_5$. Synthesis, structural, and spectroscopic elucidation of the (μ-oxido) chloridotantalates(V) $[BMPy]TaCl_6]$ $[BMPy]_4[(TaCl_6)_2(Ta_2OCl_{10})$ and $[EMIm]_2[Ta_2OCl_{10}]$. Z Anorg Allg Chem 636:2633–2640
293. Matsumoto K, Oka T, Nohira T, Hagiwara R (2013) Polymorphism of alkali bis (fluorosulfonyl)amides (M[N(SO2F)2] M = Na, K, and Cs). Inorg Chem 52:568–576
294. Marszalek M, Fei Z, Zhu D-K, Scopelliti R, Dyson PJ, Zakeeruddin SM, Grätzel M (2011) Application of ionic liquids containing tricyanomethanide [c(cn)3] or tetracyanoborate [B (CN)4] anions in dye-sensitized solar cells. Inorg Chem 50:11561–11567
295. Henderson WA, Young VG Jr, Pearson W, Passerini S, De Long HC, Trulove PC (2006) Thermal phase behaviour of N-alkyl-N-methylpyrrolidinium and piperidinium bis(trifluoromethanesulfonyl)imide salts. J Phys Condens Matter 18:10377–10390
296. Kutuniva J, Oilun-Kaniemi R, Laitinen RS, Asikkala J, Kärkkäinen J, Lajunen MK (2007) Synthesis and structural characterization of 1-butyl-2,3-dimethyl-imidazolium bromide and iodide. Z Naturforsch 62b:868–870
297. Gardas RL, Coutinho JAP (2008) Extension of the Ye and Shreeve group contribution method for density estimation of ionic liquids in a wide range of temperatures and pressures. Fluid Phase Equilib 263:26–32
298. Shannon MS, Tedstone JM, Danielsen SPO, Hindman MS, Irvin AC, Bara JE (2012) Free volume as the basis of gas solubility and selectivity in imidazolium-based ionic liquids. Ind Eng Chem Res 51:5565–5576
299. Tekin A, Safarov J, Shahverdiyev A, Hassel E (2007) (p,ρ,T) Properties of 1-butyl-3-methylimidazolium tetrafluoroborate and 1-butyl-3-methylimidazolium hexafluorophosphate at T = (298.15 to 398.15) K and pressures up to p = 40 MPa. J Mol Liq 136:177–182
300. Jaquemin J, Husson P, Mayer V, Cibulka I (2007) High-pressure volumetric properties of imidazolium-based ionic liquids: effect of the anion. J Chem Eng Data 52:2204–2211

301. Marcus Y (2013) The compressibility and surface tension product of molten salts. J Chem Phys 139:124509/1-5
302. Marcus Y (2013) Internal pressure of liquids and solutions. Chem Rev 113:6536–6551
303. Kayama Y, Ichikawa T, Ohno H (2014) Transparent colourless room temperature ionic liquids having high refractive index over 1.60. Chem Commun 50:14790–14792
304. Pereiro AB, Santamaria F, Tojo E, Rodriguez A, Tojo J (2006) Temperature dependence of physical properties of ionic liquid 1,3-dimethylimidazolium methyl sulfate. J Chem Eng Data 51:952–954
305. Gomez E, Gonzalez B, Calvar N, Tojo E, Dominguez A (2006) Physical properties of pure 1-ethyl-3-methylimidazolium ethylsulfate and its binary mixtures with ethanol and water at several temperatures. J Chem Eng Data 51:2096–2102
306. Russina O, Gontrani L, Fazio B, Lombardo C, Triolo A, Caminiti R (2010) Selected chemical-physical properties and structural heterogeneities in 1-ethyl-3-methylimidazolium alkyl sulfate room temperature ionic liquids. Chem Phys Lett 493:259–262
307. Shamsipur M, Beigi AAM, Teymouri M, Pourmortazavi SM, Irandousi M (2010) Physical electrochemical properties of ionic liquids 1-ethyl-3-methylimidazolium tetrafluoroborate, 1-butyl-3-methylimidazolium trifluoro-methanesulfonate and 1-butyl-1-methylpyrrolidinium bis(trifluoromethyl-sulfonyl)imide. J Mol Liq 157:43–50
308. Vakili-Nezhaad G, Vatani M, Asghari M, Ashour I (2012) Effect of temperature on the physical properties of 1-butyl-3-methylimidazolium based ionic liquids with thiocyanate and tetrafluoroborate anions, and 1-hexyl-3-methylimidazolium with tetrafluoroborate and hexafluorophosphate. J Chem Thermodyn 54:148–154
309. Singh T, Kumar A, Kaur M, Kaur G, Kumar H (2009) Non-ideal behaviour of imidazolium based room temperature ionic liquids in ethylene glycol at T = (298.15 to 318.15) K. J Chem Thermodyn 41:717–723
310. Kim K-S, Shin B-K, Lee H (2004) Physical electrochemical properties of 1-butyl-3-methylimidazolium bromide, 1-butyl-3-methylimidazolium iodide, and 1-butyl-3-methylimidazolium tetrafluoroborate. Korean J Chem Eng 21:1010–1014
311. Wu T-Y, Chen B-K, Hao L, Lin K-F, Sun I-W (2011) Thermophysical properties of a room temperature ionic liquid (1-methyl-3-pentyl-imidazolium hexafluoro-phosphate) with poly (ethylene glycol). J Taiwan Inst Chem Eng 42:914–921
312. Chen ZJ, Lee J-M (2014) Free volume model for the unexpected effect of c2-methylation on the properties of imidazolium ionic liquids. J Phys Chem B 118:2712–2718
313. Yan X-J, Li S-N, Zhai Q-G, Jiang Y-C, Hu M-C (2014) Physicochemical properties for the binary systems of ionic liquids [Cnmim]Cl + N,N-dimethylformamide. J Chem Eng Data 59:1411–1422
314. Lago S, Rodriguez H, Soto A, Arce A (2012) Alkylpyridinium alkylsulfate ionic liquids as solvents for the deterpenation of citrus essential oil. Sep Sci Technol 47:292–299
315. Garcia-Mardones M, Martin S, Gascon I, Lafuente C, Schröder B et al (2014) Thermophysical properties of the binary mixture 1-propylpyridinium tetrafluoroborate with methanol. J Chem Eng Data 59:1564–1571
316. Larriba M, Garcia S, Navarro P, Garcia J, Rodriguez F (2012) Physical properties of N-butylpyridinium tetrafluoroborate and N-butylpyridinium bis(trifluoro-methylsulfonyl) imide binary ionic liquid mixtures. J Chem Eng Data 57:1318–1325
317. Almeida HFD, Lopes-da-Silva JA, Freire MG, Coutinho JAP (2013) Surface tension and refractive index of pure and water-saturated tetradecyltrihexylphospho-nium-based ionic liquids. J Chem Thermodyn 57:372–379
318. Huang M-M, Jiang Y, Sasisanker P, Driver GW, Weingärtner H (2011) Static relative dielectric permittivities of ionic liquids at 25°. J Chem Eng Data 56:1494–1499
319. Hunger J, Stoppa A, Schrödle S, Hefter G, Buchner R (2009) Temperature dependence of the dielectric properties and dynamics of ionic liquids. ChemPhysChem 10:723–733

320. Mizoshiri M, Nagao T, Mizoguchi Y, Yao M (2010) Dielectric permittivity of room temperature ionic liquids: a relation to the polar and nonpolar domain structures. J Chem Phys 132:164510/1-6
321. Weingärtner H (2014) The static dielectric permittivity of ionic liquids. J Mol Liq 192:185–190
322. Bandres I, Giner B, Artigas H, Royo FM, Lafuente C (2008) Thermophysic comparative study of two isomeric pyridinium-based ionic liquids. J Phys Chem B 112:3077–3084
323. Bandres I, Giner B, Artigas H, Lafuente C, Royo FM (2009) Thermophysical properties of N-octyl-3-methylpyridinium tetrafluoroborate. J Chem Eng Data 54:236–240
324. Rocha MAA, Ribeiro FMS, Ferreira AIMCL, Coutinho JAP, Santos LMNBF (2013) Thermophysical properties of [CN-1C1iml[PF6l ionic liquids. J Mol Liq 188:196–202
325. Benito J, Garcia-Mardones M, Perez-Gregorio V, Gascon I, Lafuente C (2014) Physicochemical study of N-ethylpyridinium bis(trifluoromethylsulfonyl)imide ionic liquid. J Solution Chem 43:696–710
326. Seki S, Tsuzuki S, Hayamizu K, Serizawa N, Ono S, Takei K, Doi H, Umebayashi Y (2014) Static transport properties of alkyltrimethylammonium cation-based room-temperature ionic liquids. J Phys Chem B 118:4590–4599
327. Bhattacharjee A, Lopes-da-Silva JA, Freire MG, Coutinho JAP, Carvalho PJ (2015) Thermophysical properties of phosphonium-based ionic liquids. Fluid Phase Equilib 400:103–113
328. Montalban MG, Bolivar CL, Bnos FGD, Villora G (2015) Effect of temperature, anion, and alkyl chain length on the density and refractive index of 1-alkyl-3-methyl-imidazolium-based ionic liquids. J Chem Eng Data 60:1986–1996
329. Ma X-X, Wei J, Zhang Q-B, Tian F, Feng Y-Y, Guan W (2013) Prediction of thermophysical properties of acetate-based ionic liquids using semiempirical methods. Ind Eng Chem Res 52:9490–9496
330. Sattari M, Kamari A, Mohammadi AH, Ramjugernath D (2014) A group contribution method for estimating the refractive indices of ionic liquids. J Mol Liq 200:410–415
331. Daguenet C, Dyson PJ, Krossing I, Oleinikova A, Slattery J, Wakai C, Weingärtner H (2006) Dielectric response of imidazolium-based room-temperature ionic liquids. J Phys Chem B 110:12682–12688
332. Nakamura K, Shikata T (2010) Systematic dielectric and NMR study of the ionic liquid 1-alkyl-3-methyl imidazolium. ChemPhysChem 11:285–294
333. McHale G, Hardacre C, Ge R, Doy N, Allen RWK, MacInnes M, Bown MR, Newton MI (2008) Density-viscosity product of small-volume ionic liquid samples using quartz crystal impedance analysis. Anal Chem 80:5806–5811
334. Ghatee MH, Zare M, Moosavi F, Zolghadr AR (2010) Temperature-dependent density and viscosity of the ionic liquids 1-alkyl-3-methylimidazolium iodides: experiment and molecular dynamics simulation. J Chem Eng Data 55:3084–3088
335. Costa AJL, Esperança JMSS, Marrucho IM, Rebelo LPN (2011) Densities viscosities of 1-ethyl-3-methylimidazolium n-alkyl sulfates. J Chem Eng Data 56:3433–3331
336. Ciocirlan O, Croitoru O, Iulian O (2011) Densities viscosities for binary Mixtures of 1-butyl-3-methylimidazolium tetrafluoroborate ionic liquid with molecular solvents. J Chem Eng Data 56:1526–1534
337. Quijada-Maldonado E, van der Boogaart S, Lijbers JH, Maindersma GW, de Haan AB (2012) Experimental densities, dynamic viscosities and surface tensions of the ionic liquids series 1-ethyl-3-methylimidazolium acetate and dicyanamide and their binary and ternary mixtures with water and ethanol at T = (298.15 to 343.15 K). J Chem Thermodyn 51:51–58
338. McEwen AB, Ngo HL, LeCompte K, Goldman JI (1999) Electrochemical properties of imidazolium salt electrolytes for electrochemical capacitor applications. J Electrochem Soc 146:1687–1695

339. Kulkarni PS, Branco LC, Crespo JG, Nunes MC, Raymundo A, Afonso CAM (2007) Comparison of physicochemical properties of new ionic liquids based on imidazolium, quaternary ammonium, and guanidinium cations. Chem Eur J 13:8478–8488
340. Li J-G, Hu Y-F, Sun S-F, Liu Y-S, Liu Z-C (2010) Densities and dynamic viscosities of the binary system (water + 1-hexyl-3-methylimidazolium bromide) at different temperatures. J Chem Thermodyn 42:904–908
341. Fendt S, Padmanabhan S, Blanch HW, Prausnitz JM (2011) Viscosities of acetate or chloride-based ionic liquids and some of their mixtures with water or other common solvents. J Chem Eng Data 56:31–34
342. Domanska U, Krolikowska M (2012) Density viscosity of binary mixtures of thiocyanate ionic liquids + water as a function of temperature. J Solution Chem 41:1422–1445
343. Gomez E, Calvar N, Macedo EA, Dominguez A (2012) Effect of the temperature on the physical properties of pure 1-propyl 3-methylimidazolium bis(trifluoro-methylsulfonyl) imide and characterization of its binary mixtures with alcohols. J Chem Thermodyn 45:9–15
344. Rocha M, Neves CMSS, Freire MG, Russina O, Triolo A, Coutinho JAP, Santos LMNBF (2013) Alkylimidazolium based ionic liquids: impact of cation symmetry on their nanoscale structural organization. J Phys Chem B 117:10889–10897
345. Song D, Chen J (2014) Density viscosity data for mixtures of ionic liquids with a common anion. J Chem Eng Data 59:257–262
346. AlTuwaim MS, Alkhaldi KHAE, Al-Jimz AS, Mohammad AA (2014) Temperature dependence of physicochemical properties of imidazolium- pyridinium- and phosphonium-based ionic liquids. J Chem Eng Data 59:1955–1963
347. Zhang Q-G, Wei Y, Sun S-S, Wang C, Yang M, Liu Q-S, Gao Y-A (2012) Study on thermodynamic properties of ionic liquid n-butyl-3-methylpyridinium bis(trifluoromethylsulfonyl)imide. J Chem Eng Data 57:2185–2190
348. Oliveira FS, Freire MG, Carvalho PJ, Coutinho JAP, Canongia JN et al (2010) Structural and positional isomerism influence in the physical properties of pyridinium NTf2-based ionic liquids: pure and water-saturated mixtures. J Chem Eng Data 55:4514–4520
349. Garcia-Mardones M, Gascon I, Lopez MC, Royo FM, Lafuente C (2012) Viscosimetric study of binary mixtures containing pyridinium-based ionic liquids and alkanols. J Chem Eng Data 57:3549–3556
350. Gonzalez B, Calvar N, Gomez E, Dominguez I, Dominguez A (2009) Synthesis physical properties of 1-ethylpyridinium ethylsulfate and its binary mixtures with ethanol and 1-propanol at several temperatures. J Chem Eng Data 54:1353–1358
351. Safarov J, Kul I, El-Awady WA, Nocke J, Shahverdiyev A, Hassel E (2012) Thermophysical properties of 1-butyl-4-methylpyridinium tetrafluoroborate. J Chem Thermodyn 51:82–87
352. Tokuda H, Ishii K, Susan MABH, Tsuzuki S, Hayamizu K, Watanabe M (2006) Physicochemical properties and structures of room-temperature ionic liquids. 3. Variation of cationic structures. J Phys Chem B 110:2833–2839
353. Gonzalez EJ, Gonzalez B, Mecedo EA (2012) Thermophysical properties of the pure ionic liquid 1-butyl-1-methylpyrrolidinium dicyanamide and its binary mixtures with alcohols. J Chem Eng Data 58:1440–1448
354. Huang X-J, Rogers EL, Hardacre C, Compton RG (2009) The reduction of oxygen in various room temperature ionic liquids in the temperature range 293–318 K: exploring the applicability of the Stokes-Einstein relationship in room temperature ionic liquids. J Phys Chem B 113:8953–8959
355. Ferreira CE, Taslavera-Prieto NMC, Fonseca IMA, Portugal TC (2012) Measurements of pVT, viscosity, and surface tension of trihexyltetradecylphosphonium tris(pentafluoroethyl) trifluorophosphate ionic liquid and modelling with equations of state. J Chem Thermodyn 47:183–196
356. Deive FJ, Rivas MA, Rodriguez A (2013) Study of thermodynamic and transport properties of phosphonium-based ionic liquids. J Chem Thermodyn 62:98–103

357. Xu W, Cooper EI, Angell CA (2003) Ionic liquids: ion mobilities, glass temperatures, and fragilities. J Phys Chem B 107:6170–6178
358. Ghatee MH, Zare M, Zolghadr AR, Moosavi F (2010) Temperature dependence of viscosity and relation with the surface tension of ionic liquids. Fluid Phase Equilib 291:188–194
359. Hildebrand JH, Lamoreaux RH (1972) Fluidity. General theory. Proc Natl Acad Sci U S A 69:3428–3431
360. Marcus Y (2014) The fluidity of room temperature ionic liquids. Fluid Phase Equilib 363:66–69
361. Harris KR, Kanakubo M, Woolf LA (2006) Temperature and pressure dependence of the viscosity of the ionic liquids 1-methyl-3-octylimidazolium hexafluoro-phosphate and 1-methyl-3-octylimidazolium tetrafluoroborate. J Chem Eng Data 51:1161–1167
362. Harris KR, Kanakubo M, Woolf LA (2007) Temperature and pressure dependence of the viscosity of the ionic liquids 1-hexyl-3-methylimidazolium hexafluoro-phosphate and 1-butyl-3-methylimidazolium bis(trifluoromethyl-sulfonyl)imide. J Chem Eng Data 52:1080–1085
363. Harris KR, Kanakubo M, Woolf LA (2007) Temperature and pressure dependence of the viscosity of the ionic liquid 1-butyl-3-methylimidazolium tetrafluoroborate: viscosity and density relationships in ionic liquids. J Chem Eng Data 52:2425–2430
364. Noda A, Hayamizu K, Watanabe M (2001) Pulsed-gradient spin-echo 1H and 19F NMR ionic diffusion coefficient, viscosity, and ionic conductivity of non-chloroaluminate room-temperature ionic liquids. J Phys Chem B 105:4603–4610
365. Tsuzuki S, Shinoda W, Saito H, Mikami M, Tokuda H, Watanabe M (2009) Molecular dynamics simulations of ionic liquids: cation and anion dependence of self-diffusion coefficients of ions. J Phys Chem B 113:10641–10649
366. Harris KR, Woolf LA, Kanakubo M, Rüther T (2011) Transport properties of N-butyl-N-methylpyrrolidinium bis(trifluoromethylsulfonyl)amide. J Chem Eng Data 56:4672–4685
367. Bidikoudi M, Zubeir LF, Falaras P (2014) Low viscosity highly conductive ionic liquid blends for redox active electrolytes in efficient dye-sensitized solar cells. J Mater Chem A 2:15326–15336
368. McFarlane DR, Sun J, Golding J, Meakin P, Forsyth M (2000) High conductivity molten salts based on the imide ion. Electrochim Acta 45:1271–1278
369. Vila J, Varela LM, Cabeza O (2007) Cation and anion sizes influence in the temperature dependence of the electrical conductivity in nine imidazolium based ionic liquids. Electrochim Acta 52:7413–7417
370. Ignat'ev NV, Welz-Biermann U, Kucheryna A, Bissky G, Willner H (2005) New ionic liquids with tris(perfluoroalkyl)trifluorophosphate (FAP) anions. J Fluor Chem 126:1150–1159
371. Kanakubo M, Harris KR, Tsuchihashi N, Ibuki K, Ueno M (2007) Temperature and pressure dependence of the electrical conductivity of the ionic liquids 1-methyl-3-octylimidazolium hexafluorophosphate and 1-methyl-3-octyl-imidazolium tetrafluoroborate. Fluid Phase Equilib 261:414–420
372. Stoppa A, Hunger J, Buchner R (2009) Conductivities of binary mixtures of ionic liquids with polar solvents. J Chem Eng Data 54:472–479
373. Schreiner C, Zugmann S, Harti R, Gores HJ (2010) Fractional Walden rule for ionic liquids: examples from recent measurements and a critique of the so-called ideal KCl line for the Walden plot. J Chem Eng Data 55:1784–1788
374. Tokuda H, Hayamizu K, Ishii K, Susan MABH, Watanabe M (2004) Physicochemical properties and structures of room temperature ionic liquids. 1. Variation of anionic species. J Phys Chem B 108:16593–16600
375. Wu T-Y, Hao L, Chen P-R, Liao J-W (2013) Ionic conductivity and transporting properties in LiTFSI-doped bis(trifluoromethanesulfonyl)imide-based ionic liquid electrolyte. Int J Electrochem Sci 8:2606–2624

376. G-h S, K-x L, Sun C-g (2006) Application of 1-ethyl-3-methylimidazolium thiocyanate to the electrolyte of electrochemical double layer capacitors. J Power Sources 162:1444–1450
377. Simons TJ, Bayley PM, Zhang Z, Howlett PC, MacFarlane DR, Madsen LA, Forsyth M (2014) Influence of Zn2+ and water on the transport properties of a pyrrolidinium dicyanamide ionic liquid. J Phys Chem B 118:4895–4905
378. Ning H, Hou M-Q, Mei Q-Q, Liu Y-H, Yang D-Z, Han B-X (2013) The physicochemical properties of some imidazolium-based ionic liquids and their binary mixtures. Sci China Chem 55:1509–1518
379. Calado MS, Diogo, JCF, Correia da Mata JL, Caetano FJP, Visak ZP, Fareleira JMNA (2013) Electrolytic conductivity of four imidazolium-based ionic liquids. Int J Thermophys 34:1265–1279
380. Wang X, Chi Y, Mu T (2014) A review on the transport properties of ionic liquids. J Mol Liq 193:262–266
381. Nishida T, Tashiro Y, Yamamoto M (2003) Physical electrochemical properties of 1-alkyl-3-methylimidazolium tetrafluoroborate for electrolyte. J Fluor Chem 120:135–141
382. Sun J, Forsyth M, MacFarlane DR (1998) Room-temperature molten salts based on the quaternary ammonium ion. J Phys Chem B 102:8858–8864
383. Garcia-Mardones M, Osorio HM, Lafuente C, Gascon I (2012) Ionic conductivities of binary mixtures containing pyridinium-based ionic liquids and alkanols. J Chem Eng Data 58:1613–1620
384. Li J-G, Hu Y-F, Ling S, Zhang J-Z (2011) Physicochemical properties of [C6mim]PF6] and [C6mim]C2F5)3PF3] ionic liquids. J Chem Eng Data 56:3068–3072
385. Zech O, Stoppa A, Buchner R, Kunz W (2010) The Conductivity of imidazolium-based ionic liquids from (248 to 468) K. B. Variation of the anion. J Chem Eng Data 55:1774–1778
386. Hayamizu K, Tsuzuki S, Seki S, Fujii K, Suenaga M, Umebayashi Y (2010) Studies on the translational and rotational motions of ionic liquids composed of N-methyl-N-propyl-pyrrolidinium (P13) cation and bis(trifluoromethanesulfonyl)amide and bis(fluorosulfonyl) amide anions and their binary systems including lithium salts. J Chem Phys 133:194505/1-13
387. Borodin O (2009) Polarizable force field development and molecular dynamics simulations of ionic liquids. J Phys Chem B 113:11463–11478
388. Mondal A, Balasubramanian S (2014) A molecular dynamics study of collective transport properties of imidazolium-based room-temperature ionic liquids. J Chem Eng Data 59:3061–3068
389. Kanakubo M, Harris KR, Tsuchihashi N, Ibuki K, Ueno M (2015) Temperature and pressure dependence of the Electrica conductivity of 1-Butyl-3-methylimid-azolium Bis(trifluoro-methanesulfonyl)amide (trifluoromethanesulfonyl)amide. J Chem Eng Data 60:1495–1503
390. Kanakubo M, Harris KR, Tsuchihashi N, Ibuki K, Ueno M (2007) Effect of pressure on transport properties of the ionic liquid 1-butyl-3-methylimidazolium hexafluoro-phosphate. J Phys Chem B 111:2062–2069
391. Lopez ER, Pensado AS, Comunas MJP, Padua AAH, Fernandez J, Harris KR (2011) Density scaling of the transport properties of molecular and ionic liquids. J Chem Phys 134:144507/1-12
392. Stoppa A, Zech O, Kunz W, Buchner R (2010) The conductivity of imidazolium-based ionic liquids from (−35° to 195)C. A. Variation of cation's alkyl chain. J Chem Eng Data 55:1768–1773
393. Van Valkenburg ME, Vaughn RL, Williams M, Wilkes JS (2005) Thermochemistry of ionic liquid heat-transfer fluids. Thermochem Acta 425:181–188
394. Ge R, Hardacre C, Nancarrow P, Rooney DW (2007) Thermal conductivities of ionic liquids over the temperature range from 293 K to 353 K. J Chem Eng Data 52:1819–1823
395. Tomida D, Kenmochi S, Tsukada T, Qiao K, Yokoyama C (2007) Thermal conductivities of [bmim]PF6] [hmim]PF6] and [omim]PF6] from 294 to 335 K at pressures up to 20 MPa. Int J Thermophys 28:1147–1160

396. Tomida D, Kenmochi S, Tsukada T, Qiao K, Bao Q, Yokoyama C (2012) Viscosity thermal conductivity of 1-hexyl-3-methylimidazolium tetrafluoroborate and 1-octyl-3-methylimidazolium tetrafluoroborate at pressures up to 20 MPa. Int J Thermophys 33:959–969
397. Tomida D, Kenmochi S, Qiao K, Tsukada T, Yokoyama C (2013) Densities and thermal conductivities of N-alkylpyridinium tetrafluoroborates at high pressure. Fluid Phase Equilib 340:31–35
398. Liu H, Magino E, Visser AV, Bridges NJ, Fox EB (2012) Thermal and transport properties of six ionic liquids: an experimental and molecular dynamics study. Ind Eng Chem Res 51:7242–7254
399. Frez C, Diebold GJ, Tran CD, Yu S (2006) Determination of thermal diffusivities, thermal conductivities, and sound speeds of room-temperature ionic liquids by the transient grating technique. J Chem Eng Data 51:1250–1255
400. Fröba AP, Rausch MH, Krzeminski K, Assenbaum D, Wasserscheid P, Leipertz A (2010) Thermal conductivity of ionic liquids: measurement and prediction. Int J Thermophys 31:2059–2077
401. Nieto de Castro CA, Lourenço MJV, Ribeiro APC, Langa E, Vieira SIC (2010) Thermal properties of ionic liquids and ionanofluids of imidazolium and pyrrolidinium liquids. J Chem Eng Data 55:653–661
402. Nieto de Castro CA, Murshed SMS, Lourenço MJV, Santos FJV, Lopes MLM, França JMP (2012) Enhanced thermal conductivity and specific heat capacity of carbon nanotubes ionanofluids. Int J Therm Sci 62:34–39
403. Bhatt VD, Gohil K (2013) Ion exchange synthesis and thermal characteristics of some [N + 222]-based ionic liquids. Bull Mater Sci 36:1121–1125
404. Kozlov DN, Kiefer J, Seeger T, Fröba AP, Leipertz A (2014) Simultaneous measurement of speed of sound, thermal diffusivity, and bulk viscosity of 1-ethyl-3-methylimidazolium-based ionic liquids using laser-induced gratings. J Phys Chem B 118:14493–14501
405. Tenney CM, Massel M, Mayer JM, Sen M, Brennecke JF, Maginn EJ (2014) A computational and experimental study of the heat transfer properties of nine different ionic liquids. J Chem Eng Data 59:391–399
406. Shojaee SA, Farzam S, Hezave AZ, Lashkarbolooki M, Ayatollahi S (2013) A new correlation for estimating thermal conductivity of pure ionic liquids. Fluid Phase Equilib 354:199–206
407. Wu K-J, Chen Q-L, He C-H (2014) Speed of sound of ionic liquids: database, estimation, and its application for thermal conductivity prediction. AIChE J 60:1120–1131
408. Wu K-J, Zhao C-X, He C-H (2013) Development of a group contribution method for determination ofthermal concuctivitry of ionic liquids. Fluid Phase Equilib 339:10–14
409. Reichardt C (2002) Solvents and solvent effects in organic chemistry, 3rd edn. Wiley-VCH, Weinheim
410. Jessop PG, Jessop DA, Fu D, Phan L (2012) Solvatochromic parameters for solvents of interest in green chemistry. Green Chem 14:1245–1259
411. Marcus Y (1993) The properties of organic liquids that are relevant to their use as solvating solvents. Chem Soc Rev 22:409–416
412. Catalan J (2009) Toward a generalized treatment of the solvent effect based on four empirical scales: dipolarity (SdP, a new scale), polarizability (SP), acidity (SA), and basicity (SB) of the medium. J Phys Chem B 113:5951–5960
413. Schade A, Behme N, Spange S (2014) Dipolarity versus polarizability and acidity versus basicity of ionic liquids as a function of their molecular structures. Chem Eur J 20:2232–2243
414. Kochly ED, Citrak S, Gathondu N, Amberchan G (2014) Effect of ionic liquids in unimolecular solvolysis reactions involving carbocationic intermediates. Tetrahedron Lett 55:7181–7185
415. Moita M-LCJ, Santos AFS, Silva JFCC, Lampreia IMS (2012) Polarity of some [NR1R2R3R4]Tf2N] ionic liquids in ethanol: preferential solvation versus solvent-solvent interactions. J Chem Eng Data 57:2702–2709

416. Spange S, Lungwitz R, Schade A (2014) Correlation of molecular structure and polarity of ionic liquids. J Mol Liq 192:137–143
417. Schneider H, Migron Y, Marcus Y (1992) Hydrogen-bond donation properties of aqueous solvent mixtures from carbon-13 NMR data of dialkylbenzamides. Z Phys Chem 175:145–164
418. Schneider H, Badrieh Y, Migron Y, Marcus Y (1992) Hydrogen bond donation properties of organic solvents and their aqueous mixtures from carbon-13 NMR data of pyridine-N-oxide. Z Phys Chem 177:143–156
419. Seddon KR, Stark A, Torres M-J (2000) Influence of chloride, water, and organic solvents on the physical properties of ionic liquids. Pure Appl Chem 72:2275–2287
420. Chapeaux A, Simini LD, Stadtherr MA, Brennecke J (2007) Liquid phase behavior of ionic liquids with water and 1-octanol and modeling of 1-octanolwater partition coefficients. J Chem Eng Data 52:2462–2467
421. Cho C-W, Preiss U, Jungnickel C, Stolte S, Arning J, Ranke J, Klamt A, Krossing I, Thoming J (2011) Ionic liquids: predictions of physicochemical properties with experimental andor DFT-calculated LFER parameters to understand molecular interactions in solution. J Phys Chem B 115:6040–6050
422. Bonhote P, Dias A-P, Papageorgiou N, Kalyanasundaram K, Grätzel M (1996) Hydrophobic, highly conductive ambient-temperature molten salts. Inorg Chem 35:1168–1178
423. Anthony JL, Maginn EJ, Brennecke JF (2001) Solution thermodynamics of imidazolium-based ionic liquids and water. J Phys Chem B 105:10942–10949
424. Chun S, Dzyuba SV, Bartsch RA (2004) Influence of structural variation in room-temperature ionic liquids on the selectivity and efficiency of competitive alkali metal salt extraction by a crown ether. Anal Chem 73:3737–3741
425. Wong DSH, Chen JP, Chang JM, Chou CH (2002) Phase equilibria of water and ionic liquids [emim]PF6] and [bmim]PF6]. Fluid Phase Equilib 194–197:1089–1095
426. Luo H, Dai S, Bonnesen PV (2004) Solvent extraction of Sr2+ and Cs + based on room-temperature ionic liquids containing monoaza-substituted crown ethers. Anal Chem 76:2773–2779
427. Shimojo K, Goto M (2004) Solvent extraction and stripping of silver ions in room-temperature ionic liquids containing calixarenes. Anal Chem 76:5039–5044
428. McFarlane J, Ridenour WB, Luo H, Hunt RD, DePaoli DW, Ren EX (2005) Room temperature ionic liquids for separating organics from produced water. Sep Sci Technol 40:1245–1265
429. Chen P-Y (2007) The assessment of removing strontium and cesium cations from aqueous solutions based on the combined methods of ionic liquid extraction and electrodeposition. Electrochim Acta 52:5484–5492
430. Freire MG, Neves CMSS, Carvalho PJ, Gardas RL, Fernandes AM, Marrucho IM, Santos LMNBF, Coutinho JAP (2007) Mutual solubilities of water and hydrophobic ionic liquids. J Phys Chem B 111:13082–13089
431. Salminen J, Papaiconomou N, Kumar RA, Lee J-M, Kerr J, Newman J, Prausnitz JM (2007) Physicochemical properties and toxicities of hydrophobic piperidinium and pyrrolidinium ionic liquids. Fluid Phase Equilib 261:421–426
432. Deng Y, Besse-Hogan P, Husson P, Sancelme M, Delort A-M, Stepnowski P, Paszkiewicz M, Gołebiowski M, Gomes MFC (2012) Relevant parameters for assessing the environmental impact of some pyridinium, ammonium and pyrrolidinium based ionic liquids. Chemosphere 89:327–333
433. Gonzalez EJ, Dominguez A, Macedo EA (2012) physical and excess properties of eight binary mixtures containing water and ionic liquids. J Chem Eng Data 57:2165–2176
434. Gonzalez EJ, Macedo EA (2014) Influence of the number, position and length of the alkyl-substituents on the solubility of water in pyridinium-based ionic liquids. Fluid Phase Equilib 383:72–77

435. Kaar JL, Jesionowski AM, Berberich JA, Moulton R, Russell AJ (2003) Impact of ionic liquid physical properties on lipase activity and stability. J Am Chem Soc 125:4125–4131
436. Lee SH, Lee SB (2009) Octanol/water partition coefficients of ionic liquids. J Chem Technol Biotechnol 84:202–207
437. Ropel L, Belveze L, Aki SNVK, Stadtherr MA, Brennecke JF (2005) Octanol-water partition coefficients of imidazolium-based ionic liquids. Green Chem 7:83–90
438. Lee B-S, Lin S-T (2014) A priori prediction of the octanol-water partition coefficient (KOW) of ionic liquids. Fluid Phase Equilib 363:233–238
439. Zhao H, Baker GA, Song Z, Olubajo C, Zanders L, Campbell SM (2009) Effect of ionic liquid properties on lipase stabilization under microwave irradiation. J Mol Catal B Enzym 57:149–157
440. Deng Y, Besse-Hogan P, Sancelme M, Delort A-M, Husson P, Costa Gomes MF (2011) Influence of oxygen functionalities on the environmental impact of imidazolium based ionic liquids. J Hazard Mater 198:165–178
441. Hassan S, Duclaux L, Leveque J-M, Reinert L, Farooq A, Yasin T (2014) Effect of cation type, alkyl chain length, adsorbate size on adsorption kinetics and isotherms of bromide ionic liquids from aqueous solutions onto microporous fabric and granulated activated carbons. J Environ Manag 144:108–117
442. Domanska U, Bogel-Lukasik E, Bogel-Lukasik R (2003) 1-Octanol/water partition coefficients of 1-alkyl-3-methylimidazolium chloride. Chem Eur J 9:3033–3041
443. Chun S, Dzyuba SV, Bartsch RA (2001) Influence of structural variation in room-temperature ionic liquids on the selectivity and efficiency of competitive alkali metal salt extraction by a crown ether. Anal Chem 73:3737–3741
444. Rickert PG, Stepinski DC, Rausch DJ, Bergeron RM, Jakab S, Dietz ML (2007) Solute-induced dissolution of hydrophobic ionic liquids in water. Talanta 72:315–320
445. Kolarik Z (2013) Ionic liquids: how far do they extend the potential of solvent extraction of f-elements? Ion Exch Solvent Extract 31:24–60
446. Padro JM, Ponzinibbio A, Agudelo Mesa LB, Reta M (2011) Predicting the partitioning of biological compounds between room-temperature ionic liquids and water by means of the solvation-parameter model. Anal Bioanal Chem 399:2807–2820
447. Galan-Sanchez, M (2008) Functionalised ionic liquids, absorption solvents for CO_2 and olefin separation. Ph. D. thesis
448. Torralba-Calleja E, Skinner J, Dutierrez-Tauste D (2013) CO2 capture in ionic liquids: a review of solubilities and experimental methods. J Chem 473584:1–16
449. Yokozeki A, Shiflett MB, Junk CLM, Foo T (2008) Physical chemical absorptions of carbon dioxide in room-temperature ionic liquids. J Phys Chem B 112:16654–16663
450. Zhang X, Liu Z, Wang W (2008) Screening of ionic liquids to capture CO2 by COSMO-RS and experiments. AIChE J 54:2717–2027
451. Karadas F, Atilhan M, Aparicio S (2010) Review on the use of ionic liquids (ILs) as alternative fluids for CO2 capture and natural gas sweetening. Energy Fuels 24:5817–5828
452. Baltus RE, Culbertson BH, Dai S, Luo H, DePaoli DW (2004) Low-pressure solubility of carbon dioxide in room-temperature ionic liquids measured with a quartz crystal microbalance. J Phys Chem B 108:721–727
453. Carvalho PJ, Alvarez VH, Marrucho IM, Aznar M, Coutinho JAP (2010) High carbon dioxide solubilities in trihexyltetradecylphosphonium-based ionic liquids. J Supercrit Fluids 52:258–265
454. Zhang S, Chen Y, Ren RX-F, Zhang Y, Zhang J, Zhang X (2005) Solubility of CO2 in sulfonate ionic liquids at high pressure. J Chem Eng Data 50:230–233
455. Soriano AN, Doma BT Jr, Li M-H (2009) Carbon dioxide solubility in some ionic liquids at moderate pressures. J Taiwan Inst Chem Eng 40:387–393
456. Jalili AH, Mehdizadeh A, Shokouhi M, Ahmadi AN, Hosseine-Jenab M, Fateminassab F (2010) Solubility diffusion of CO2 and H2S in the ionic liquid 1-ethyl-3-methylimidazolium ethylsulfate. J Chem Thermodyn 42:1298–1303

457. Cadena C, Anthony JL, Shah JK, Morrow TI, Brennecke JF, Maginn EJ (2004) Why is CO2 so soluble in imidazolium-based ionic liquids? J Am Chem Soc 126:5300–5308
458. Blanchard LA, Gu Z, Brennecke JF (2001) High-pressure phase behavior of ionic liquid/CO2 systems. J Phys Chem B 105:2437–2444
459. Mejia I, Stanley K, Canales R, Brennecke JF (2013) On the high-pressure solubilities of carbon dioxide in several ionic liquids. J Chem Eng Data 58:2642–2653
460. Kim YS, Choi WT, Jang JH, Yoo K-P, Lee CS (2005) Solubility measurement and prediction of carbon dioxide in ionic liquids. Fluid Phase Equilib 228–229:439–445
461. Shin E-K, Lee R-C (2008) High-pressure phase behavior of carbon dioxide with ionic liquids: 1-alkyl-3-methylimidazolium trifluoromethanesulfonate. J Chem Eng Data 53:2728–2734
462. Huang J, Rüther T (2009) Why are ionic liquids attractive for CO2 absorption? An overview. Aust J Chem 62:298–308
463. Wang C, Luo X, Zhu X, Cui G, Jiang D, Deng D, Li H, Dai S (2013) The strategies for improving carbon dioxide chemisorption by functionalized ionic liquids. RSC Adv 3:15518–15527
464. Seo S, DeSilva MA, Brennecke JF (2014) Physical properties and CO2 reaction pathway of 1-ethyl-3-methylimidazolium ionic liquids with aprotic heterocyclic anions. J Phys Chem B 118:14870–14879

Author Index

Y. Marcus, *Ionic Liquid Properties*, DOI 10.1007/978-3-319-30313-0

K

Z

Subject Index

Y. Marcus, *Ionic Liquid Properties*, DOI 10.1007/978-3-319-30313-0

MIX
Papier aus verantwortungsvollen Quellen
Paper from responsible sources
FSC® C105338

If you have any concerns about our products,
you can contact us on
ProductSafety@springernature.com

In case Publisher is established outside the EU,
the EU authorized representative is:
Springer Nature Customer Service Center GmbH
Europaplatz 3, 69115 Heidelberg, Germany

Printed by Libri Plureos GmbH
in Hamburg, Germany